The Animal Kingdom

Volume 7:
The Class Aves 2

Georges Cuvier
Edited and translated by
Edward Griffith

CAMBRIDGE UNIVERSITY PRESS

Cambridge, New York, Melbourne, Madrid, Cape Town,
Singapore, São Paolo, Delhi, Mexico City

Published in the United States of America by Cambridge University Press, New York

www.cambridge.org
Information on this title: www.cambridge.org/9781108049603

This edition first published 1829
This digitally printed version 2012

ISBN 978-1-108-04960-3 Paperback

CAMBRIDGE LIBRARY COLLECTION

Books of enduring scholarly value

Life Sciences

Until the nineteenth century, the various subjects now known as the life sciences were regarded either as arcane studies which had little impact on ordinary daily life, or as a genteel hobby for the leisured classes. The increasing academic rigour and systematisation brought to the study of botany, zoology and other disciplines, and their adoption in university curricula, are reflected in the books reissued in this series.

The Animal Kingdom

Georges Cuvier (1769–1832), made a peer of France in 1819 in recognition of his work, was perhaps the most important European scientist of his day. His most famous work, Le Règne Animal, was published in French in 1817; Edward Griffith (1790–1858), a solicitor and amateur naturalist, embarked in 1824, with a team of colleagues, on an English version which resulted in this illustrated sixteen-volume edition with additional material, published between 1827 and 1835. Cuvier was the first biologist to compare the anatomy of fossil animals with living species, and he named the now familiar 'mastodon' and 'megatherium'. However, his studies convinced him that the evolutionary theories of Lamarck and St Hilaire were wrong, and his influence on the scientific world was such that the possibility of evolution was widely discounted by many scholars both before and after Darwin. Volume 7 is the second of three books on birds.

Cambridge University Press has long been a pioneer in the reissuing of out-of-print titles from its own backlist, producing digital reprints of books that are still sought after by scholars and students but could not be reprinted economically using traditional technology. The Cambridge Library Collection extends this activity to a wider range of books which are still of importance to researchers and professionals, either for the source material they contain, or as landmarks in the history of their academic discipline.

Drawing from the world-renowned collections in the Cambridge University Library and other partner libraries, and guided by the advice of experts in each subject area, Cambridge University Press is using state-of-the-art scanning machines in its own Printing House to capture the content of each book selected for inclusion. The files are processed to give a consistently clear, crisp image, and the books finished to the high quality standard for which the Press is recognised around the world. The latest print-on-demand technology ensures that the books will remain available indefinitely, and that orders for single or multiple copies can quickly be supplied.

The Cambridge Library Collection brings back to life books of enduring scholarly value (including out-of-copyright works originally issued by other publishers) across a wide range of disciplines in the humanities and social sciences and in science and technology.

THE

ANIMAL KINGDOM

ARRANGED IN CONFORMITY WITH ITS
ORGANIZATION,

BY THE BARON CUVIER,

MEMBER OF THE INSTITUTE OF FRANCE, &c. &c. &c.

WITH

ADDITIONAL DESCRIPTIONS

OF

ALL THE SPECIES HITHERTO NAMED, AND OF
MANY NOT BEFORE NOTICED,

BY

EDWARD GRIFFITH, F.L.S., A.S.,

CORRESPONDING MEMBER OF THE ACADEMY OF NATURAL SCIÉNCES
OF PHILADELPHIA, &c.

AND OTHERS.

VOLUME THE SEVENTH.

LONDON:

PRINTED FOR WHITTAKER, TREACHER, AND CO.
AVE-MARIA-LANE.

MDCCCXXIX.

LONDON:

SHACKELL AND BAYLIS, JOHNSON'S COURT, FLEET-STREET.

THE

CLASS AVES

ARRANGED BY THE

BARON CUVIER,

WITH

SPECIFIC DESCRIPTIONS

BY

EDWARD GRIFFITH, F.L.S., A.S., *&c.*

AND

EDWARD PIDGEON, Esq.

THE ADDITIONAL SPECIES INSERTED IN THE TEXT OF CUVIER

BY

JOHN EDWARD GRAY, Esq., F.G.S., *&c.*

VOLUME THE SECOND.

LONDON:

PRINTED FOR WHITTAKER, TREACHER, AND CO

AVE-MARIA-LANE.

MDCCCXXIX.

LONDON:
SHACKELL AND BAYLIS, JOHNSON'S COURT, FLEET-STREET.

LIST OF PLATES IN THE SEVENTH VOLUME.

LIST OF PLATES IN THE SEVENTH VOLUME.

Errata in the Plates.

The Satin Gracle, both Plates, for *Macleyii*, read *Macleayii.*
Double-crested Lark, for *Cilopha*, read *Bilopha.*
Natterer's Nutchatch, for *Anabatiodes*, read *Anabatoides.*
Sumatra Bee-eater, for *Merors*, read *Merops.*

Errata in Vol. VII.

Page	*Line*						
50	22	*for*	Throythorus	..	..	*read*	Thryotherus
62	15	..	Holm ..	..	..	..	Stockh
67	26	..	Fancica ..	..	..	..	Francica
72	11	..	Isabellanus	..	..	..	Isabellinus
73	10	..	climaturus	..	..	..	climacurus
75	3	..	macrurus	..	..	..	macrourus
75	26	..	in ..	..	..	..	n.
77	15	..	subtulated	..	..	..	subulated
77	21	..	gerrated	..	..	..	serrated
115	24	dele	It may possibly be the female of Dr. Latham.				
122	23	..	Oranor ..	..	..	..	Oranoir
334		..	Bennet's Magpie		..	..	Bernet's Magpie*
352	note	..	amgustirostus	..	..	..	angustirostris
354	note	..	Malomimes	..	..	..	Malouines
365	9	..	brassy ..	..	..	..	grass
370	11	..	Taumautias	..	..	..	Thaumantias
374	7	..	mesolencos	..	..	..	mesolencus
381	13	..	Poriss ..	..	..	..	Briss
381	16	..	sonimanga and calibris	..		..	souimanga and colibris
383	14	..	Praner ..	..	..	..	Promer.
579	3	..	Matous ..	..	..	..	Matons

* This bird is described by Mr. Vigors in Vol. IV. of the Zoological Journal, under the name ***Pica Colliei.***

THE

SECOND ORDER

OF

BIRDS.

SUPPLEMENT ON THE DENTIROSTRES.

(CONTINUED.)

THIS bird is of a shining metallic black—with feathers of a silky and glistening character, and the point of the bill and the feet inclining to yellow. The female is green above, paler underneath, and variegated with white spots, and fuscous lunules. The wing-coverts, remiges and rectrices, are of a rufous brown.

Mr. Vigors and Dr. Horsfield have designated this bird (which hitherto had received no scientific name) from the late respected Secretary of the Linnæan Society, and called it *Ptilonorhynchus Macleayii*.

We insert a figure of both male and female from the Museum of the Linnæan Society.

LYRES (MÆNURA) are birds of New Holland, whose description will be found in our text. Of their habits, unluckily, little or nothing can be said with any certainty. They are said generally to remain on trees, and never to descend except for the purpose of seeking food.

The only notions we have concerning the manners of the MANAKINS (PIPRA) are owing to Sonnini, who has observed them in Guiana, where they prefer humid and cool woods to the hot and dry lands. They do not, however, frequent marshes or the borders of streams; their flight is rapid, but short, and low; they only perch on the middle branches of trees in the woods, which they never quit to proceed into the open plains or to the neighbourhood of habitations. They assemble in the morning, in small troops from eight to ten in number, of the same species, and frequently join company with other small birds. At such times, their chirping is rather agreeable. But, about nine or ten o'clock, they separate, and retire alone, until the following day, into the most shady recesses of the forests. Their food consists of little wild fruits, and they also eat insects.

The *Rupicola*, *Coq de Roche*, and *Rock Manakin*, of Latham, has been ranged with the manakins, but with little propriety, being a bird differing from them considerably in size and habits. It inhabits in Guiana and other parts of South America, the deep clefts of rocks, and still more frequently large and obscure caverns, where the light of day cannot penetrate. This circumstance has caused many to believe that it was a nocturnal bird, but as it flies during the day, others have placed it among the diurnal. It is, however, now well known that many of the howlers, and other birds, reputed nocturnal, do the same thing. There is also another point of approximation between the rupicolæ and the nocturnal birds. The former are found more frequently in their obscure habitations, than in open places. Great numbers of them are discovered in caverns, where one cannot enter but with torches. The male and female are equally lively, and extremely wild. It is impossible to shoot them, but by remaining concealed behind some rock: here it is necessary to wait many hours before they come within range of shot; for the instant they perceive the hunter, they fly off very rapidly, but their flight is short and low. The males come forth from the caverns more frequently than the females, which seldom show themselves, and probably only come out during the night. They place their nest in the hole of a

THE SATIN GRACKLE. *Female*

PTILORYNCHUS MAC LEYII. *Vigors & Horsfield*

London Published by G.B. Whittaker, Oct. 1828

THE SATIN GRACKLE. *Male.*

PTILORHYNCHUS MAC LEYII. Vigors & Horsfield.

London Published by G. B. Whittaker, Oct. 1. 1828.

rock; it is rudely constructed of dry branches, and the female usually lays two spherical and white eggs, about the size of a pigeon's egg. They feed on small wild fruits: they have the habit of scratching the ground, and clapping and shaking their wings like cocks and hens; but this is the only point of relation between them and the latter birds, for they have neither the cry of the hen, nor the crowing of the cock: they are easily tamed, and sometimes left at liberty to live and run about with the poultry.

We now come to our author's great family of MOTACILLA, which he terms in French, "*bec-fins*," from the general tenuity of the beak: they are most of them comprehended under Latham's genus of the WARBLERS. The first division are the *traquets*, a word which we must preserve.*

BECHSTEIN and Meyer term these birds *saxicola*, M. Vieillot, after Gesner, Willoughby, and Ray, has preserved to them the denomination of *œnanthe*.

They inhabit during the fine season, dry and stoney places; those, called in French, *tariers*, are more partial to herbage, whether on the mountains or in the plains; all nestle on the ground: some in a tuft of grass, others in a

* We may remark the very serious inconvenience to which a translator of French works in natural history is subjected, by the names given to subdivisions. It is always difficult, and often impossible, to find a proper English equivalent. The names of species are given to genera, and subgenera; and, in many cases, were we to translate these names, we should convey to the reader a wrong idea of the animals comprehended under the section; for, in several instances, the animal, whose specific name is thus generalized, actually belongs to a division wholly different. Even when this is not the case, the inconvenience is far from being inconsiderable. The generic name is followed by the same, specifically applied, with some clumsy and circuitous appendage. It is much to be wished that naturalists would agree to speak one and the same language: at present their confusion of tongues occasions a most perplexing and vexatious Babel. For all divisions, except specific ones, scientific names should alone be employed.

E. P.

hole in the earth, under stones, at the foot of bushes, and sometimes under roots; they rarely perch on the summit of trees, some remain always on the ground, and others on low bushes, &c.

Motacilla Œnanthe, the *Wheat-ear Warbler* of Latham, is a migratory species; they pass only the fine season in France. Those found in our northern climates arrive towards the end of March, and spread themselves through the fields; they are to be seen in lands newly ploughed, always perched upon clods of earth, in French, *mottes*, whence they are named *motteux*. The white colour of the under part of their bodies is discovered when they fly, and causes them to be easily distinguished in the air from all other birds. They seek in the furrows of cultivated lands for insects and small worms, which constitute their principal nourishment; such are the places which they inhabit during the first days of their arrival, and after the hatching. But in spring, they seek untilled or fallow lands, preferring the borders of hills, the rocky slopes of the mountains, and dry places in general; they jump from stone to stone, and avoid the hedges, bushes, and trees, where they are rarely seen to perch except when greatly disturbed; their usual flight is short and rapid, but more elevated during their migrations; they pursue insects on the ground.

The cry of these wheat-ears when disturbed, resembles the syllables, *far*, *far*, pronounced rapidly. The male has also another cry which may be thus expressed, *titreu*, *titreu*, which appears to be one of alarm, and is never heard except at hatching time. Clumps of turf, heaps of stones in untilled grounds, little old dry walls, are the places in which these birds generally conceal their nests. Moss and fine plants compose the exterior; feathers and wool line them within. These nests, constructed with great care, are moreover remarkable for a kind of shed placed above them, and cemented against the stone or clump under which they are situated; the stone or clump is usually turned to the south side, and the entrance of the nest is to the north. Four or five eggs are laid, of an undulating pale blue, and an elongated form; the females are so absorbed in the care of hatching, that they are often taken on the nest. The male, towards the middle of the day, assists her in this care, and, moreover, provides for her subsistence. He is cunning and dextrous in misleading those who might disturb her. If he sees one passing, he runs or flies before him, makes short pauses to draw him to the opposite side, and when he sees him sufficiently remote, he takes a circular flight, and regains the nest. At such moments he is heard more frequently to repeat his cry of alarm.

These birds seldom have, in our parts of the world, more than one brood every year, and as soon as the little ones are feathered, and even some time previously, they quit uncultivated lands, and frequent grounds recently tilled, where they continue until September and October, when the family assembles and sets out on the southern voyage. Some of them are occasionally caught at sea, when, at considerable distances from land, being much fatigued, they drop on the masts of vessels to repose.

These birds inhabit all Europe, Africa, and Southern Asia. Great numbers of them are caught here, especially in Sussex, towards the commencement of autumn, at which time they are fat, and of a delicate favour.

The *Stone-chat Warbler*, (" *traquet proprement dit*,") *Sylvia rubicola*, delights in dry, arid, sloping grounds, and in bushes and briars. It is distinguished by its vivacity and agility. It is always seen hopping from bush to bush, never perching but at the extremity of the most elevated branches of hedges, and shrubs, or at the summit of the highest vine-poles. It rises in the air by little springs, and comes down in a sort of pirouette, appearing and disappearing every moment, and continually agitating its wings and tail; this movement has been compared to that of the clapper of a mill, from which its French name is taken.

According to Buffon, its cry resembles the word *ouistrata*, frequently repeated, but Dr. Latham thinks that it is more like the clicking of stones together, whence, probably, its English name.

This bird is of a solitary character, being always alone, except during the season of love. It migrates, and arrives in spring alone; it quits this part of Europe in autumn, usually in September; though if the season be mild it remains longer, and has been seen in France in December. As soon as t has chosen a companion, the couple proceed to the construction of the nest, which is placed at the foot of a bush under the roots, or under the covert of a stone, and tolerably deep in the earth. This nest is concealed so well, that it is very difficult to find it, which difficulty is much increased by the many circuits which the bird makes before entering, or after departing. If it wants to enter, it always first crosses different bushes, and when it comes forth it threads cautiously through the brambles, to some distance; so that when we see this bird with a worm or insect in the bill for the young ones, enter a particular bush, it is not there that we must look for the nest, but at the foot of some neighbouring bushes.

The eggs are from five to six, of a greenish white, with some spots of reddish-yellow. The little ones are born covered with down. The parents feed them with worms and insects, which constitute their habitual aliment. Their solicitude is so great that they never cease to cry when they are approached; this even seems to increase when the young ones quit the nest; they call and cry to them continually, and never quit them until they are entirely capable of providing for themselves.

This species is extended throughout Europe, from England and Scotland, to Italy and Greece. It is even found farther north, and is said to have been seen in Siberia. The flesh in autumn is fat and delicate, not inferior to that of the beccafico in flavour.

We shall now proceed to consider the most remarkable of the family of the Warblers, contained under our author's subgenera, RUBIETTES, (*Sylvia,*) FAUVETTES, (*Curruca,*) ACCENTOR, ROITELET, &c.

These birds are all comprehended by M. Vieillot under the general term FAUVETTE, *Sylvia* of Latham, and *Motacilla* of Linnæus. With the exception of the red-breast, they all leave our climates at that season when they are left destitute of food by the defoliation of the trees, and the death or lethargy of the insect tribe. But as soon as the flowers begin to blow, and the woods put on the livery of spring, we find the numerous family of the warblers re-appear, and disperse themselves through our groves and fields. Many establish themselves in gardens and shrubberies, others prefer the borders of coppices or the deep recesses of the woods; and some, in fine, are partial to the neighbourhood of water, where they fix their domicile. They all spread animation through the places which they inhabit, by the gaiety of their songs, the variety, the vivacity of their movements, their sportive tricks, and amorous combats.

Some of these birds feed on insects only; others unite with these, berries and succulent fruits. When they are nourished with grapes, figs, mulberries, &c. they grow fat, and their flesh becomes as savoury as that of the beccafico, with which bird they are frequently confounded in the south of Europe. Thickets, groves, shrubberies, &c. are the places where the majority of them choose to establish their nests, while others prefer reeds and rushes. Some confide their progeny to the hollows of walls, rocks, and trees. Some of the *reguli* nestle on the ground, and give to their nest the form of a little oven. The usual number of eggs among the majority is from four to five; a few of the *reguli* lay from six to eight.

The nomenclature of these birds, in various authors, presents a chaos, which we may say, we trust, without profaneness, would almost require the voice of a Deity to reduce to order—a confused, undigested accumulation, where we may well say,

> "Nulli sua forma manebat,
> Obstabatque aliis aliud."

It is most surprising that this should be the case with European birds, familiar to every observer, yet such is the fact. Authors have separated what they should have united, and united what they should have separated. The figures, too, given of these birds have, for the most part, contributed to increase the embarrassment. They have been either defective, or if correct, not in accordance either with text, or synonime. Linnæus, the grand guide of almost all naturalists, has sometimes occasioned mistakes, by indicating specific characters incompletely, or in a manner which rendered them susceptible of application to different species. Brisson has described very well such species as he had seen, but has not been equally happy in the use of synonimes. Buffon, who flattered himself that he had thrown some light on this subject, has only proved how contagious is error when supported by an eminent name. His mistakes, repeated by other naturalists, have passed with the great majority for incontestible truths. It is easy, however, to perceive, that in his illuminated plates, many figures of warblers of Europe are far from being in accordance with the text: so little so, indeed, that one cannot help believing that he did not always compare his descriptions with the figures. Even his history of many species is wanting in exactitude. He often attributes to one bird, the manners, the song, the nest, and the eggs of another. Subsequent naturalists have much increased this confusion, by the erroneous citations of synonimes, and specific periphrases, by no means analogous.

Our business, here, however, is not with classification or specific description; we shall, therefore, confine ourselves to the notice of the more distinguished species in whose natural history there is any thing remarkable or interesting to record.

We shall begin with that well known bird, the *Red-breast* (*Sylvia rubecula*).

It is found all over Europe, from Spain and Italy, to Sweden. It inhabits this country at all seasons, but seems more numerous in winter, as it then approaches the habitations of man. In France, the red-breasts are more numerous in Lorraine and Burgundy than elsewhere. They are very much sought after there, and their flesh acquires an exquisite fat, which renders it a very delicate meat, and which is doubtless owing to the abundance of fruits and tender berries, which constitute the principal nutriment of all the insectivorous birds in autumn. It is not so in the other provinces, and accordingly its flesh is little esteemed there. Of the birds of this species some do not quit their native country, while others, and those constitute the greatest number, prepare for their departure at the epoch when the red colour commences to appear in the breast of the young, whose plumage, during the moulting, presents an agreeable variety, from the mixture of the tints of infancy, and the colours of more advanced age.

In all seasons the red-breast preserves its solitary character. It voyages alone, while on such occasions almost all other birds seek the society of their species, and unite in flocks more or less numerous. In fine weather, the most shady woods and groves, and humid places, constitute its favourite sojourn. They seem to possess so much attraction for it, that it appears to quit them with regret when the frost destroys its sources of subsistence. It is only then that it approaches habitations, and remains in hedges, gardens, and orchards, and enlivens such places, when all other birds are silent, by a song which is far from disagreeable. The more rigorous the season grows, the more familiar this bird becomes; and when the snow deprives it of all nourishment abroad, or renders it difficult to be procured, it comes into houses, to seek the crumbs from the table, fibres of meat, and various grains, for at this period it is omnivorous. It exhibits so much desire to enter the houses, that if every avenue is closed, it will strike at the windows with its bill for admission, which is always willingly granted. It soon grows so familiar that it will remain there the whole winter, if undisturbed, without showing the slightest inclination to depart. On the approach of spring, it will give the same signal, to indicate its desire of returning into its solitude.

The red-breast which remains in the forest becomes the companion of the woodman, warms itself at his fire, picks his bread, and never ceases the entire day to flutter around him. It always shows an affection for man, and seems to delight in his company. It follows or precedes travellers in the forests, and that for a very long time. Less wild than other birds, it will suffer itself to be approached so nearly, as to make one imagine that it can be caught with the hand, but as soon as it is within reach, it hops on further, where it remains for the purpose of again removing in the same manner.

In spring, when the red-breasts which migrate return, they are seen in greater numbers in orchards and gardens, though but for a short time: they soon hasten to the distant woods to enjoy the sweets of solitude and love, beneath the embowering foliage. Forests of great extent, especially such as are provided with springs of living water, are the haunts which they prefer, and where they are found in the greatest numbers.

Of all birds, the red-breast is one of the earliest, its song being heard from the dawn of day. It is also the last which is seen to flutter after the setting of the sun. Its song, composed of light, delicate, and tender sounds, is but a chirping, during the winter season. But at the time of reproduction, it receives its full extent, is terminated by louder modulations, and interspersed with soft and touching tones. It has different cries, one which is heard afar, especially in the morning and evening, and when it is moved by any new object. This is loud and abrupt; the other is a sort of chirp which seems to be a note of call, for it is sufficient to imitate it by sucking the finger, to assemble all the red-breasts in the neighbourhood.

This lover of solitude chooses obscure places, in which to build its nest. It conceals it more or less near to the earth, in the roots of trees, in tufts of ivy, or in very thick bushes. It is composed externally of moss, mingled with hair and oak leaves, and furnished within with the hair of cattle and feathers. There are some red-breasts, says Willoughby, which, after constructing this nest, cover it with accumulated leaves, and under this mass leave a narrow oblique entrance, which they stop up with a leaf on going out. From five to seven whitish eggs, spotted with reddish, constitute the lay. The male covers them in the middle of the day, while the female goes in search of food. Like the nightingale, the male red-breast will suffer no other bird of its own species in the particular district which it has adopted. It pursues it violently the moment it appears, and soon forces it to retire. The female has two or three broods during the year, and the young ones are fed with worms and insects, which the parents hunt with much dexterity. Taken adult in the after season, the red-breast may be easily retained in captivity, and will sing for some time after the loss of liberty. It may be preserved, by giving it the same food as the nightingale receives.

In the back season, this bird joins to its natural insectivorous regimen, tender berries and fruits of different kinds. Its flesh is then delicate, and in estimation.

The Blue-throated Warbler (Sylvia Suecica) has a similar

mode of living, and the same familiarity as the red-breast, but differs in some of its habits from that bird. The latter, during summer, seeks for solitude in the depths of the forest, but the blue-throat confines itself to their borders, seeks marshes, humid places, osieries, and even reeds. After the fine season it quits them, and, previously to its departure, visits hedges and gardens, where it may be taken without much difficulty. These birds, like the red-breasts, are not encountered in flocks, and rarely more than two are ever seen together. Towards the end of summer, they frequent cultivated fields, and especially fields of pease, whither they are no doubt attracted by the numerous insects which are found there. But in autumn, when they voyage to the south, they eat various berries. When this bird is on the ground, it carries its tail elevated.

This species is much less numerous than the red-breasts every where.

It places its nest commonly on willows, briars, and other shrubs which grow on humid places, and constructs it of grass, interlaced at the origin of the boughs or branches.

We now come to that most celebrated of birds, in both ancient and modern times, the *Nightingale (Sylvia Luscinia).*

This species appertains to the old world. It inhabits Europe from Italy and Spain, as far as Sweden, and is also found in Siberia and a part of Asia. We are also informed that it has been seen in China and Japan; but there are some countries in which it delights more than others. There are some where it does not even remain.

The nightingales quit the temperate countries of Europe in autumn. As there was no certainty of any being found in Africa, it was imagined that they retired into Asia. But at present, it is known that those of Europe do take refuge in Africa, to pass the winter. Sonnini informs us that there are nightingales in the most eastern part of that great continent, and that in autumn, they arrive in Lower Egypt. This traveller has seen several during the winter, on the fresh and smiling plains of the Delta, and has also witnessed their passage in the islands of the Archipelago. In some parts of Asia Minor, particularly Natolia, the nightingale is common, and never quits the forests or woods which it has chosen for its abode. During their passage through the islands of the Levant, and their sojourn in climates which are foreign to them, as they are not employed in the reproduction of their species, these delightful songsters never pour forth their enchanting melody. There seems no doubt but that some of them retire into Barbary, as they are found in greater numbers in the countries bordering on the Mediterranean, at that time than at any other. They are to be seen there, when they have totally disappeared from our more northern climates, and nearly a month sooner than they reappear in the north. They advance thither only in proportion as the cold relaxes: they set out and return with the common warblers, the fig-eaters, the becaficos, and other insectivorous birds. So powerful is the instinct of migration in the nightingale, that those which are retained in captivity are observed to be particularly agitated, especially during the night, at the usual periods when the species migrate These birds change places, not only to avoid the cold, but also to seek out the countries in which they can procure a suitable aliment.

The nightingale, naturally timid and solitary, migrates, arrives, and departs alone. It appears in England in the middle of April, or at farthest in the beginning of May. In France, a little sooner. It remains at first among the hedges which border cultivated lands and gardens, where it finds a more abundant nutriment. But it remains there but for a short time, for as soon as the forests begin to be covered with verdure, it retires into the woods and thickets where it delights in the thick foliage. The shelter of a hill-side, the neighbourhood of some purling stream, the proximity of an echo, are the haunts which it usually prefers. The male has always two or three favourite trees, on which it delights to sing, and seldom will it give any where else to its voice all the compass of which it is capable. The one it most particularly prefers is that which is nearest to the nest, on which it ever keeps a watchful eye. Once mated, it will never suffer any of its own species to approach the district which it has chosen, the extent of which depends on the greater or less abundance of nutriment which it affords for the subsistence of the family. We find that where food is abundant, the distance between the nests is considerably less. Jealousy, however, has something to do with the extent of their mutual distance, for the males fight outrageously for the choice of a companion. These combats are frequently reiterated at the period of the arrival of the nightingales. It is a common opinion that the females are much less numerous in this species, but this may partly arise, as Dr. Latham well observes, from the males migrating sooner than the females, in consequence of which more of them are caught at such periods.

They commence the construction of their nests about the beginning of May. Coarse weeds and dried oak leaves are employed in great quantities without. Horse-hair, little roots, and cow's hair furnish the inside. The whole is bound together, but in so fragile a manner that as soon as the nest is displaced the whole edifice crumbles to pieces. It is usually constructed near the ground, in brush-wood, at the foot of a hedge, &c. or on the lowest branches of some tufted shrub. The eggs are four or five in number, and of a greenish brown. It is said that the male does not partake the incubation, which would be an exception to the established order among insectivorous birds, all of which relieve their females in this care towards the middle of the day. Such a supposed exception, however, demands, before it can be admitted, the best authenticated observations, confirmatory of its existence.

It is also reported that the female quits the nest but once

in the four-and-twenty hours to take food; this would be a long time for a bird to fast that feeds on insects: but the truth is, that the male supplies her with food in the course of the day. As soon as the young have broken the shell, both parents take equal care of them, but they do not disgorge the food for them, like canaries. In this point they do not differ from the other insectivorous birds, for, like them, they have no crop. They fill their beak, as far as the œsophagus, with small worms, naked caterpillars, ant's eggs, and those of other insects, which they distribute equally to the young. When the food is abundant near the nest, they content themselves with carrying it at the end of their beak, as they do when they bring up their young in aviaries. The little ones have the body covered with feathers in less than fifteen days, and quit the nest before they are able to fly. They then are observed to follow their parents, jumping from branch to branch. As soon as they can flutter, the male takes care of the rest of their education, while the female is occupied in constructing a new nest for the second brood; she has generally two in the year, seldom three, at least in our part of the world, except the first has been destroyed, which often occurs, after the placing of the nest.

After the birth of the first brood the nightingale ceases to sing, and is seldom heard during the second hatching, however late it may take place. But it often utters a piercing cry, especially in the evening, which is heard at a great distance, and a kind of low, hoarse, croaking, which the father and mother repeat incessantly, whenever the nest is approached. These are cries of inquietude and alarm, which far from proving a security, serve only to reveal the place of concealment, and expose the young to danger; still, at this signal, the young family remains motionless, squats down on the branches, or conceals itself in the bushes, and especially preserves a most profound silence.

Towards the end of August, or even sooner, if their

habitual nourishment grows rare in the woods, they quit them, both young and old, to approach green hedges, lands newly ploughed, and gardens where it is more abundant, and they then unite to it tender berries and fruits. Their flesh then grows fat and delicate, and is much esteemed in some places, especially in Gascony in France. But neither there nor elsewhere, are they fattened for table as some naturalists have asserted.

The nightingale, has, of all other birds, a voice of the greatest harmony, variety, and compass. Even the mocking thrush, which some have set far above it, has been universally pronounced by all, who have heard both birds, to be much inferior, to hold at least but the second rank, among the feathered songsters. The nightingale unites the talents of all the singing birds, and succeeds in every style ; sixteen different burdens may be reckoned in its song, well determined by the first and last notes. It can sustain the song uninterrupted during twenty seconds, and the sphere which its voice can fill is at least a mile in diameter. Song is so peculiarly the attribute of this species, that even the female possesses it, less strong and varied, it is true, than that of the male, but as to the rest entirely resembling it ; even in its dreaming sleep the nightingale warbles. What peculiarly constitutes the charm of this bird is, that it never repeats itself, like other birds ; it creates at each burden, or passage, and even if it ever resumes the same, it is always with new accents and added embellishments. In calm weather, in the fine nights of spring, when its voice is heard alone, undisturbed by any other sound, nothing can be more ravishing and delightful ; then it developes, in their utmost plenitude, all the resources of its incomparable organ ; but from the setting in of the summer solstice, it grows more sparing of its song, it is seldom heard, and when it is, there is neither animation nor constancy in its tones. In a few days, at this time, the song altogether ceases, and we hear nothing but hoarse cries and a croaking sound, in which we would in vain endeavour to recognize the melodious Philomela.

It has long been an object of desire to find the means of enjoying the song of this bird; but to preserve to its voice the charm which in the free bird disappears with the season of love, it is necessary to retain it in captivity. Great patience, attention and care, are necessary in the management of the nightingale, far more than in that of other birds, for it is a captive of a temper very difficult to manage, which renders the desired service only in proportion as it is well treated.

Nightingales are procured in three ways: in the nest, in autumn, or in spring, on their arrival. To find the nest of a nightingale, where there are young ones, it is necessary to go in the morning at sun-rise, and in the evening at sun-set, near the place where the male has been heard to sing, which is generally at no great distance from the nest; it is necessary to remain quiet, without making the least noise. The entrances and exits of the father and mother, and the cries of the young will assuredly indicate the proper place. They should not be taken from the nest until they are well covered with feathers; those of the first brood are preferable, they are always more vigorous and will sing sooner; besides, the moulting, which generally causes some of them to perish, will then come on in the warm weather, when they are better able to support it. They should be put with the nest and some moss in a basket, with an open work lid, and which during the night should be covered with some warm stuff. Care must especially be taken that they do not get out of the basket after receiving their food, as they may catch the gout, which in them is an incurable malady. They must be kept very clean in this basket, until they are able to sustain themselves well upon their limbs; they then are put into a cage, the bottom of which is furnished with moss. It is necessary to be particularly careful in the proper proportioning of their food, for they are so delicate that the slightest excess will destroy them; no regard should be paid to their reiterated demands for food, for they open the bill every instant, whether one approaches them or touches the nest. To succeed properly with them, the following system should be strictly adhered to.

They should receive their first bill-full, half an hour after sun-rise, the second an hour after, and so on from hour to hour, until the last, which is towards the setting of the sun. After which they must be refused, although they demand food; but the last allowance should be a little more than the preceding on account of the night. They should be fed with a small wooden skewer, very smooth, rather slender at the end, and about the breadth of one's little finger. After three weeks or more, they will eat of themselves, and the males begin to warble; they are then separated, and put into different cages, for these birds are fond of living alone.

As these birds are very delicate, the attempt to bring them up in the manner we have described is not always successful; on which account, if it be possible, they should be tended by the father and mother. These may be easily caught by laying nets near the nest, which nets are furnished with the lava of an insect, which abounds in flour, the meal worm *(tenebrio molitor)*. As soon as they are taken, they are put with the nest and the young ones into a closet, where very little light is permitted to enter; three delft pots not very deep are provided for them to eat and drink out of; in one of these is water, in the other sixty or seventy of the above-mentioned worms, and in the third the usual aliment given to them, (which we shall explain by and by,) to which are joined the eggs of ants. It is necessary to familiarize them with their new domicile; branches covered with leaves should be put into it, and the floor spread with moss; trees in boxes, evergreens, such as laurels, &c., are still better, as there will then be no necessity to enter the place and disturb the birds, for the purpose of changing the verdure. By managing in this manner, those who bring up these birds will soon have the satisfaction of seeing the father and mother take the food which is provided, and give it to the young. These birds have so much affection for their young, that they speedily forget the loss of liberty, lavish on them the same care which they do in the woods, and exhibit for them the same solicitude; they also utter those cries of alarm, when any thing occurs to annoy them, on which signal the young conceal themselves in the moss and foliage.

Those who are desirous of making them nestle the following spring, preserve the old ones, and either put them into separate cages, or give the female and the young of her sex their liberty. The females are easily recognised by their silence, and on the contrary, the young males commence their song as soon as they are able to eat alone. A young one, which a month after this, does not warble, may to a certainty be pronounced a female.

The young ones may receive such instruction as their owners may think proper, but the best is that given by the old nightingale, and he that has the best voice should be chosen, for all do not sing equally well. In proportion as the young male advances in age, his voice is formed by degrees, and is in its full power towards the end of December. It will easily learn different airs, whistled by the mouth, or played on the flageolet, if it be made to hear them constantly for some months. It can even be taught, it is said, to sing alternately with a chorus, to repeat the couplets, and to speak; but it is necessary in this case, to sacrifice its natural song, which is lost altogether, or greatly injured by these foreign acquisitions. This is often a very just source of regret, for as variety constitutes its principal charm, this gives place to a monotony, which in the long run becomes tiresome. Another inconvenience which frequently occurs, is, that it forgets a portion of its native notes, and acquires only a part of the foreign, from which results an interrupted and imperfect song.

Still if it be deemed necessary to teach it some airs, the mode is this: as soon as it commences to warble, it must be placed in a cage covered with green serge, which is put into a chamber entirely remote from the hearing of every other bird. It must also be free from disturbance of every kind, no one must enter except the person who has charge of the bird, for the utmost tranquillity is indispensably necessary. His cage, during the first eight days, is fastened near the window, after which it is gradually removed thence into the darkest part of the room, where the bird should remain during the whole period of tuition. Six lessons a day will suffice; two in the morning, two in the middle of the day, and two in the evening; those in the morning and evening should be the longest, for then the bird is most attentive. The air designed to be taught, should be repeated ten times at each lesson; the bird will be able to execute two airs with facility, but if more are attempted to be taught, it will generally confound them, and know nothing perfectly.

A deep toned flageolet is the best instrument for teaching the nightingale by. The instructor too, must not abandon his pupil, because he hears it warble on in the usual manner, and appear to derive no profit from his lessons. This will often be the case after the moulting, and even during the entire winter, and yet the bird will repeat the airs which it has heard, in the commencement of spring. Among the young birds so brought up, some will sing during the night, but the major part commence in the morning, more or less early according to the season.

The nightingale taken adult, may be made to sing, if properly treated, if its prison be made as much as possible like its native woods, freed from intrusion, and secured against cold. It should also receive an abundant and agreeable nutriment. On these conditions it will begin to sing about the end of eight days, and even sooner, it it be not yet mated, that is, before the 25th of April, otherwise it generally dies in consequence of the loss of the female; if taken after the 15th of May, it will not sing at all.

The race which our author notices, of a larger size, in the eastern parts of Europe (*Syl. Philomela*, Meyer), differs from the common nightingale, not only in dimensions, but in depth of colour, and extent of voice. The latter is so powerful, that the bird can hardly be kept in a room; it sings more slowly than the other, and its style is altogether inferior. It can be heard at a much greater distance than the common nightingale, particularly at night.

The nightingale though a timid and solitary bird, is capable, in the long run, of an attachment to the person who tends it. Some have been known to die of regret at the change of masters, and others having been left free to fly into the woods, have voluntarily returned to their captivity. It is very certain that the bird does not like change. It grows sad, unquiet, and ceases to sing, if transported from one situation to another, or even if it changes place in the same apartment; therefore, not to interrupt its song, it should remain in the same spot during the whole season.

Nightingales which are kept in cages, have a habit of bathing themselves after they have sung; they should therefore be provided every day with fresh water. This bird, when not tamed, grows desperate at the sight of any strange object; it would infallibly perish, if placed, like other birds, in a cage open to the light on all sides; it would dash itself against the bars until it was killed. But when the light is excluded, it remains quiet, and consoles itself by singing and eating.

These birds have naturally but two seasons for singing, May and December. But in a state of captivity this order may be changed; for this purpose, the old male is put in a proper cage, and shut up in a closet rendered obscure by degrees; it is kept there until the end of May, and the light is gradually restored, as it was withdrawn; this produces on the bird the effects of spring, and it will sing in June. The same experiment may then be tried with another bird, and so on to the end of November. Thus with two males, a person may always enjoy the song of one, while the other is silent; but care must be taken, that one shall never hear the song of the other, and that during the winter the cold be carefully excluded. Some deprive the nightingale of sight, for the purpose of making it sing almost continually; but we shall spare our readers the details of this barbarous and disgraceful operation.

Notwithstanding the great obstacle presented by the love of liberty, which is stronger in these birds than almost any others, the means have been found to make them nestle in their prison, and rear their offspring. The best for this purpose, are those which have been caught the preceding spring, and made to bring up their young: the male and female should have been kept in separate cages during the winter, and then placed together (in their cages) in an apartment destined to the desired object, that they may grow accustomed to each other by degrees. They should be suffered to come out of their cages from time to time. This alliance will be all the better, the more it is owing to nature, for the attempt to pair them by force is rarely successful. At the commencement of April their cages are left continually open. They are then furnished with such materials as they are in the habit of employing for their nests: such as oak-leaves, moss, hair, &c. Four fagots of dry and small wood are placed in a corner of the chamber, near the window, one against the other, connected together but loosely, and fixed by the gross end. They are furnished with oak leaves on the top, sides, and between the branches, leaving no opening to facilitate an entrance except one, through which the hand can be passed. There is placed there, moreover, a little bucket of wood, about two inches deep and three feet in diameter, and filled with earth, and a vase about an inch in depth and filled with water, so that the birds can bathe; this must be renewed every day; but the vase should be withdrawn when the female hatches. This nest should be exposed to a southern aspect, but very close and guarded against the north wind; some curiosi procure themselves a more agreeable enjoyment, by placing the couple in a large aviary, planted with yews, lilacs, &c., or rather in the corner of a garden furnished with these trees, and converted into an aviary by being surrounded with nets. This is the most favourable and certain method. It has been many times observed, that the father and mother may be let go with perfect security as long as the young ones are not able to fly; there is no danger of losing them then. It is only necessary, for the first few days, not to let them both go out at once, but to free the male first, alone, then the female alone, and then both together. But the aperture through which they go out and return, must be near the nest. They will take advantage of this liberty to catch many insects which cannot be procured for them, and which are very necessary to the rearing of their young. It is necessary to avoid entering frequently into the aviary, while the parents have this liberty, and to take especial care that no dog or cat approach, which would be sufficient to make the birds abandon it directly.

There is no man who would not desire to possess, in a garden ornamented with groves, a bird of such extensive, various, and melodious powers of voice, as the nightingale. The means of fixing this charming musician near one's dwelling are these: when the little ones have been for about eight days out of the shell, the father and mother must be taken with nets, as we mentioned before, when the object was to make them rear the young. They must be taken very early in the morning, which may be done in about one hour; as soon as they are caught, they are shut up separately, each in a silk bag; after which the nest is removed, without touching the young, all the branches upon which it is placed being removed with it. If it is on a shrub, that should be taken away altogether; the whole is transported to the destined spot, and placed in a situation the most similar to that in which the nest was found; then the male is put into one cage, and the female into another. These cages must be covered with green serge, tolerably thick, with a door in front, arranged so that it can be opened from a distance by a packthread attached to it. The nest being fixed, the cages are placed one on each side, at a distance of from twenty-five to thirty paces, so that the young ones shall be nearly in the same line, and between both. The doors should face them. All being thus prepared, the young should be suffered to cry for a certain time until their note of appeal is completely heard, both by father and mother; then the cage of the female should be opened, the person who does so, not showing himself; when she has flown forth, the cage of the male should be opened; the movements of nature will carry them directly to the place where they have heard their little ones cry; they will administer food to them, and continue the same cares until they are brought up. The young family will, it is positively asserted, return thither the following year, and people the adjoining groves; for they are in the habit of returning every year into the places in which they have been reared, if they find abundant nutriment and proper accommodations for nestling. If this be not the case, all the labour we have described will be thrown away.

It is necessary to say a few words, in conclusion, on the food proper for the nightingale in a state of captivity.

This bird, being naturally voracious, will accommodate itself to aliment of every description, provided it be intermixed with meat. Some people feed their nightingales with equal parts of bruised hemp-seed, crumbs of bread, parsley and boiled beef, all hashed small and mixed together; others use the heart of beef or mutton in the same way, with some farinaceous appendage. These mixtures agree very well with the bird, but must be made fresh every day, as otherwise the meat speedily corrupts, disgusts the bird, and makes it lean and silent. Pastes, however, can be composed for it, that will keep for years, and preserve the bird in excellent health.

The first is made with two pounds of rolled beef, a pound of grey peas, a pound of sweet almonds, an ounce and a half of saffron, in powder, and twelve fresh eggs. The peas must be pounded and sifted; the almonds peeled in hot water, and pounded as fine as possible; the beef should be hashed exceedingly fine, and cleansed carefully from skin, fat, and fibres; the saffron infused in half a goblet of boiling water; all being thus disposed, the eggs are taken and broken in a plate, and then all these ingredients are successively mixed, finishing with the saffron. Round cakes are then formed of it, about the thickness of one's finger, which are dried in the oven, after the bread is withdrawn, or in a large baking pan, rubbed with fresh butter, and placed at a very gentle fire; these cakes are sufficiently dressed, when they have acquired the consistence of new-made biscuits. They are broken and crumbled in the hand for the use of the bird.

The second paste is made like the first, but with half-a-pound of poppy seeds in addition, as much of roasted millet, two ounces of flour, a pound of white honey, and two or three ounces of fresh butter. The peas and the millet are pulverized and sifted; the poppy seeds well pounded, as also the sweet almonds; these each, indeed, should be reduced into a perfect paste, as otherwise the bird cannot digest them. There should be no lumps left; the meat should be prepared as for the first paste, then the eggs should be broken, the yolks of which, alone, should be put into an earthen plate, adding the honey and saffron; when these ingredients are well mixed, the meat, sweet almonds, and flour are successively incorporated, stirring the whole well with a wooden spatula, so as to make a sort of soup of even consistence and without lumps; then the whole is poured into another large dish of japanned earthenware, the bottom of which must be greased with butter, and placed upon a very mild fire, stirring it continually to prevent the paste from attaching. This process must be continued until it is quite done, which will be known by its no longer sticking to the fingers. It is then taken from the fire and left to cool, completely, in the dish; after which, it is put into a box of white iron, covered with its lid, and kept in a dry place for use.

This paste is difficult to prepare, and its goodness depends upon the exact degree to which it is dried, and which is very much a matter of chance. When too dry, it loses its substance, and it is often necessary to join sheep's heart to it to keep the nightingales in good condition. If, on the contrary, it be not sufficiently dry, it will corrupt, and must be immediately used.

These two preparations are very proper for the nightingale, because they are stimulating, and, it is said, excite the bird to sing. Perfumes are said to have a similar effect. But these pastes are improper for the generality of the warblers, and other small birds with slender bills. For the first few months they agree with them extremely well, and the birds grow fat upon the regimen: in the end they dry them up, and produce consumption.

A provision of the larvæ of the *tenebrio molitor*, or meal-worm, formerly mentioned, should always be kept for the use of the nightingale. They constitute a strengthening food for the bird, during the season in which it sings; they should be laid in during summer, as they are difficult to be got in the commencement of spring. They are preserved in pots of delft or varnished earthenware, very wide bottomed, and fed on bran; some pieces of cork or rotten wood are thrown into this vessel, into which the insects retire, and where they get fat very speedily. The pots should be of the ware above-mentioned, because these insects would soon escape, if they were put into a box or vessel against which they could climb. For this reason, also, there should be two or three inches difference between the bran and the edges of the opening. This precaution is indispensable, because they could otherwise escape, and also because being exceedingly voracious, they would destroy both books and furniture. The vessel should be kept in a dry plaee; the bran must be renewed from time to time; it is easy to observe when it is consumed, as it is then reduced to a sort of grey dust; it should be sifted twice a year, and then renewed altogether, for otherwise it contracts a bad odour, and a humidity occasioned by the mixture of the excrements of these insects, which causes them to become emaciated and perish.

To understand the symptoms of illness in the nightingale, it is necessary to be acquainted with the signs of its good health: when in this latter state, it will sing frequently during the season, which is from December to the end of June. We must except, however, the first year of its captivity, in which it is seldom heard before February. It will also frequently pick and cleanse its plumes, especially on the back. It will evince much gaiety and alertness, stirring about in its cage, shaking its wings, &c. Finally, it will sleep on one foot, eat well, and be particularly eager after the insects above mentioned.

Such are the signs of health, but when the nightingale remains during the night at the bottom of its cage, it is a sign of illness, unless its toes should be embarrassed by the attachment of dung, which will happen, if proper attention be not paid to cleanliness. The dung will harden to such a degree, as to prevent the bird from remaining on the perch. In this case, he should be taken out by the hand, and his feet steeped in warm water, to cleanse them. He will also find much difficulty in perching, if his claws are too long, but it will be sufficient to pare them from time to time.

If the nightingale is attacked with an abscess on the crupper, it should be cut with the point of a scissars, pressed a little with the end of the finger, and the bird should receive some of the insects above mentioned, some wood-lice, and spiders. This malady may be avoided by purging the bird at times, especially in the month of March, with half a dozen of the latter.

When, by dint of singing, the nightingale gets dried up and thin, poppy seed in the paste is excellent for tranquillizing, refreshing, and procuring sleep. Sheep's heart, with the skin, fibres, and veins removed, hashed very fine, and mixed with the paste, fattens it very quickly, as also do figs and elder berries. The poppy seed should be omitted after the moulting, for then the bird grows exceedingly fat, and is exposed to perish of the disease called molten grease.

Constipation may be removed by four or five of the larvæ of the tenebrio, or a large black spider, which is most efficacious.

For diarrhæa, which may be observed by the dung being more liquid than usual, by the continued shaking of the tail, and bristling of the feathers, sheep's heart, prepared as above mentioned, is an excellent remedy.

These birds are subject to the gout, particularly the young ones brought up by hand. Those which have it before they can eat alone, infallibly perish. As soon as they begin to limp, it is loss of time to attempt to rear them. When the old ones taken in the net, are attacked, which rarely happens, it proceeds from the cage being exposed to some wind, through crevices which the bird cannot avoid. It is then sufficient for the purposes of cure, to put it in some warm place. To prevent this malady, the bottom of the cage should be furnished with moss and sand. Of all these complaints, of which the bird in a state of freedom experiences nothing, the most dangerous is the falling sickness: it will immediately destroy it, if prompt assistance be not rendered.

When the nightingales have swallowed any thing indigestible, they reject it in the form of pills, or little pellets, like the birds of prey. This is not a disease, but proceeds from their having no crop, only a single canal conducting to the stomach.

It is necessary to examine the nightingale twice a year, to see if he be too fat, or too lean, for his external appearance is often deceitful. Sometimes he is ill, without appearing to be so; and sometimes the reverse is the case, from derangement of the feathers. This examination should take place in the months of March and October.

The River Nightingale (*Turdus Arundinaceus*) was placed among the thrushes by former naturalists. Our author places it at the head of the warblers, which immediately follow the nightingale. It inhabits marshes, the borders of ponds and rivers, generally remaining in reeds and hedges, from which it has received its scientific appellation. It climbs along reeds and willows of no great elevation, like the creepers, and lives on the insects which it finds there. The male sings during the night, as well as day, in the season of re-production. Its song, and the habit of remaining in humid places, have obtained it the denomination of river nightingale. Its voice, however, though of considerable compass, has none of the charms of that of the songster of the woods. This bird accompanies its song with lively action, and a tremulous motion of the entire body. It flies heavily, clapping its wings, places its nest on precipitous banks, and in places furnished with moss. It lays about five eggs, of a yellowish white, spotted with brown, and a little larger than those of a sparrow. It is abundant in the South of France, and is also found in the North, but not so frequently. It inhabits the Southern provinces of Russia, and the islands at the mouth of the Vistula. It does not inhabit England; it is found in Asia, for Sonnerat brought back an individual from the Philippine islands.

The *Reed-wren (Mot. Arundinacea)* is smaller than the last, but very analogous to it in form and habits. These birds frequent the edges of rivers, lakes, and ponds, where they remain in the reeds, and generally in those watered places where these plants cross. They seize their stalks across with their toes, and traverse them jumping along. The male, during the day, and of calm nights, utters a song which may be expressed by the syllables *tran, tran, tran,* repeated a dozen or fifteen times running. The nest is found in the same places which the bird frequents, within about a foot of the water. It is constructed of the same materials as that of the preceding species: leaves and little stalks of aquatic plants, constitute the bed, on which the female lays four or five greenish eggs, irregularly spotted with olive green; these spots are confluent towards the large end. This nest is attached to several reeds, so that it is suspended in air. It is pretended, that by means of three or four loose rings, composed of moss and horse hair, it can be raised or lowered according to the height of the water. This assertion, however, is combated by many naturalists, who affirm that these nests, although suspended in this way, cannot be raised above two or three inches at most, the rings being stopped by the knots of the reeds: therefore if the increase of the water should be considerable, the nests will be submerged. It may be mentioned here, that the inhabitants of Lorraine judge of the height to which the waters will arrive, by the elevation of the nest of the river nightingale.

The *Spotted Warbler* (*Sylvia Nævia*) is the smallest of the aquatic species. It is more frequent in this country than in France, remains usually in thick hedges, briars, and even reeds. The eggs of this species are of a pale blue.

The *Black-cap* (*Sylvia Atricapilla*, Lath.). Of all the

warblers, there is none so affectionate to its female as the male of this species. It shews the same affection for its young, and its song is very agreeable and prolonged. It is of this bird that Buffon speaks, when he says, "The warbler *(fauvette)* was the emblem of transitory love, as the turtle was of fidelity; nevertheless, this same warbler, lively and gay as it is, is not the less loving, or the less faithfully attached on that account; nor is the turtle dove, in spite of its melancholy and plaintive character, the less a libertine. The male of the atricapilla lavishes on the female a thousand little cares during the time of incubation. It partakes of her solicitude for the little ones, which have just burst the shell, and does not quit her even after the education of the young, for its love appears to remain after its desires are satisfied."

Nothing can alter its tender affection, not even the loss of liberty, if it is deprived of it with its family. It will then feed the young and the female, even forcing the latter to eat, when the chagrin occasioned by captivity would lead her to refuse all sustenance.

The males of this species arrive in the early days of April, but the females do not appear until towards the 15th. If, at this epoch, any return of cold should deprive them of insects, they will feed upon the berries of the laurel, the ivy, the privet, and the hawthorn. It is the same way with such of these birds as a late brood, or other accidents, compel to remain during the winter in our climates, which, however, is a circumstance of rare occurrence.

Immediately after the arrival of the females, these birds employ themselves in the construction of the nest. The male seeks out the most favourable position, and when his choice is made, he appears to announce it to the female, by a sweeter and more tender song. It is almost always in the small bushes of eglantine and hawthorn, at an elevation of two or three feet from the ground, that the female fixes the nest. It is small and not very deep, composed of dried plants without, and a deal of horsehair within. Four or five eggs are laid, marbled with a deep moronne, on a clearer ground of the same colour. If the eggs are touched, the female will abandon them, though not so speedily as the other warblers. The male relieves her, during the labour of incubation, from ten in the morning to four or five in the evening. The little ones are born without down, covered with feathers in a few days, and quit the nest very soon, especially if they are disturbed: it is often enough only to approach them, to drive them away. They will then follow their parents, hopping from branch to branch, and they all assemble in the evening to pass the night together. The whole family perch on one branch, the male at one end, and the female at the other, and the little ones in the middle, all closely pressed against each other. After the first brood these warblers will make a second, and sometimes more, if interrupted.

The male of this species is in great request for the cage. Its song partakes something of that of the nightingale, and its modulations, though not very extended, are agreeable, flexible, and varied, and its tones sprightly and clear. To this it adds an amiability of disposition, not at all common. It is most peculiarly affectionate to the person who has the care of it, and will call him in a particular tone. At his approach, its voice grows more expressive of affection. It will dart towards him, against the wires of its cage, as if to break through this obstacle, for the purpose of joining him, and by a continual clapping of the wings, accompanied with little cries, it seems to express eagerness and recognition. Such is the picture drawn of it by Olina, and it is of this warbler that Mademoiselle Descartes has said,

"N'en déplaise à mon oncle, elle a du sentiment."

These birds are procured in various ways. The young ones taken towards the months of August and September are generally preferred. Their song is said to have more melody, and a greater analogy with that of males in a state of freedom. To accustom them to the cage, the extremities of their wings are tied, and they receive the same aliment as the nightingale, with tender fruits, and even with apples and pears. When it is desired to bring up the young ones from the nest, they must be taken when half fledged, that is, about eight or nine days after their birth, and fed like the young nightingales. They must be kept extremely clean, on dry moss, regularly renewed twice a day. They may, moreover, receive a liquid paste, composed of yolk of egg, bruised hemp-seed, and crumbs of bread. When they can eat alone, parsley, hashed very fine, is added to this, and the whole receives a greater consistence. This diet, however, is sometimes apt to fatten them too rapidly, and thus occasion death. The hemp-seed has peculiarly this quality. This may be corrected, by giving them pears or apples, cut in two, figs, grapes, and other little fruits to which they are partial. During winter they must be kept in a warm place; it is sufficiently so if their meat and drink be not frozen. It is said that their song may be improved, by placing them within hearing of the nightingale. At the epoch of migration, in autumn, these captured warblers are very much agitated during the night, especially at the full of the moon, which causes a great number of them to perish. This torment continues until November, after which they are tranquil until the same season the following year. This desire of voyaging does not quit them till after some years of captivity.

They have been preserved in cage for ten years; but the ordinary duration of their lives is from five to six. With care they may be brought to nestle in captivity. For this purpose they should be kept in a garden, and the aviary should be provided with evergreens. During winter they must be kept in an apartment.

This warbler is common enough in Europe, both south and north, but is rather rare in this country.

The *Orphæa*, or warbler proper, is found in almost all the temperate countries of Europe, inhabits gardens, and the borders of woods, and makes a nest in the bushes, in which the female lays from five to six whitish eggs, with points and spots of a greenish grey.

The *Babbling Warbler* (*Mot. Curruca*, Lin.) does not frequent gardens, unless there are very thick woods in their proximity. It delights in coppices of three and four years old, and prefers the thicket and most solitary places, where the male, without quitting his favourite retreat, sings in a manner that has some analogy with the style of the reed wren. As there are no intervals between the burthens of the song, it is probable that the epithet attached to this warbler has been derived from this circumstance.

The babbling warbler constructs its nest in the middle of the thickest bushes it can find, places it at about three or four feet from the ground, employs many more materials, and gives the nest more depth and thickness than do the other warblers. It first of all puts stalks of coarse plants negligently at the base and sides, to which are added finer plants, interlaced with a little wool. The eggs are from four to six in number, white, glazed with a clear grey, punctated with olive and black on the middle, with spots of the former colour numerous and irregular towards the gross end.

The *Passerine Warbler* inhabits Lombardy, Sardinia, France, &c.; comes later into the latter country than other warblers. These birds sojourn in coppices, groves, orchards, and often in gardens, even in the midst of the most populous cities, provided there be there hedges of yoke-elm, with trees of a certain elevation. On this account the name of *hortensis* has been applied to it, which, however, has also been given to another warbler by Latham and Gmelin. It is called *aedonia* by M. Vieillot, from the beauty of its song. The fowlers of Paris call it *bretonne*. It does not frequent thickets, but is fond of perching on the top of middle-sized trees. It is always lively, perpetually in motion, and sings even while in search of food. Its voice has less brilliancy of tone than that of the atricapilla, but it is equally melodious, and the burthens of its song seem to possess greater variety. Its cry, when it is disturbed, is the same, and it frequently repeats it when alarmed for its offspring.

The nest is usually exposed on the hedges above-mentioned, or on large shrubs, and is rarely found elsewhere. It is of no great consistence, loosely constructed, so that the light appears through the interstices, composed externally of the stalks of plants, and furnished with horsehair within. It is neither very large nor deep, the eggs are four in number usually, marbled with two shades of brown, on a ground of sombre and dirty white.

The Hawk-like Warbler (S. Nisoria) is found in Germany, and in Piedmont, on its passage. It frequents coppices on plains, hedges, and bosquets which surround meadows, or border on them. Its nest is usually in the thickest bushes; it is composed externally of plants and small roots, internally of horn and cattle-hair. The eggs are four, of a whitish gray, sown with irregular and confluent spots, of a reddish ash, and leaden gray. The cry of this warbler resembles the sound with which the nightingale preludes its song; it is the least agile of all the warblers; its motions are awkward and heavy, but its flight is extremely rapid. During the season of re-production, it is observed to elevate itself in a right line in the air to the height of fifteen or twenty feet, the head being raised, and the tail perpendicular; it then stops, descends slowly, clapping its wings, and flutters for a moment above the bush which it has just quitted.

The ACCENTOR, or Alpine Warbler, made the type of a new genus, by Bechstein, was long ranged under the head of Sylvia. It differs, however, from the birds of that genus, in habits, manners, and entire mode of life. This bird, which Buffon termed, "*fauvette des Alpes*," is called in the mountains of Upper Comminge, *pegot*. The word *pee*, in the vulgar language of the country, means imbecile; it inhabits the Pyrenees and the Alps; it ordinarily chooses the most elevated and solitary points of those arid mountains; its nest is circular, and composed of moss and grass; it fixes it in the sheltered hollow of a rock, for it seems to be afraid of the north wind, and always remains in a southern aspect; the eggs are green and five or six in number.

These birds never quit their beloved mountains; but when, in winter, tempests and hurricanes arise, then they precipitate themselves in flocks into the vallies, take refuge in the infractuosities of rocks, or retire behind the shrubs which grow in their clefts and hollows; they are either so terrified, or so dull, that they give into snares of every kind, and even serve as a sport to the children, who kill them continually with stones. Travellers frequently meet them on the summits of the mountains, either perched on the ground, two by two, or climbing along the rocks with the assistance of their wings; they are either so confident or so stupid, that the sight of man does not frighten them; they will suffer themselves to be approached very nearly.

It does not appear that these birds have been retained in a state of captivity, or even that any one has ever heard their song. M. Picot La Peyrouse, who has observed them much, never even heard them utter a cry; they are both granivorous and insectivorous, but more especially the latter. Buffon says, that they remain commonly on the ground, where they run very fast, after the manner of quails and partridges, and not hopping like the other warblers; that they also perch on stones, but rarely on trees; that they go in small troops, and have a rallying cry like that of the water-wagtail.

This species inhabits not only the Alps and Pyrenees, but

also the high mountains of Persia. It seems necessary to notice that it has been three times repeated by Gmelin and Latham, as distinct species. By the former under the names, *motacilla alpina*, *sturnus collaris*, and *sturnus mauritanus* ; by the latter as *collared stare*, *Persian starling*, and *Alpine warbler*.

The Winter Warbler, or *Accentor Modularis*, is classed in this genus. This bird has been known under a variety of denominations, which it is needless to recapitulate ; it seems to be remote from the warblers in its mode of life and habits, to have less gaiety and vivacity, and a song more feeble, plaintive, and monotonous. It is heard most usually, in the morning and evening ; it then perches on a middle-size tree, or the top of a shrub ; its song is agreeable in a season of the year, when all other birds are silent. It has also a little trembling cry.

Nature, always provident, having destined this bird to pass the winter in our northern climates, has clothed it much better than the other warblers, and given it a better furnished plumage. In autumn these birds appear in great numbers near habitations ; all at this epoch, quit the woods, their summer domicile, and spread themselves in hedges and bosquets which neighbour on gardens. When the cold grows rigorous, they approach houses and particularly barns and lofts where grain is thrashed, seeking in the straw for little insects and small grains ; every aliment suits them at this time, for corn has been found in their crop, though they are usually insectivorous. They may be fed with hemp-seed, which they will swallow entire, as pigeons do ; but it is only necessity which makes them granivorous. As soon as the cold relaxes they remove from houses and barns, and remain in hedges and bushes, seeking the chrysalides on the branches, and the little insects which are benumbed under the moss ; at the approach of the fine days they remove still farther, retire to the borders of woods, and eventually penetrate into the thickest recesses.

Such is the mode of life of these birds in the more southern provinces of France. But in the more northern, as Normandy, for instance, some of them always remain near habitations, if there be hedges of green trees, &c., where they can nestle. In the woods they prefer the thickest bushes; they are so little wild that they will make their nests in an orangery if they can enter, and will even hatch in an aviary, if it be furnished with tufted shrubs.

This is one of the earliest of the sedentary birds to announce the return of spring, and commence its amorous music. From the earliest days of March, the male and female are observed to be engaged in composing the cradle of their young. They usually place the nest at a moderate elevation, but always in a secret situation. It is formed with a considerable quantity of moss, especially at the base and sides, and it is furnished within with wool, horse-hair, and feathers, softly arranged. On this bed the female deposits four or five eggs, of a handsome clear blue, without spot; the male remains in the neighbourhood, and cheers his companion by his song, at such moments as he does not relieve her from the cares of incubation. The young are born, covered with down; and do not abandon the nest until they are well feathered. When taken from the nest, they are easily reared; and when caught in nets in their youth, are tamed without difficulty. The mother does not abandon her eggs, although they may be touched, and she shows much attachment for the young; she understands how to lead the enemy astray which seeks for their destruction. Like the partridge before a dog, she throws herself before a cat when it approaches, and hovers on the ground from place to place, until it is sufficiently remote. The cuckow, according to Latham, frequently lays in the nest of this bird.

This species is found in all parts of Europe, but more frequently in northern climates. It is easily taken in snares of every description.

C. Hamilton Smith Esq. del.

Griffith sc.

BLUE THROATED MALURUS *of Vieillot.*

MALURUS GULARIS —— Vieil.

London Published by G.B. Whittaker, March 1828.

1 DWARF WARBLER

ACANTHIZA NANA. Vigors.

2 EXILE WARBLER.

MALURUS EXILIS.

London. Published by G.B. Whittaker, July, 1826.

The *Accentor Montanellus* of M. Temminck inhabits the eastern parts of the south of Europe, and under the same latitude in Asia. It was found by Pallas in eastern Siberia, and in the Crimea. It is rare in the Neapolitan States, Dalmatia, and the South of Hungary. It always lives in the mountains, and never shews itself in the plains except in winter. In summer it is insectivorous, and probably will eat grain in winter. Its mode of propagation is unknown.

The word *malurus* is used by Vieillot to designate as a genus some species of the warblers distinguished by a gauze-like tail. The opposite figure is of a bird which seems referable to this group. The whole upper part is dusky brown, the throat brown, the tail and streak over the eye are azure blue, the remainder of the bird is bright reddish brown; the tail feathers, only five, are spread and very thin, the middle is the longest, and the other two on each side decrease in length successively.

The figures of the Dwarf Warbler, and Exile Warbler, are from specimens in the Linnæan Society, and are described at pages 469 and 470.

The first bird in our author's division of Regulus, is the *common Gold-crested Wren. Motacilla Regulus* of Lin. This, the smallest of our European birds, must not be confounded with the common wren, which belongs to the baron's division Troglodites. The latter is a little larger and more bulky, and as it seldom quits our rural habitations, is much better known than the other, which only inhabits the woods, and is seen only in autumn and winter. The gold-crested wren is so small, that it passes through the meshes of common nets, and easily escapes from all cages. A leaf is sufficient to conceal it from the most piercing sight, which may probably be the reason of its appearing more rare in summer than it really is, for at the fall of the leaf these birds are seen in tolerable numbers; and when the trees are totally defoliated, they are found in small troops of from ten to twelve, which join titmice and other little birds for the purpose of migratioa.

The gold-crested wrens are discoverable by a small sharp cry, which has much resemblance to that of the grasshopper. They are fond of oaks, elms, lofty pines, fir-trees, and willows. They are not at all distrustful, and may be approached and killed with great ease.

These birds live on similar food with the titmice, and are very analogous with them in their habits. Their mobility is extreme; they fly incessantly from branch to branch, climb trees, and hold themselves indifferently in all positions, and often have the feet upwards. The smallest insects constitute their ordinary nutriment; sometimes they take them on the wing, at others seek them in the clefts of the bark, or the heaps of dead leaves which remain at the end of the branches; they also eat larvæ and all kinds of small worms; it is said that they will eat the berries of evergreens; be that as it may, it is quite certain, that they are fonder of those trees than any others, probably because they find there a greater abundance of insects which constitute the basis of their aliment. They grow fat in autumn, and their flesh is then good eating; notwithstanding their smallness, Montbeillard tells us that the markets of Nuremberg abound with these little birds; many are taken in the environs of that city, and usually by the bird-call.

These wrens are seldom seen in France, but in the after season. They scarcely ever nestle in any of the French provinces; but in summer they are abundant in the woods of Germany and England. Their nest, curiously constructed, and suspended to the extremity of the little branches of pines and other trees, is tissued without with moss, wool, and spiders' webs, and furnished within with the softest down; this nest is spherical, and its aperture is at top. The female lays from six to eight eggs, about the size of peas, of a yellowish brown, without any spot, according to some naturalists, but of a pale flesh-colour, undulated with a deeper shade, according to Meyer. M. Temminck declares them to be a rose-coloured white; but Dr. Latham says they are of a brownish white! The nest is also found sometimes suspended to the extremity of a bundle of ivy, which escapes from the branches of a tree, or from a wall; but it is always concealed in the foliage. When the female hatches, the male sings a short song, not without agreeableness; he also relieves her in the middle of the day from the cares of incubation.

This species is extended throughout all Europe, from Sweden to Italy. It is also found in Asia, as far as Bengal, and even in the United States of America, where it principally frequents the northern provinces, not advancing southwards, except in the autumnal season of the year.

M. Vieillot says that there are two races of this bird, one, which has been described, and another which he calls crested wren, with mustachios; the cry of the latter, according to this naturalist, is stronger than that of the other. It remains at the summit of the highest trees, and is seen only in couples, male and female. It is also more distrustful and approached with greater difficulty; neither is it met with precisely at the same period as the other.

The Yellow Wren Warbler (M. Trochilus, et Pouillo of Buffon) is a very small bird, spread universally throughout Europe, as far north as Sweden. It is with us a bird of passage, coming in spring and departing in autumn, proceeding doubtless towards more southern climates. It is known by various names in the different countries which it inhabits, derived either from its habits or its cry, which is nothing but the frequent repetition of the monosyllables *tuit, tuit.*

During the fine season, this bird remains in the woods, where it lives on small insects, such as gnats, &c. It fixes its nest in tufted plants or bushes. This nest is composed with much art and care. Externally it is constructed of moss, and internally furnished with horsehair and wool. It has the form of a little ball, and no other aperture than a hole in the side, which the female takes care to close when she is obliged to quit the eggs or the young ones, to provide for her own sustenance, or any other wants. The brood consists of four or five eggs, a little larger than peas. They are white, picked out with reddish. The young ones do not quit this cradle of their infancy, until they are able to fly as well as the father and mother, and to follow them in their wandering courses.

In autumn these little families quit the woods, and extend themselves in our orchards and gardens, where they find subsistence until their departure, which now draws near. Then these birds are seen in perpetual motion, and even when they do take a moment of repose, a sort of trembling is observed in the tail.

This bird is not more bulky than the gold-crested wren, but it has a more variegated and elegant shape.

The opposite figure is from a specimen brought from Chili by Lord Byron, and presented by him to the British Museum. Mr. Gray has named it from the noble donor, *Byron's Golden Crested Wren*, to distinguish it from the European species.

The upper part of this bird is dark green; on the top of the head is a bright red streak, and over the eye is a broadish lunated yellowish white streak; the breast and belly is yellow; the quills of the tail and wings are black, and a band of that colour, broader at the sides than the middle, passes round the breast; the smaller wing coverts and the throat are white.

The *Common-Wren* is placed in the subdivision TROGLODITES, by our author, a name derived from the Greek, and signifying an inhabitant of clefts and caverns.

After the gold-crested wren, this is the smallest of our European birds, being little more than four inches long. It is known under as many different names as it inhabits different countries.

The wren lives on worms, on flies, and other small insects.

GOLDEN CRESTED WREN.

REGULUS BYRONENSIS.

London. Published by G.B. Whittaker. May. 1828

In the summer it remains in the woods where it constructs its nest, near the ground, sometimes on the ground, and sometimes in the shelter of some rock, to which it attaches it. This nest, of a spherical form, seems externally nothing but an unformed heap of moss, which causes it easily to escape search; but internally, it is arranged with great neatness. It has but one narrow entrance, situated on one side, and always diametrically opposite to the wind which most usually prevails in the adjacent mountains. The female lays nine or ten eggs, of a tarnished white, with a zone of reddish points towards the gross end. It is not uncommon, in spring, to find in their nests the young of mice or field-mice, who have taken possession of them.

At the approach of winter this pretty little bird quits the woods, and draws near to the habitations of men. It introduces itself into the clefts of walls, where it may be seen entering and coming out with precipitation, incessantly agitating its wings with a rapid tremulous motion, and always holding its tail in an elevated position. It accompanies these movements with a little cry constantly repeated. Its song is also soft and piping, and the more agreeable from the general rarity of feathered music during winter. The wren gives more animation to his song in proportion as there is a greater abundance of wind, and accompanies his singing with a little vibration of the tail from right to left.

The wren, as well as the red-breast, is one of the least distrustful of our birds. It is naturally very curious and inquisitive. The sight of man causes in it no fear. It suffers itself to be approached very nearly, and will flutter for some time along the hedges, at some paces in front of the traveller, which would create a belief, that it was fond of preceding him. It is true it is rarely pursued, and in many places the people scruple, not only to kill it, but even to touch its nest; for the wren, as well as the red-breast, is held in a sort of veneration by the lower classes, highly favourable to their preservation, and "it must be a most wicked and mischievous boy," says Dr. Latham, "who will not pay some sort of deference to a very trite proverb, viz.—'The robin and the wren are God Almighty's cock and hen.'"

This species is tolerably extended throughout Europe. But the winters of the north are too rigorous for its constitution. It is seldom seen, says Linnæus, in Sweden and the North of Russia. It is, however, reported to have been found at Oonalashka; but it is more than doubtful if the species be the same.

There is a bird in North America, called the *Winter Wren*, (*T. Hyemalis*), which if not of the same species as the last, has certainly very striking analogies with it. Its plumage, conformation, song, and mode of life approximate very closely to those of our European wren. It arrives in the central parts of the United States in autumn, and remains there during the mild winter. It frequents the backs of ditches, ravines, old deracinated stumps, small bushes, and the brambles in aquatic places. It is often observed in rural habitations, where it conceals itself in piles of wood; but at the end of the bad season it returns northward. According to Wilson, this wren nestles in the mountain-forests of upper Pennsylvania; others imagine that it does so in countries still more northward, as does the ortolan, and several other birds, which are only seen in winter in the United States.

The *Brown Warbler* of our text, *House-wren* of Wilson, *Troglodytes Œdon.* Vieil., is another North American bird. Like our common wren, it seems to take great delight in approaching the habitations of man. It is sufficient to procure it the advantages which the building of its nest requires, to be certain of attracting it into a garden, and making it nestle there every year if its brood remains untouched. It merits, in all respects, the attention bestowed upon it by the Americans, for it is in no wise hurtful, living only on larvæ, chrysalides, and small insects, and is the only singing bird which establishes itself in cities. Its song is as strong and sonorous as that of the chaffinch, but more mellow, varied, and extended. In consequence of this it is known by the name of the nightingale of North America. The Americans, who have not this bird near their dwelling, and are desirous of fixing it there, attach a calabash against the house, or at the end of a perch in the middle of their garden. Others construct, for the same purpose, a little house, attached in like manner to the end of a perch. This little nook rarely remains untenanted; for the young couples, on their return from the south, being forced to seek a district where they may be isolated from their fellows, immediately take possession of it. In default of these artificial retreats, they make their nest in the hollow of a tree. Every place which is close and obscure suits them best. Filaments of roots, cattle-hair, moss, fine plants, &c. constitute the materials which this species employs without much art, and heaps together without much order, as do most of the birds that nestle in hollow trees. Its first brood consists usually of from six to eight eggs, white, or flesh coloured, and spotted with a purple red; the second is less numerous, One takes place at the arrival of the bird in the month of May, the other in July.

This species is extended through all North America, from Canada to Louisiana.

The *Arada* has been placed at the end of the ant-eaters, by Buffon, and in our tabular view; but as it has great analogies with the wrens, we chuse to notice its habits here. In these it differs from the ant-eaters; it is solitary, perches on trees, and never descends to the earth, but for the purpose of catching ants, and other insects, of which it also eats. It differs from them still more by its song, which is peculiarly fine; whereas all the ant-eaters utter nothing but cries or sounds, totally destitute of all modulation.

The traveller, who wanders through the immense and

solitary forests which cover almost the entire soil of Guiana, is at first struck with the gloomy silence which prevails in the depths of these sombre retreats, which are nevertheless peopled by a crowd of animals of every class, and every genus. The more he penetrates into their interior, the more general this silence becomes; animated nature appears mute, or, if the monotonous uniformity should be interrupted by any sound echoing from afar, the impression produced on the senses and the mind is disagreeable and painful. Sometimes the ear is struck by the horrible howlings of the alouatta; sometimes by the alarum of the chiming thrush; sometimes by the sudden stroke given with the tail by the great adder; and sometimes by the startling and reiterated crash of many falling trees, which, tumbling one over the other, break in rapid succession, causing an instantaneous clearance in the midst of the most magnificent plantations of nature. One sound, however, more singular than the others, will occasionally arrest the attention of the wanderer through these mighty woods. Removed to the distance of many leagues from every human habitation, his ear will be suddenly saluted by a whistle like that of a bandit calling to his brothers of spoil. This whistle will be repeated, and the traveller will believe that he is approaching one of those wild settlements, which the desire of liberty or the tyranny of the colonist has forced the fugitive negro to form in the depth of almost impenetrable forests, or in the distant solitude of nearly inaccessible mountains. Advancing towards the point from which the sounds appear to issue, he will find them to recede; but should he approach unperceived within sufficient distance he will discover, to his astonishment, that this whistling is not produced by a man, but by a bird, though nothing can be more perfect than the resemblance. Neither will he be long in perceiving that the same bird has a most melodious song, and that the whistler is also a most agreeable musician. Its song is less varied and less brilliant than that of the nightingale, but more grave, touching, and tender, and more resembling the mellow sounds of a soft-toned flute. It is modulated on different keys and accents, to which the seven notes of the octave, which the bird delights to repeat, serve, in some sort, as a prelude. In those warm climates, where the young broods take place several times in the year, the song, which is only the expression of love, lasts longer than in cold or temperate regions. This constitutes a very decided advantage in favour of the *arada* over the nightingale. From its song, the name of *musician-thrush* has been given to the *arada;* but it certainly does not belong to the thrushes: and as the epithet musician has been applied to other birds of different genera, the native name of *arada* was preferred by Buffon. The species is rare, and avoids the neighbourhood of inhabited places.

The *Buenos Ayres Wren* (*Sylvia Platensis*) has great analogy with the Œdon, already described; it is known by the Guaranis under the name of *Basacaraguay.* But at Buenos Ayres, they called it a mouse, from its cry, and its habit of gliding, especially in winter, under roofs and into the crevices of walls, and holes of trees, and of entering sometimes into houses, to catch spiders and other small insects. This bird never frequents the plains or forests; it remains in thickets, on the edge of woods, in enclosures and rural habitations. It hops lightly along the ground, holding the tail almost always erect, and appearing by no means frightened at the approach of men. The male sings all the year, and in the season of love accompanies its song with a clapping of the wings; the female replies to the male, by a single cry, thus, *chi,* low and tender. The voice of the male is always elevated, clear, and agreeable; its song consists of eight or ten syllables, pronounced quickly, and repeated at intervals, and often for a long time together; its rhythm is not unlike that of the nightingale, but its phrases are neither so varied nor so expressive.

It nestles in holes of walls, but in inhabited places, which it constantly frequents. It places its nest on the beams of the scaffolding of houses, and, more usually, in the apertures which they leave in the walls. M. Azara discovered a nest of one of these birds on the ground, in the cranium of a dead cow, and another under a roof. It is composed of feathers, and blades of straw, and furnished internally with abundance of horsehair. The eggs are four, red at the gross end, and speckled in the remainder, with the same colour, on a white ground.

We insert a figure of a wren from a specimen in the Langsdorf Collection, drawn by Major Hamilton Smith, which though it appears to have some affinity to the Great Carolina Wren of the American Ornithology, the Barred-tailed Wren of Latham, may nevertheless be distinct.

The bill is very long, and nearly straight and slight; the top of the head, neck, small wing coverts, and upper half of the back, are of a dirty brown colour; the larger wing coverts tail, and lower part of the back, reddish brown, or brick-dust colour; the forehead and vent are ashy, but a white streak passes from the anterior corner of the eye over it, and is continued to the bottom of the side of the neck; the throat is white, the breast is light yellow, and the belly reddish brown; the large wing coverts and tail feathers have several black bars passing on each side from the shaft, not quite rectangular in their direction, but inclining towards the tip of the feathers.

The genus THRYOTHORUS of Vieillot, consists of a small group of species which inhabit only North and South America. There is so great an analogy between them, that description is scarcely sufficient for the purposes of accurate distinction. The differences, however, which characterize them are easily observable in nature.

The Great Carolina Wren (*Thryothorus Littoralis*, Vieill.) is found constantly in the month of May on the banks of the

C. Hamilton Smith Esq.e del.t

Langsdorff Col.

THE LONG-BEAKED WARBLER.

TROGLODYTES LONGIROSTRIS.

London. Published by G. B. Whittaker, March 1828.

Delaware. Yet it is seen rarely in Pennsylvania, and still more rarely in the state of New York. It is very frequently met with on the banks of James river. It seems to delight in the obscurity of the cypresses that border the marshes, in profound caverns, and in piles of timber fallen together from age, in the neighbourhood of rivers, and small streams. It has all the habits of our wren. It conceals itself in holes, in crevices of the earth, and is perpetually in motion. It appears and disappears every moment. Its cry, which it utters from time to time, is loud and strong, not unlike a burst of laughter, and seems to express, according to Wilson, the word *chirr-up*, the first syllable being lengthened and strongly dwelt on. It has another song, but much more soft and musical. It sounds something like our English words, *sweet William, sweet William.*

The Marsh Wren of Wilson, is another bird of this division (*Certhia Palustris*). It inhabits marshy places ; sojourns in reeds, and prefers those whose roots are bathed by water. It is continually jumping over their stalks, like the *reed*-wren, already described, with which it has additional relation from its song and continual babble. It has not been observed to fix on trees or shrubs. It even seems to avoid fixing on the brambles or bushes which are on the edges or in the centre of its usual haunts. Its song, if indeed such a name can be given to an assemblage of various cries, repeated twenty times in succession, without interruption, and in the same key, is as hoarse and disagreeable as the croaking of the frogs which are its habitual companions, and as troublesome, from its long duration. The slight cracking, says Wilson, made by globules of air forcing their way through a marshy soil, as you walk along, may give you a tolerable notion of this delightful warble. Many couples of this species are found in the same district, and the males seem to take pleasure, like the frogs, in vying with each other in uttering the loudest cries. This work continues, during hatching time, from dawn until midday, recommences some time before the setting of the sun, and continues one or two hours after.

This bird is very common in the marshes near the city of New York. It visits them in the month of May, and quits them on the approach of autumn. As if in compensation for its disagreeable song, nature has endowed this bird with a rare industry, in making a shelter for its young, against every inclemency of the atmosphere. It attaches its nest to several stalks of reeds, and always above the highest waters. The attachments are so solid, that the most violent winds cannot remove them. The form of the nest is that of an elongated melon. Stalks of plants, small roots, and dried leaves, are at its exterior. All these materials are intermingled with mud, forming a sort of wall, which the water cannot penetrate, when it is dried by the sun. The inside of this cradle is furnished with feathers, cattle-hair, and other softer materials. The entry is on one side towards the middle, and is surmounted by a little roof, which advancing a little over it, prevents the rain from coming in. The eggs are five or six in number, very small, and of a deep pewter-colour.

M. Vieillot has enumerated some other species under the genus Throythorus, which other naturalists, with greater propriety, have placed under different divisions of the dentirostral family.

We now come to the division of the Wagtails.

Linnæus comprised, under the denomination of *Motacilla*, a great number of birds with slender beaks, which have subsequently been divided into many genera. Bechstein has restrained the name to the wagtails proper, and budytes, which have more elevated limbs, and a longer tail than the rest, which they are continually lowering and raising. To such the name is more suitable than to any of the others. These birds have, moreover, as distinctive marks, certain scapulary feathers, which, extending to the end of the wing, give them some relation with the majority of the grallæ, and a tail composed of twelve rectrices nearly equal, with the two lateral, however, shorter than the eight intermediate quills.

M. Cuvier has separated the wagtails proper from the budytes, a name derived from these latter birds being frequently seen amongst cattle. There is, however, very great analogy between the two sections. Perhaps our popular name of wagtail is the best to apply to both.

The majority of the wagtails proper, and all the yellow wagtails (*Bergeronnette de Printemps*), migrate from our northern countries at the approach of winter. The *boarula*, on the contrary, comes to pass the winter with us, and quits us when the others return. It is said to nestle in the German districts, which border on the French territories.

All these birds frequent meadows, and humid and marshy places, delighting in the borders of rivulets and rivers. Most of them have an undulating flight. They all run rather than walk; seldom perch, sing, or cry, during their flight; and construct their nest on the ground. That of the white wagtail is, however, sometimes found in a pile of wood, along side of the banks, or in the hole of some wall whose base is washed by waters. Insects and small worms are their only aliment.

The *White Wagtails* (*M. Alba*) have a mode of life peculiar to themselves, and habits which distinguish them from budytes. They more readily approach man and his habitations, being fond of nestling in our neighbourhood. The others, more wild, inhabit the vicinity of the meadows and isolated herbaceous tracts. The former prefer stagnant waters, and the latter delight more in the borders of springs and running streams. Both run with the cattle, fly about the labourer, accompany him in his rural labours, and follow the plough in pursuit of small worms and larvæ, of which the newly turned furrows present a vast abundance. These insectivora, as useful as the fly-catchers and swallows, sometimes in the flight, but more frequently on the ground, amidst the herbage, seize upon the flies and gnats which have escaped the murderous bills of their other pursuers in the air. All the insect population of ponds and marshes constitute the nutriment of these volatiles. Their slight forms, little head, delicate feet, and long tail, perpetually balanced, cause them to be at once distinguished from all other birds with slender bills. They are therefore with great propriety formed into a small distinct family.

The *Motacilla Alba* is spread throughout Europe. It is even seen in Siberia, Kamschatka, Iceland, and the Feroe islands. It also inhabits Africa and India.

They form in autumn numerous flocks, which extend themselves through the fields, and withdraw, on the approach of evening, into osieries and willows which border canals and rivers. There they perform a noisy concert until night-fall. They depart in October, and often at this period they are heard passing in the air, sometimes at a very considerable height, and clamouring to each other incessantly. They do not, however, all migrate at this season, for some, though a very few, are occasionally to be met with. They then abound in Egypt, where the people, says Maillet, dry them in the sand, to preserve them for the purpose of food. They are also to be seen in Senegal at the same season; but, like the swallows and quails, they disappear from thence in spring to return to our climates, where they arrive at the end of March.

These birds possess the most astonishing gaiety and lightness. They appear in flying to rest upon their long outspread tail, as upon a broad oar which assists them to balance, spring, and perform a variety of evolutions in the air. During such sports, they are frequently heard to utter a little cry, lively, clear, and redoubled, which sounds like the syllables *guit, guit, guit, guit, guit*. They have also a soft and delicate song, which, in autumn, is reduced nearly to a murmur. "Encore," says Belon, in his old French, "savent *rossignoller* du gosier melodieusement, chose qu'on peut souvente fois ouïr sur le commencement de l'hiver." The motion of their tail in flying is horizontal, but on the ground its position is perpendicular. As they delight in being upon the edge of the water, and often approach the washerwomen that are there, and seem to imitate with their tails the beating of the linen, the French have given them the name of *lavandières*. They run lightly, with very nimble steps, upon the strand, and their long legs enable them even at times to enter the water to a small depth; but they are usually seen placed upon the stones and other little elevations about it.

The wagtail fixes its nest on the ground, under some roots, or below the turf; more frequently at the edge of waters, under some hollow bank, in elevated piles of wood alongside of rivers, and sometimes in heaps of stones. It is composed of dried herbs, small roots, and moss, connected carelessly together, and it is furnished inside with horse-hair, and feathers in abundance. The eggs are from four to six in number, of a bluish white, spotted with brown. There are usually two broods in the year. The male relieves the female during some hours in the day from the labour of incubation. The little ones are born covered with down. The father and mother defend them with much courage when they are approached. They meet the enemy, fly about to lead him to a distance, and often succeed in deceiving him by their manœuvres. If their young family is carried off, they fly about the head of the ravisher, turn incessantly, and continually utter piercing cries. It has been remarked that they attend very scrupulously to their young, keeping the nest extremely neat, and cleansing it carefully from all kinds of filth and ordure. They fling these out, and even carry them to a certain distance. This last precaution seems to be the result of a different instinct from that of mere cleanliness. It would seem to be done rather with the view of removing every indication of the proximity of their nest. Many other birds use a similar precaution, especially during the first ten or twelve days after the birth of the young. They even carry off the eggshells when the young are evolved, and take them to a considerable distance. This habit is so innate in birds, that even canaries, which, in a long lapse of captivity, one would imagine, would leave it off, take the shell, the moment the little one comes out, and either transport it to the dung which is in that part of their cage the farthest from the nest, and conceal it there, or else break it to pieces and swallow it.

When the young family is in a state to fly, the parents still conduct and feed it, for three weeks or a month. This is a period in which they wage incessant war with the insect tribe, seizing and devouring them with the most extraordinary quickness, without appearing even to give themselves time to swallow them. They collect the little worms on the ground, gorge themselves with the eggs of ants, and often make turns in the air, to catch the flies and gnats.

The wagtails are not distrustful, and are less fearful of man than of the birds of prey. They are not even much frightened by fire-arms, for, on being aimed at, they do not fly far, and frequently return and place themselves within a short distance of the fowler. They give into all kinds of snares which are laid for them, quite easily; but if taken when adult, they cannot be preserved in cages, but will die in four-and-twenty hours. For this purpose, they must be taken from the nest, and reared like the nightingales.

The *Motacilla Flava* is, of migrating birds, one of the earliest which re-appear in spring, and one of the latest which depart in autumn. In the southern provinces of France, many remain during the winter. In autumn, they assemble in numerous troops. They more willingly frequent elevated and cultivated soils, where they seek a more abundant nutriment, and find it more readily in the track of herds and flocks, with which they love to associate. From this last circumstance, these birds are called, in French, *bergeronnettes*, to which, from their early appearance in spring, the epithet *de printemps* is added. In autumn, their flesh acquires a delicacy which causes them to be objects of research; but it is far inferior to that of the becafico.

These birds, which do not avoid man, but rather seem to take a pleasure in his society, cannot, nevertheless, support a state of slavery. They die as soon as they are shut up, but in a large enclosure, they will soon familiarize themselves, and afford some amusement by their activity and dexterity, in seizing flies and other insects. They will not live in a state of captivity, even when taken from the nest, more than three or four years.

This species is spread through Europe. They fix their nests in meadows, and sometimes at the edge of the water under the root of a tree. It is composed of dry herbs and moss without, feathers in abundance; horse-hair and wool inside. Six or eight rounded eggs compose the brood, of a dirty white, shaded with green olive, clear brown, and flesh-colour. The male partakes with the female the construction of the nest, and the hatching of the young.

The *Pipits*, or *Field Larks*, have been separated from the larks by M. M. Bechstein and Meyer, and formed into a peculiar genus, under the name of ANTHUS. Their habits have much analogy with those of the larks proper, though they differ in certain details of conformation. Like the larks, they sing in flying, and elevate themselves to a certain height in the air. They seek their nutriment, nestle, and sleep on the ground. Some frequent cultivated fields and meadows; others delight, during the summer season, in the borders of woods, in glades, in furze, and brushwood, thinly scattered; many prefer mountains, steep shores, rocks, and maritime pastures. Some few, in fine, inhabit, during summer, the little hills in sandy and stony situations, and during the after season, sojourn on the banks of rivers, and seek their food upon the strand. A very small number have the power of perching constantly upon trees. There is considerable trouble in distinguishing them specifically.

The *Anthus Sepiarius*, called by M. Vieillot, *pipi des buissons*, (bush pipit) is the smallest of the genus. The male sings when flying, and uses much action. He erects himself, half opens the bill, spreads the wings, and every thing announces that his is a song of love. The song is simple, but soft, harmonious, and clear. Both the male and female send forth a cry when flying, and when disturbed, which very well expresses the syllables, *pi*, *pi*, *pi*, *pi*, repeated three or four times successively. From this, this bird, and the entire genus, derive their name, which, in French, is *pipi*, in English, *pipit*, in German, *piep*, and in Danish, *pibe*.

There is nothing more in the detail of the habits of the other species, sufficiently interesting or sufficiently known, to detain us any longer upon them.

We shall now resume so much of the text of Cuvier, as relates to the Fissirostral Family of the passerine order of birds, and shall, as heretofore, insert therein such additional species as have been named, without pledging ourselves, and much less the Baron, for the propriety of treating them respectively as distinct.

The field larks would conduct us directly to the larks; but we are obliged first to speak of a small family allied to this by the fly-catchers. It is that of

THE FISSIROSTRES,

A family not numerous, but very distinct from all others, by the short broad beak, flatted horizontally, slightly crooked, without indention, and very deeply cleft, so that the opening of their mouth is very wide, and they easily swallow the insects which they pursue upon the wing.

They are most closely allied to the tribe of the fly-catchers, especially to the procnias, whose bill scarcely differs from theirs except in its indention.

Their regimen, exclusively insectivorous, eminently constitutes them migrating birds which quit us in the winter.

These birds are divided into diurnal and nocturnal, like the birds of prey.

THE SWALLOWS. HIRUNDO. *Lin.*

Comprehend the diurnal species, all remarkable for their close plumage, the extreme length of their wings, and the rapidity of their flight.

Among them, we distinguish

THE MARTENS. APUS. CUV. CYPSELUS. Ill.

Of all birds, those which have the longest wings in

proportion, and which fly with the greatest rapidity. Their tail is forked; their feet, very short, have this very peculiar character, that the thumb is directed forward almost like the other toes, and that the middle and external toes have each but three phalanges like the internal.

The shortness of their humerus, the breadth of its apophyses, their oval furca, their sternum without indention towards the bottom, indicate, even in the skeleton, to what a degree these birds are organized for a vigorous flight; but the shortness of their feet, joined to the length of their wings, prevents them when they are on the ground from taking their spring. Accordingly, they pass, so to say, their entire lives in the air, pursuing, in flocks, and with vociferous cries, the insects in the highest regions. They nestle in the holes of walls and rocks, and climb with rapidity along the smoothest surfaces.

The common species, *Hirundo Apus*: L. En. 542. 1;

is black, with a white throat.

The species of the high mountains. *Hirundo Melba*. L. Edw. 27. Vail. Afric. 243.

is the largest; brown above, white beneath, with a brown collar under the neck.

This is the *Swift* of Latham. The *Micropus Murarius* and *Brachypus Murarius* of Meyer, and the *Cypselus Murarius* of Temminck.

The white bellied Swift of Latham. The greatest Martin. Edw. F. 27. The *Hir. scopol. Micropus Alpinus* of Meyer, and *Cyp. Alpinus* of Tem. forms the variety *C. Alpinus Africanus* of Vail. O. A. 243.

Chinese Swift, Hirundo Sinensis. Gm.

Brown; beneath, greyish red; crown red; throat and orbits white. China.

H. Leucothea. Vaill. Ap. 244. 1.

Black; sides of rump white. Cape of Good Hope.

H. Velox. Vaill. Afr. 244. 2.

Blue black, beneath white; tail forked. South Africa.

Cypselus mystaceus. Less. et Garn. Zool. de la Coquille. F. 22.

Head, tail, and back, blue black; whiskers, above and below the eye, white; neck, chest, back, and abdomen, brownish slate; wing and tail coverts white. New Guinea.

Cyp. comatus. Tem. Pl. col. 268.

Green bronze; wings and tail blue, cheeks ochraceous red, with two white whiskers above and below the eye; crest of long feathers. Sumatra.

Cyp. longipennis. Tem. Pl. col. 83. f. 1. *Hir. Klecho.* Horsf.

Green black; beneath, ash coloured. Java.

The above species are admitted by the Baron. The following have been named.

Cyp. gutturalis. Vaill. Ap. 243. f. 1.

Brown; throat white; lower tail coverts white edged; feet and toes feathered. South Africa.

Cyp. Caffer. Licht.

Sooty; forehead and eyebrows white; back and belly black; throat white. Africa.

Cyp. parvus. Licht.

Mouse coloured; throat whitish. Tail forked; outer tail feather very long. Nubia.

Balasian Swift. Lath. *Cyp. Balasiensis.*

Dull brown, with the outer toe versatile. India.

THE SWALLOWS PROPER. HIRUNDO. CUV.

Have the toes of the feet and the sternum arranged like the majority of the passeres.

Some have the feet clothed with feathers even to the claws; their thumb too shows a slight disposition to turn forwards. Their tail is forked, and of moderate size.

The Window Swallow. *Hirundo urbica.* Lin.
Enl. 542. 2.

Black above, white beneath, and on the crupper. Every one is acquainted with the solid nests which they construct of clay in the angles of windows, under the ledges of roofs, &c.

White coloured Swift. *Hir. Cayenensis.* Gm.
Pl. Eul. 752.

Blackish violet; head black; eye-band and thighs white. Cayenne.

Hir. Ludoviciana. Cuv. Pl. Eul. 725. Catesby I. F. 57. *H. Cayanensis,* var. Vieil. *H. Cinerea.* B. Lash.

Grayish black. Louisiana.

Crag or Rock Swallow. Hir. rupestris. H. Montana. Gin. Stor. Deg. IV 409, young Vail. O. A. 246.

Pale brown; beneath whitish, quills and tail blackish; under edge of tail feathers with an oval white spot.

Others have the toes naked, the tail forked, and the bifurcations often very long.

The Chimney Swallow. Hirundo rustica. Lin. En. 543. 1.

Black above; forehead, eyebrows and throat red; The rest beneath white. Its name is derived from the habitation it usually selects.

The *H. Smithii* of Leach, *H. corhinca,* is like it but smaller.

The River Swallow. Hirundo riparia. L. Enl. 543. 2.

Brown above, and on the chest; the throat and beneath white. It lays its eggs in holes by the water-side. It appears well authenticated that it falls into a lethargic state during the winter, and even that it passes that state at the bottom of marshy waters.

The type of the genus *cotyle* of Borè. Also found in America. Wilson, Orn. v. p. 38. fa. Vaillant, O. A. t. 246. f. 2., is also a variety according to Temmink.

Among the foreign swallows should be remarked

The Salangane. Hir. esculenta. Lin.

A very small species, of the Indian Archipelago, with a forked tail, brown above, and at the end of the tail and beneath, whitish. Celebrated for its nests, of a whitish gelatinous substance, disposed in layers, which it makes with a peculiar species of fucus, which it macerates and bruises for the purpose. The restorative virtues attributed to these nests have made them an important article of commerce in China. They are prepared for eating like mushrooms.

According to Dr. Horsfield, the tips of the tail are sometimes black.

The *Hir. fuciphaga* of the *Act. Holm.* seems allied to this.

Barn Swallow. Hir. rufa. Gm. *H. Americana.* Wilson. Pl. Enl. 724. Wilson, V. 38.

Above, and band on breast, steel blue; forehead, and beneath, rufous; tail forked, with a white spot on side of feathers; outer tail feathers long and narrow. North America.

Fulvous, or Cliff Swallow. Hir fulva. Vieil. Bonap. A. O. 7.

Blue black; beneath, brownish white; throat, front, and rump ferruginous; tail even. America.

The black-bellied Swallow, Hir. melanogaster of Swainson, seems allied to this.

White-bellied Swallow. *Hir. fasciata.* Gin. Pl. Eul. 724.

Black; abdominal band and femoral spot white. Cayenne.

Hir. violacea. Pl. Enl. 722. *H. purpurea.* Wilson, v. 39.

Glossy purple, with a copper, or blue reflection.

Chalybeate Swallow. *Hir. chalybea.* Gm. Pl. Enl. 545.

Steel black; beneath, brownish gray.

Senegal Swallow. *Hir. Senegalensis.* Lin. Pl. Enl. 301.

Shining bluish black; beneath, and rump red, quills and tail black. Senegal.

Hir. Capensis. Gm. Pl. Enl. 723.

Bluish black; beneath, yellowish black, streaked; crown red, side tail feathers with a white spot. South Africa.

Rufous-headed Swallow. *Hir. Indica.* Gm. Latham. Syn. 56.

Brown; crown red; body beneath whitish. India.

Panayan Swallow. *Hir. Panayensis.* Gm. Sonnerat. Voy. 76.

Silky black; beneath, white; forehead and throat, yellow ferruginous; throat, margined by a black collar. Phillipine Islands.

Hir. subis. Lin. Edw. 120

This species, according to Lichtenstein, is a variety of *H. purpurea.*

Ambergris Swallow. Hir. ambrosiaca. Gm. Briss. ii. 45.

Gray brown, beneath paler; tail much furcated.

Brazilian Swallow. Hir. tapera. Lin. *Hir. Americana.* Briss. ii. 45.—3.

Brown, beneath grayish; belly white; tail slightly forked. Brazil.

Black Swallow. Hir. nigra. Gm. Briss. ii. 46. f. 3.

Black; wings very long. Cayenne. Sometimes found with a white frontal band.

Daurian Swallow. Hir. Daurica. Lin. *Hir. Alpestris.* Pall.

Blue; beneath white; temples and rump ferruginous; outer tail feathers very long, with a white spot on the inner edge. Siberia.

Hir. rufifrons. Vieil. Vail. O. A. 245.

Bluish black, belly white, forehead red, bill and feet black. South Africa.

Hir. paludicola. Vieil. Vail. A. O. 246. 2.

Ash brown; beneath paler; tail short. South Africa.

Hir. cristata. Vieil. Vail. O. A. 247.—1.

Pale gray, beneath ashy white; crested. South Africa.

Hir. fuscata. Pl. Col. 161.

Brown, beneath white; head, neck and breast rufous; tail very slightly forked. Brazil.

Hir. jugularis. Pr. Max. Pl. Col. 161.

Red brown; throat rufous; breast and sides ashy yellow. Brazil.

Hir. melanoleuca. Pr. Max. Pl. Col. 209.

Lustrous black above; throat, belly and abdomen very pure white. Length five inches. Brazil.

Hir. Minuta. Pr. Max. Pl. Col. 209.

Fine blue above, beneath white; wings and tail black, the former slightly forked, and shorter than the tail. Brazil.

Green, blue or *white-bellied Swallow. Hir. bicolor.* Vieil. *Hir. viridis.* Wil. A. O. V. 38.

Dark greenish; beneath blue; tarsi naked, white. North America.

The following species, not noticed in the "Règne Animal," have been named.

Hir. domestica. Vieil. Azara. n. 300.

Shining blue; throat and crop whitish, varied with brown; chest and belly white; tail forked. Paraguay.

Hir. Cyanoleuca. Vieil. *Hir. melampiga.* Lich. Azara. 303.

Above steel colour; wings, and tail, which is forked, sooty above, white beneath. Paraguay.

H. Pyrrhonota. Vieil. not. Lath. Azara. n. 305.

Crown blue; forehead, reddish brown; cheeks and throat reddish; back blue, edged with whitish; rump brown. Paraguay.

Ooonalaschka Swallow. Lath. *H. Ooonalaschikensis.* Lath.

Blackish; beneath ash; rump whitish. Ooonalaschka; length, 4½ inches.

Peruvian Swallow. H. Peruviana. Brisson.

Black; beneath white; abdominal band ash; wings and tail pale grey.

Otaheitan Swallow. H. Tahitica. Gm. Lath. Syn. iv. title page.

Blackish brown, with bluish reflection above the forehead; neck and beneath purplish; tail, slightly forked, black. Otaheite; length, 5 inches.

H. nigricans. Vieil.

Blackish brown; beneath pale white; throat and crop brown streaked; tail slightly forked. Australasia.

Hir. thalassinus. Swainson.

Above changeable green, with lilac reflection; beneath snowy white; wings and tail violet brown; tail slightly forked.

* *Tail even; two outer feathers very long, and half filiform.*

Wire-tailed Swallow. Lath. *H. filifera.*

Bluish; beneath white; wing and tail black; tail with a white subterminal spot; brown nape; upper part of neck rufous. India.—British Museum.

Hirundo ruficeps. Licht.

Crown and forehead chesnut; upper part of body and eye-streak steel black; beneath white; chest-band ferrugineous; tail notched shorter than the wings; two outer tail feathers of male very long. Nubia.

*** Two middle tail feathers different from the rest.*

H. Albicollis. Vieil. *cypselus collaris.* Pr. Mag. Pl. Col. p. 195.

Black; neck above and chest white. Brazils. A *chætura* of Stephens.

Hir. Rutila. Vieil.

Blackish; forehead, cheeks, throat, and neck red; two middle tail feathers sharp tipped, rest rounded.

Foreign countries have some swallows with a tail almost square, and others, whose square and short tail has its quills terminating in a point.

St. Domingo Swallow. H. Dominicensis. Briss. *H. albiventer.* Vieil. Pl. Enl. 545.

Steel-black; belly white; tail blackish.

Brown-collared Swallow. H. torquata. Gm. Pl. Enl. 723.

Brown; beneath white; pectoral band and thighs brown. South Africa.

White-winged Swallow. H. leucoptera. Gm. Pl. Enl. 546.

Shining bluish; ash beneath; rump and wings varied with white. Cayenne.

Gray-rumped Swallow. H. Fancica. Enl. 544.

Blackish with white rump. India.

Wheat Swallow. *H. Borbonica.* Gm. *H. virescens,* Vieil.

Blackish brown; beneath gray, spotted with brown. Isle of France.

Rufous-rumped Swallow. *Hir. Americana.* Gm.

Blackish brown; with a shining green tint; beneath white; rump and vent red. South America.

Hir. fulva. Vieil. Bonap. Am. Ornith. F. 7 *H. lunifrons.* Sav.

Blue-black; beneath brownish; throat white; forehead and rump ferrugineous. North America.

Also have been named

H. rupicollis. Vieil.

Body above and chest grey brown; throat red; belly yellowish red. Brazil.

H. Javanica. Lath. Sparmann Mus. 100. Pl. col. 83.

Blackish shining blue; beneath ash; forehead, throat, and crop, ferrugineous; side-tail feathers spotted white at tip. India and New Holland.

H. Cyanoptyrha. Vieil.

Head and body, above, blue; forehead, and beneath throat red; quills and tail brown. South America.

H. pyrrhonota, *H. Americana.* B. Lath.

Blackish brown; beneath whitish; rump and vent red. India.

H. Hortensis. Lich. *H. flavigastra,* Vieil. Azara, 306.

Body above blackish brown; throat and crop red; chest and belly yellowish white. Paraguay.

Red-headed Swallow. Lath. *Hir. Erythrocephala.* Gm.

Blackish above; white beneath; head red; wings and tail brown. India.

Ash-bellied Swallow. Lath. *H. cinerea.* Gm.

Shining black; beneath, ash; tail edged with yellowish gray. Peru and Otaheite.

H. leucorhœa. Vieil.

Blue, with violet tint over the eyes and body; beneath, white; quills and tail, black. Paraguay.

H. fusca. Vieil. Azara. 301.

Brown; throat and belly white, with a brown and white band on chest; quills and tail brown. Paraguay.

Others have square, short tails, and the tail feathers ending in a point.

Temminck has referred these sharp-tailed swallows to the Cypseli, or Swifts. And Mr. Stephens separates them generically under the name of *Chœtura.*

Sharp-tailed Swallow. Lath. *H. acuta.* Gm. *H. martinicana.* Briss. Pl. Enl. 544. *Chœtura martinica.* Stephens.

Black; beneath, brown; throat grey Martinique.

Aculeated Swallow. Lath. *H Pelasgia.* Lin. Pl. Enl. 726. Catesby's Carolina, 8.

Brown; throat whitish.

Cypselus giganteus. Pl. Col. 364.

Body black; neck brown; tail and wing-coverts and side of belly, bottle-green; back and scapulars, dull ash; sides of belly and vent, white; tarsi naked.

H. albicollis. Vieil. *Cypselus collaris.* Pr. Max. col. 195.

Sooty black in general; deeper on the body than wings or head; a white collar. Brazil.

H. caudacuta. Lath. *Chætura australis.* Step.

Dusky, tinged with shining green; forehead and throat white. New Holland.

H. Pacifica. Lath.

Dull brown; throat and rump whitish. New South Wales.

H. Oxyura. Vieil.

Throat whitish, tail coverts blackish, varied with reddish brown. South America.

Chætura Sabini. Gray. MSS.

Bluish black; belly and rump white. Africa. Capt. Sabine.

THE GOATSUCKERS. CAPRIMULGUS. Linn.

Have the same light soft plumage, and shaded with gray and brown, which characterizes the nocturnal birds. Their eyes are large; their beak is still more cleft than that of the swallows, furnished with strong mustachios, and capable of engulphing the largest insects, which it retains by means of a glutinous saliva. On the base are the nostrils, in the form of small tubes; their wings are long, the tail squared, the feet short, with feathered tarsi, and toes united at the base by a short membrane; the thumb itself is thus united to the external toe, and can be directed forward; the claw of the middle toe is often indented at its internal edge, and the external toe has but four phalanges, a conformation rare among the birds. These birds live in an isolated manner; fly only during twilight or on fine nights, pursuing the phalenæ, and other nocturnal insects. They nestle on the ground inartificially, and lay a small number of eggs. The air, which is engulphed in their large bill, as they fly along, produces a peculiar humming sound.

We have but one species of them in Europe.

Caprimulgus Europeus. L. Enl. 195.

As large as a thrush; of a gray brown, undulated and spotted with a blackish brown; a whitish band proceeding from the bill to the nape. It nestles in the furze, and lays only two eggs.

America produces many of these birds, with a round or square tail, one which is as large as an owl. *Caprim. grandis.* Enl. 325. and another, *C. vociferus,* Wils. v. 41, celebrated by the loud noise which it makes during spring.

Another European species has been discovered

Cap. Rufitorquis. Vieil.

Above gray and black-banded, and black and gray streaked; beneath yellowish black, cross lined; throat white; collar reddish; three primaries internally white spotted; two outer tail-feathers white tipped; Marseiles. M. Baillon.

The species of this genus are much involved in obscurity.

Some have the tail rounded or wedge-shaped, and the middle claw serrated.

C. Egyptius. Licht. *C. Isabellanus.* Temm. Pl. Col. 379.

Tail equal; a little longer than the wings; Isabella very finely marbled with thin ziz-zag lines; outer quill well banded with black; inner spotted with white; half collar white. Egypt.

Caprimulgus Nubicus. Licht.

Tail equal, a little longer than the wings; pale ash; crown, back, and coverts striated with black, and spotted with Isabella; gula-band white; chest and belly, black waved. Length, eight inches. Nubia.

Caprimulgus Icteropus. Vieil.

Blackish gray, varied with ferrugineous and white; beneath reddish white; crown banded; wings blackish and yellowish, transversely streaked. China.

Cap. Cyanus. Lath. *Cap. Cayennenis.* Gm. *C. Leucurus.* Vieil. Pl. Enl. t. 760.

Red and gray, variegated, and lined with black; throat and wing-band white; temples red, with five black streaks: side-tail feather white edged; tail equal. South America.

Cap. Griseus. Gm.

Gray; wings blackish, gray banded; tail rather longer than the closed wings, black banded. South America.

Cap. Nattereri. Temm. Pl. Col. t. 107.

Breast, wing, tail, and above black, red spotted; throat dusky, with a white lunule spot; belly, vent, and under tail-coverts rufous, black barred. Brazils.

Long tailed Goatsucker. Lath. H. vii. t. 14. *C. climaturus.* Vieil. Gal. t. 112. *Cap. longicaudus.* Steph.

Variegated brown; ferrugineous and black; crown, ash, spotted with rust; throat with a white spot; tail longer than the body, barred with dusky. Senegal.

Gold coloured Goatsucker. Capr. torquatus. Gm.

Two middle feathers of tail longest; ash brown, varied with dull yellow, and whitish spots; collar golden. Brazils.

Capr. gracilis. Lath.

Tail lengthened; above ash; varied with brown and white; beneath white striated, and spotted with ferrugineous yellow; middle toe. New Holland.

Caprimulgus Hirundinaceus. Spix. Braz. ijt. 2. f. 1.

Small; fuscous brown; throat, and beneath, red black banded; tarsi woolly; outer tail feather banded, and internally tipped with white; middle claw serrated; tail equal; Brazil.

Chuck Will's Widow. Capr. Carolinensis, Gm. Wils. A. O. t. 54. f. 2. pl. Enl. t. 735. *C. rufus.* Vieil. C. *brachypterus.* Shaw.

Bristles shorter than the bill; tail rounded, reaching an inch beyond the wings; three outer-tail feathers on the inner web at tip white; *in fem.* ochraceous. North America. Twelve inches.

Whip-poor-Will. Cap. Virginianus. Vieillot. O. A. t. 23. not Lin. *C. Vociferus.* Wils. A. O. v. t. 41. f. 1, 2, 3.

Bristles much longer than the bill; tail much rounded, reaching one-half beyond the wings; primaries mottled; outer-tail feathers tipped with white; in *fem.*; ochreous. North America.

Capri. infuscatus. Ruppel. Atlas 16.

Body reddish brown; feathers all finely waved with black; throat, chin-band, quills, and two outer tail feathers white; whiskers large, strong.

White-throated Goatsucker. Cap. Albicollis. Lath. Vieillot. Azara. n. 310.

Above reddish brown, varied with pale black; beneath red, transversely black lined; throat white; primaries white banded. S. America. Allied to *Virginianus.*

Caprimulgus Sphenurus. Vieil. Azara.

Head and scapulars black and brown; wings black, varied with white and red; throat reddish white; body, beneath, and back, blackish; white streaked.

Cap. Torquatus. Vieil.

Ash brown, varied with dull yellow and whitish spots;

collar golden; two middle tail feathers longest; feet black. S. America.

Javan Goatsucker. Lath. *Capr. Macrurus.* Horsf.

Clouded, ferrugineous, and blackish; vertical streak and band on the wings quite black; tail longer than the body; wedge-shaped; throat band white. Java. Length, ten inches.

Chuppa Goatsucker. Lath. *Cap. Affinis.* Horsf.

Variegated, black, brown, and ferrugineous; quills brown; three outer white banded; rest variegated, ferrugineous, and gray; two outer tail feathers internally white. Java. Length nine inches. Allied to C. *Asiaticus.* Lath.

Bombay Goatsucker. *Cap. Asiaticus.* Lath. *C. Pectoralis.* Vieil. Vail. O. A. t. 49. *C. Pectoralis.* Cuv.

Ashy, clouded, with black and ferrugineous; pectoral band ash; vertical streak blackish; maxilla and gular spot pale; middle claw toothed. India. South Africa.

American Goatsucker. *Cap. Americanus.* Gm. Sloane. Jam. t. 255. f. 1.

Body gray, variegated with yellowish brown; nostrils cylindrical. America.

Varied Goatsucker. *Caprimulgus Variegatus.* Vieil. Azara. in. 313.

White, varied with black and brown; beneath, cross lined, with blackish; throat white. Paraguay.

Guiana Goatsucker. Cap. Guyanensis. Gmel.

Fulvous, spotted and streaked with reddish; a white subgular moon.

Rufous Goatsucker. Cap. Rufus. Gmel. *Cap. Guyanensis.* Var. Vieil. Pl. Enl. t. 735.

Rufous, varied with black; wing coverts, and body beneath blackish-banded; quills red and black; tail black-banded. Cayenne.

Brazilian Goatsucker. Cap. Brasilianus.

Blackish, varied with yellowish, and punctated with white; beneath, variegated white and black; tail as long as the wings.

Indian Goatsucker. C. Indicus. Lath. *Cap. Cinerescens.* Vieil.

Ashy, with black cross lines; cheeks, chest, and wings spotted with ferrugineous; tail ashy, black banded; outermost feathers variegated, feruginous and black. India.

Caprimulgus Nacunda. Vieil.

C. Diurnus, Pr. Max. Pl. Col. 182. *C. Campestris.* Licht. *Nacunda.* Azara. n. 312.

Red-black; dotted throat, with a white lunule spot beneath; white chest, brown-lined; middle toe said to be serrated. Paraguay.

White-collared Goatsucker. *Cap. Semitorquatus.* Lath. Pl. Enl. 734.

Blackish, spotted with red and grey; neck with a white crescent. Cayenne.

Jamaica Goatsucker. Lath. *Capri. Jamaicensis.* Gml. Enl. 98.

Ferrugineous black, streaked; wings white, variegated; quills brown, white spotted; tail black-banded; middle claw entire. Jamaica.

Caprimulgus Leucopygus. Spix. I. 3. f. 2,

Small: olive-brownish black; beneath black; throat white waved; and wings spotless; tail equal, inner base white marked; feet naked, not serrated. Brazil.

Caprimulgus Longicaudatus. Spix.

Ferrugineous; spotted, with black chest; beneath with white lunule spot; tail longer than the wings; sub-graduated; posthumeral coverts whitish; quills and tail reddish, black banded; tarsi scarcely naked; middle claw not serrated. Brazil.

One half of the end of tail-feather acutely subtulated.

Sharp-tailed Goatsucker. Caprimulgus Acutus. Lath. Pl. Enl. t. 7. 32.

Olive-grey; beneath red, rayed all over with blackish; head and neck red-brown. South America.

One has tail forked, and the middle claw gerrated.

Cap. Enicurus. Vieil. Azara. 315.

Crown whitish, varied with blackish lines and dots; throat with a white crescent; body olive-brown; beneath red, cross-lined with blackish. Brazil.

Africa has also some of them; and, among the number, some with pointed tails, and others with forked tails approaching Hirundo. There is one also in America, the forks of whose tail are longer than the body. The middle claws of these species with forked tails is not indented.

Scissor-tailed Goatsucker. Lath. Caprimulgus Cornutris, Vieil. *Urutau* Azara, n. 308. *C. Grandis*, Gml. *C. psalurus*. Tem. Pl. Col. F. 157-158. Pl. Enl. 325.

Body olive-red, varied with black and brown; beneath red, brown, and black; belly whitish-brown; throat reddish; side of the face with egrets.

Fork-tailed Goatsucker Cap. Forficatus. Vieil. Vail. O. A. p. 47-48. *C. Furcatus*. Cuv.

Black, brown, and white variegated; tail very much forked. Cape of Good Hope.

Some have the tail forked, and the middle toe serrated.

Cap. Furcifer. Vail. Azara. 309.

Blackish; chest whitish, lined with reddish and blackish; belly pale red; quills blackish, cross-lined with red and white; feet feathered.

Caprimulgus Rupestris. Spix. Braz.

Throat whitish; abdomen, middle of the wings, and the sides of the tail, which is shortly furcated, white; chest brownish; tarsi half naked. Brazil.

Night Hawk. Capri. Virginianus. Gm. *C Americanus*. Wils. A. O. fi. 40. f. 12. *C. Popetue*. Vieil. O. A. f. 29.

Bill without bristles; tail forked, not reaching to the

tips of the wings; primaries plain blackish with a white spot. In the male, the triangular throat spot and tail band white. North America.

Tail forked, but of which the middle claw has not been observed.

Banded Goatsucker. Cap. Vittatus. Lath. Supp.

Under part bluish ;, back bluish-clouded ; crown and nuchal band black ; quills and tail, ferrugineous brown. New Holland.

Strigoïd Goatsucker. Cap. Strigoïdes. Lath.

Ferrugineous brown; above varied with dull streaks and spots; beneath brown; wing-coverts with three pale lines ; eyebrows white; tail slightly forked. New Holland.

Cap. Manurus. Vieil.

Silvery grey, spotted with black ; smaller wing-coverts white spotted; tail, two outer feathers five inches longer than the middle ones, the third and fourth very short. Brazil.

A species, also from Africa, but with the tail round, is very remarkable for having a feather twice the length of the body, arising near the bend of each wing, which is only bearded near its extremity. *Cap. Longipennis* of Shaw. Nat. Misc. t. 265.

C. Macropteropus of Afzelius. Hist. Sierra Leone.
White-throat Goatsucker. C. *Albo Gularis.* Vigors, and Horsfield, from specimen in possession of M. Leadbeater.

Brown, varied with black, grey, and yellow; underneath yellowish ; spot on throat white.

Spotted Goatsucker. C. Guttatus. Vigors and Horsfield. Lin. Trans.

Ferrugineous, spotted with brown; quills spotted with yellow. New Holland.

New Holland Goatsucker. Caprimulgus Novæ Hollandiæ. Lath. Philip. New South Wales.

Olive-brown, clouded with black and white; beneath whitish; neck and chest, obscurely banded; tail rounded; middle claw entire. New Holland.

This forms the genus ÆGOTHELES of Horsf. and Vigors.

THE PODARGES. PODARGUS CUV.

Have the form, colour, and habits of the goat-suckers, but their bill is slender, and they have neither interdigital membranes or indentation of the middle claw.

Mr. Vigors considers them as connecting the goat-suckers with the owls.

The Ash-coloured Goatsucker. P. Cuvierii. Vieill. Gal. 123.

Coloured, varied tints of ash, whitish and blackish; as big as a crow.

The Red Podargus. P. Javanensis. Horsf.

Red, varied with brown; a white band along the scapulars.

The Horned Podargus. P. Cornutus. Pl. Col. 159.

Red, varied with white; and with great tufts of feathers at the ears.

M. Temminck identifies this species with the last.

Cold-river Goatsucker. Lath. vii. 369. *P. Humeralis.* Vig. Hors. *P. Australis.* Stephens.

Dark, varied with black stripes, and black upper feathers; outer web of quills with white: tail slightly banded.

The Great-headed Podargus. Caprimulgus Megacephalus. Lath.

Pale brown; head and neck very large, and full of feathers; belly pale-ash. New South Wales.

Trinidad Goatsucker. Steatornis Caripennis Humb.

Gray-brown, varied with small streaks and dots of black; quills and tail white. South America.

M. Humboldt separates this remarkable species generically from Podargus, under the name Steatornis.

SUPPLEMENT ON THE FISSIROSTRES.

THE habitat of the swallow is said to extend itself to all those parts of the earth's surface in which insects, its principal food, are to be found. In climates which are exposed to considerable changes of temperature, these birds indeed are only seen periodically; but in intertropical countries, especially those which border on the temperate zones, as Egypt, Ethiopia, Lybia, &c. they remain during the whole year. The object attained by this phenomenon is principally a continued supply of food, and has probably but little reference to any given degree of atmospheric temperature necessary to these birds, whose country seems almost to be the world at large.

The cleft beak has furnished systematists with a character whence to name this family of birds *Fissirostres*, but the rapidity and dexterity of their flight would have afforded a still more striking distinction. They may, with peculiar propriety, be called tenants of the air, since they hunt their prey, and eat it, drink, and occasionally even feed their young, on the wing; indeed when not under the migratory influence, they may be said to live entirely on the wing and in the nest. Their rapidity is well known, and the "murder aiming eye" of the most experienced sportsman will seldom avail against the swallow, unless it be at the instant the bird makes an angle in flight, which it is remarkable for doing, probably in the pursuit of some insect out of its direct course; hence they themselves seldom fall a prey to the raptorial birds, while they commit the greatest havoc among the insect tribes, particularly those in the winged state, which they pursue with unabated vigour and success. This pursuit is either close along the surface of the ground, or more or less elevated above it, according to the state of the atmosphere, which seems to govern the elevation of flight of insects. The relative state of humidity in the air may thus be inferred from the elevation of the swallow's flight, while the vulgar regard it as actually anticipating the weather.

That swallows are serviceable to mankind, by the destruction of hosts of insects, injurious to our industry and convenience, cannot be doubted; but there is certainly more of enthusiasm than of sound philosophy in asserting, that this was the object of their creation; and there is perhaps too great a proneness in many well meaning persons, of warm imaginations, to conclude, that man is as it were the sole object of creation, and that every thing beneath him is wholly subservient to his existence. Man has indeed enough to be grateful for without derogating from the value of inferior creatures, or concluding that one set of beings was created solely for the destruction of another, that the balance of power might be thus preserved exclusively for his convenience. That this balance of power is a subject of the greatest admiration is perfectly true; but that it is attributable to a much higher and more direct agency than that alluded to, is most probable.

The emigration of birds is a subject so entirely independent of all physical principles, as far at least as we can discover, that it may, without any figure of speech, be considered a standing miracle. We are generally ready rather to run into absurdities in endeavouring to account for phenomena by material sensible agency, than to admit that our own powers of intellect are too finite to conconceive the *modus operandi* of Providence.

Hence, with reference to the particular subject of the migration of Swallows, has probably originated the absurd systems incompatible with all the principles of physiology, which have been invented to account for the periodical disappearance of these birds, although the actual fact of emigration is more easily observable in them than in any other of the class.

A bishop of Upsal in 1555, *Olaus Magnus* (the epithet could not have been applied to him as a naturalist), asserted that the fishermen in northern countries frequently caught in their nets clusters of swallows attached closely to each other, which, if kept sufficiently warm, were soon restored to animation, and that those left in the water, on the approach of warm weather, rose to the surface, and then took to flight, resuming their accustomed habits. However incredible this tale, it was generally received, and even supported by real observers of nature, among whom must be reckoned even the great Linnæus; a position, however, emanating from such a source, and recommended by such authority, was certainly, in respect not to itself but its supporters, deserving serious investigation, which it has therefore meritedly undergone.

Montbeillard, who has examined this story, observes,'that if it were true that all the swallows of any particular place or country, voluntarily plunged into the water yearly in October, and quitted it again in April, there must have been frequent opportunities of verifying the fact, either at the period of their immersion, or their subsequent emersion: yet no credible witness, whose veracity and judgment could be relied on, has vouched for having seen it; and notwithstanding the rewards offered at different times, no specimens of swallows found in a state of suspended animation under water, have ever been forthcoming. The certificates produced by Klein, (another supporter of this doctrine) in his dissertation *De Hybernaculis Hirundinum*, speak only of isolated facts, not recent, or else founded on mere hearsay. Testimonies of this sort cannot avail against the established physiological principle, that a bird to whose being, air in abundance is so essential, could exist even by suspended animation, in the medium of water which would prevent the access of all air to the lungs.

Another hypothesis to account for the disappearance of swallows has been invented, which is, that they retire to caverns and holes in the ground, and there remain torpid, like some of the mammalia, during the winter. This has been insisted on principally from the story published by Achard, who states, that while travelling down the Rhine about the end of March, in the year 1791, he saw children pulling birds out of holes on the banks, and having purchased some of them, which he found to be dormant, and, as it were, inanimate, he put one in his bosom, which, in about half an hour, awaked and flew away.

Montbeillard, and especially Spallanzani, who was never stayed in the pursuit of natural knowledge by any feeling for the victims of his experiments, have examined the probability, or rather possibility, of these birds falling into a torpid state. The latter has ascertained, by the use of artificial cold, that swallows do not appear to suffer by cold at the freezing point; that at 8° or 9° below it, they are sensibly affected but not killed, and that at 13° or 14° they speedily die.

To ascertain the effect of a continuance of low temperature, Spallanzani enclosed some swallows in wicker baskets, covered with waxed silk to preserve them from humidity, and buried them in snow, through which he made a hole for the admission of fresh air. At the end of thirty-five hours some of them were dead, and others were greatly debilitated, but shewed no signs of lethargy; ten hours after they had all died: and this experiment, first practised in the month of May, was repeated in the following July with the same effect. To assure himself that the death of these birds was not caused by want of food, the experimentalist placed other swallows in his own room, and without food; some of them resisted death till the fifth day, and none died till after three days and an half. There remained, therefore, no doubt, that the buried birds died from cold and not from starvation, and did not pass into a lethargic state; and hence we may conclude they never do, a piece of knowledge, like many of those of the Italian naturalist, purchased at far too high a price,—no less than the sacrifice of the best feelings of humanity, when the end proposed was merely one of curiosity inapplicable to all useful purposes.

So prevalent, even still, is the notion, that swallows take spontaneously to the beds of rivers, or pass the winter in a torpid state, that the re-recital of these experiments, and matter of fact to refute them, is not, as yet, needless.

Swallows have been kept in cages during the whole winter, and have been found to moult during that season, which could not happen if the birds were under water. In Mr. Bewick's History of British Birds is the following account, from the pen of Mr. Pearson:—

"Five or six of these birds were taken about the latter end of August, 1784, in a bat fowling-net at night; they were put separately into small cages and fed with nightingale's food; in about a week or ten days they took food of themselves; they were then put all together into a deep cage four feet long, with gravel at the bottom; a broad shallow pan, with water, was placed in it, in which they sometimes washed themselves, and seemed much strengthened by it. One day Mr. Pearson observed that they went into the water with unusual eagerness, hurrying in and out again repeatedly with such swiftness as if they had been suddenly seized with a frenzy. Being anxious to see the result, he left them to themselves about half an hour, and on going to the cage, found them all huddled together in a corner apparently dead; the cage was then placed at a proper distance from the fire, when only two of them recovered, and were as healthy as before; the rest died. The two remaining were allowed to wash themselves occasionally for a short time only, but their feet soon after became swelled and inflamed, which Mr. P. attributed to their perching, and they died about Christmas; thus the first year's experiment was in some measure lost. Not discouraged by the failure of this, Mr. P. determined to make a second trial the succeeding year, from a strong desire of being convinced of the truth respecting their going into a state of torpidity. Accordingly, the next season, having taken some more birds, he put them into the cage, and in every respect pursued the same method as with the last; but to guard their feet from the bad effects of the damp and cold, he covered the perches with flannel, and had the pleasure to observe that the birds throve extremely well; they sung their song through the winter, and soon after Christmas began to moult, which they got through without any difficulty, and lived three or four years, regularly moulting every year at the usual time. On the renewal of their feathers, it appeared that their tails were forked exactly the same as in those birds which return hither in the spring, and in every respect their appearance was the same."

M. Natterer has also kept a number of swallows in cages for eight or nine years together, and has observed that they all moulted about the month of February, the period when these birds having quitted Europe, are found in the warmer countries of Africa and Asia. It is obvious that their moulting at this period is irreconcilable with the idea that they hibernate in this country under water.

The activity of these birds, as the period of their emigration approaches, is very amusing, and we have a fine opportunity, especially on the southern shores of our own island, of witnessing their departure in many and numerous flocks. For some days previously to their quitting our shores, they may be seen in clusters so thick, sticking to the walls and sides of houses and elevated buildings, as to hide, even at a small distance, all appearance of that on which they rest, till at last they all quit together, and fly for a time east or west, possibly in wait for stragglers not yet arrived from the interior, or take directly to the south, and are soon lost sight of altogether for the allotted period of their absence.

We have already observed that the primary object of emigration is a supply of food; and as in temperate countries, occasionally exposed almost to the extremes of cold as well as heat, the air abounds with insects only at particular seasons, it is natural enough that these birds, whose locomotive powers, to almost any distance, are so striking, should quit such countries when the cold begins to deprive them of a sufficient supply of food. The fact abstractedly is less surprising than the mode, the regularity, and the punctuality with which it is performed. The whole of them quit our shores and the whole continent of Europe, and pass simultaneously in organized bodies over the Mediterranean into Africa; they arrive in Senegal about the ninth of October, and quit it for the north, in spring; and however difficult to be credited, it seems ascertained beyond doubt, that the same pair which quitted their nest, and thelimited circle of their residence here, return to the very same nest again; and this, for several successive years, in all probability, therefore, for their whole lives. The simple experiment of tying a silk thread round the leg of one of these birds (which has been repeatedly done,) will convince any one of the truth of this astonishing fact. Others have been caught, and carried in covered cages to very great distances, and then turned out. These have been observed immediately to fly to a great elevation, and then to describe a large circle several times, until having decided on the right course, they have proceeded, almost with the rapidity of an arrow, undeviatingly to their nest.

Many anecdotes are related of the sociability and mutual readiness of these birds to assist each other; but the story that these birds will unite to close the entrance of any nest which may have been taken possession of, during their absence, by a sparrow, seems to want confirmation. That this intruder will sometimes get possession of their nest, seems ascertained; but the possibility of any number of swallows keeping the sparrow in while they compound, agglutinate, and harden a sufficient barricado against his escape, may be well doubted. Montbeillard has observed, that in such cases, the swallows will associate and threaten, if not attack, the intruder; but he seems to think, that the sparrow is left eventually in peaceable possession of his usurped domicile, which, according to that naturalist, even if abandoned by the sparrow, would be spoiled by his temporary occupation, for the uses of its original builder.

The swallows, properly speaking, are divided into many sections in the "Règne Animal," all of them having the toes and the sternum disposed as in the passerine birds in general; but in some, the thumb is nearly capable of being turned in front, and these have the feet feathered down to the toes, and the tail forked. Others have the toes naked, and the forks of the tail very long. Some have the tail nearly square, &c.; but we shall proceed to the habits, &c., of some of the species, by which the general character of the whole genus may be estimated.

The *Marten (Hirundo urbica*, Lin.*)*, says Daubenton, is domestic by instinct, seeking the society of man in spite of all its inconveniences, in preference to all others. This opinion, maintained also by Montbeillard, and on which many sentimental reflections have been expressed, is opposed by Spallanzani, who would refer the nidification of these birds in our chimneys and window corners, more to the convenience afforded by this angle to the bird for the purpose, than to any predilection for human society. The dog, and some other quadrupeds, teach us, certainly, that *they* are capable of sentiment and preference; but there are many circumstances of parity in the existence of man and the quadrupeds, which are not to be found in creatures of two distinct classes. Nor does there seem to be any other circumstance from which we may reasonably conclude in favour of the instinctive propensity of this bird for the society of man, than that of its attaching its nest to our dwellings, and this must be limited in its extent. The architecture of Africa cannot afford similar conveniences to these birds during their absence from Europe.

This species is thought to be less familiar than the swallow, properly speaking, and arrives here about the middle of April, eight or ten days later than the other. They build in the corners of windows, or under the eaves of houses. These are composed exteriorly of earth or clay, especially worm casts, which they moisten and agglutinate more effectually than the best prepared wall of a mud cabin. Within, they are lined principally with the feathers of other birds, dry grass, &c.; the nest is larger than that of the swallow, but the opening, which is on the side at top, is smaller. They are said to have three broods during their stay here, viz., in May, June, and July, probably four or five at a time.

Each incubation continues about fifteen days, and so well do the parents supply their young with food, that the crop may be observed to be enormously distended, and the young, as it is said, to weigh even more than the old birds; nor do they quit the nest immediately on being able to fly, but continue some time longer to nestle with the parent bird.

It appears, that in the south of Europe, these birds do not confine themselves to human habitations, but frequent also marshy spots, where they build and rear their young; and it is observed of these by Montbeillard, that they do not, like those which frequent houses, return several seasons to the same nest, but build a new one every year.

The *Common Swallow*, or *Chimney Swallow*, so called to distinguish it from the marten, or window swallow, is best distinguished by the feet, which are naked, while those of its congeners are downy. This species visits us a few days earlier in the spring than the marten; nor does it migrate, as is said, so far to the south as the other species above named. Once arrived in this country, they are not observed to desert it again until the usual period of their return; but in the south of Europe, according to Spallanzani, they will retire to the south even after their arrival, in the event of a prevalence of sharp and cold winds, and return again in settled fine weather. The experiments alluded to in the former part of these observations, on the genus, evince that some degrees below freezing will not destroy these birds, even when left altogether exposed.

The swallow builds in general in the inside of chimnies, in which no fires are kept, and sometimes in other covered shelters about buildings. The female is said to have but two broods here: the first, about five, is probably more numerous than the second: the eggs are white, with reddish spots, while those of the marten are simply white. The experiment of marking a pair of these birds, by tying a silken thread to the leg or otherwise, has evinced, by the return of the same individuals to the same spot, in the following season, that they are monogamous. The parental tie appears to be equally strong as in the marten.

In their habits, instincts, persons, migration, &c. the swallow, the marten, and the swift are greatly assimilated. Their physical specific differences are mentioned in the text; and their instincts, equally disposed in favour of human habitations, differ only in the preference of one for the chimney, and of the other for the window; a difference which, however unimportant, is nevertheless curious for its undeviating nature in both.

The *River Swallow* (*H. riparia.* Lin.) does not arrive in this country until a few days after the swallow and marten, and it is thought to return to the south about the same time before them. They nestle in holes made principally in sand-banks which are two or three feet in length horizontally. The eggs, from four to six, are white. They may be seen flying continually to and from near the surface of the water, without departing any distance from their nests in the neighbouring banks. They seem, like their congeners, to have or three broods in a season.

Spallanzani tried a similar experiment on this species as he had previously done on the swallow and marten. Having drawn a male and female from the nest at the extremity of the hole, by means of a hooked wire, he carried them from Pavia to Milan; and having arranged with a friend the precise moment when they were to be set at liberty at the latter place, they were seen to enter the hole again exactly thirteen minutes after their departure; and, by means of silken threads, the same pair were observed to return the following season to the same hole.

It has been thought that this species is more hardy than the rest, and hibernates in the hole in which it breeds; but many of these holes have been inspected in the winter, and no birds have been found: nor does there seem any good reason to conclude that they differ from the swallow and marten in anything more essential, than those two species differ from each other; the sand-bank distinguishes these from them, as the window and the chimney distinguish them from one another. It will be observed, however, by his language in the text, that the Baron is inclined to a contrary opinion.

The Purple Swallow (*Hirunda purpurea. Lath.*) is a beautiful species. At first sight it appears altogether black, but its plumage, far from being uniform, varies according to the position of the eye which observes it, and the movements of the bird, and casts the most brilliant reflections of blue, violet, and purple.

This swallow is protected by the Americans with great justice. It not only diminishes the number of winged insects which cause so much inconvenience, but it warns the poultry of the approach of birds of prey. The instant one of the latter makes his appearance, these swallows all assemble, and soon put him to the rout, by the pertinacity of their attacks and their clamorous cries. They nestle in holes made expressly for that purpose, about the houses or under cornices, like the window swallow. During the entire summer they are found in such places, but retire on the approach of winter.

This swallow is an inhabitant of Louisiana, of Carolina, and of all North America, as far as Hudson's Bay.

The *Hirundo Fusca, Vieill.*, is a native of Paraguay. It rarely enters towns or villages. It is usually observed singly, or in pairs; but when winter approaches, which is the period of its departure, it forms troops, sometimes of an hundred individuals. It is much more rare and more wild than the domestic swallow of the same country. It is imagined generally to nestle in holes, and is said to dispute sometimes for the nest of the rufous bee-eater, or the dwarf-parrot.

Hirundo Cayanensis (the *White-collared Swift* of Latham) is distinguished, as its name imports, by a collar of the purest white, finely contrasting with a velvet black with violet reflections, which predominate in the rest of the plumage. This bird makes its nest in the houses. It is very large, and constructed with dog's-bane, in the form of a truncated cone, one of the bases of which is five inches in diameter, and the other three; its length is nine. The larger base is composed of a sort of pasteboard made of the same material. The cavity is obliquely divided about half-way its length, by a partition, which extends over that part of the nest where the eggs are, that is pretty near the base, and over this is a sort of plug of the soft down of the dog's-bane, to shield the young ones from the external air,

The *White-rumped Swallow* of Paraguay (*Hirundo Leucorrhoa, Vieillot*) is not wild. It flies very near the ground in the open country, and does not usually enter inhabited places. It is fond of accompanying travellers and seizing the flies and butterflies which they disturb. It nestles in the holes of palm trees, &c.; but near the river of La Plata, where there are no trees, it lays its eggs in holes in the ground. The nest is constructed with nothing but leaves and plenty of horse-hair. The entrance of this nest is so narrow, that it is impossible to draw out the young ones. This swallow is sedentary in Paraguay, rather common, and chirrups much in the spring season.

The *Domestic Swallow* of Paraguay is called by the Guaranis, *Mbiyui.* This name expresses its cry, which consists in repeating several times the syllables of this word, and they have applied the same name to all the species. It inhabits Paraguay and the river La Plata, and nestles on the cabins and country houses. But in cities and towns it chooses, in preference, churches and large buildings, where it fixes its nest to the beams or walls, but always in such a manner, that it is scarcely apparent; sometimes this nest is found under the tiles. It is said to be composed of clay without and a little straw within. The eggs are three or four in number. This species frequently perches on the crosses of weather-cocks, or the ridges of roofs, and on the barriers of enclosures. During summer, these swallows sleep in the interior of orangeries, or of tufted trees; but if ever so little cold should occur, they pass the night in holes, or under the tiles. They are birds of passage, but their periods of absence and return are not so strictly regulated as in Europe. They are determined by the greater or less duration of the cold, and consequent abundance of insects;. so that if the winter be mild, they are scarcely two months out of Paraguay, but, in the contrary case, they are absent four months. They pass the winter in the 20th degree of south latitude.

The *Gray Rock-Swallow* (*H. Montana*), from which it is probable that *H. rupestris* ought not to be separated, notwithstanding some little differences of colour, nestles in the Alps, and does not descend into the plains except for the purposes of hunting. Its flight is slower than that of the other species, and it pursues its prey in company with the marten swallow. These birds arrive in Savoy towards the middle of April, and depart about the 15th of August.

The *Hirundo Javanica* has been sometimes confounded with our *rustica*, from which, however, it differs (according to Mr. Vigors and Dr. Horsfield) chiefly in its inferior size; the side-feathers of the tail are also shorter and less slender; the frontal band is wider, and the ferrugineous colour extends over the breast, instead of the broad black band which distinguishes ours. It is a native of Australia as well as Java, and the specimens from the former region, in the collection of the Linnæan Society, accord perfectly with MM. Temminck's and Sparman's figures and descriptions of the Javanese species. To find the same birds in countries so widely remote from each other, is only to be accounted for by their migratory disposition.

From Mr. Caley's MSS. we extract the following observations.

"The resting-places of these swallows are on the dead boughs of large trees, where I have seen several of them gathered together, in the same manner as European swallows on the roof of a house. I apprehend, however, that it is when their young have taken to flight that this occurs.

"The earliest period of the year that I noticed the appearance of swallows, was on the 12th of July, 1803, when I saw two; but I remarked several towards the end of the same month, the following year (1804). The latest period I observed them was on the 30th of May, 1806, when a number of them were twittering, and flying high in the air. When I have missed them at Paramatta, I have sometimes met with them among the North Rocks, a romantic spot about two miles northward of the former place.

"The natives call the swallow Berrin'-nin; they told me it built its nest in the hollow limbs of white gum-trees, using bark, grass, hair, or similar substances; but when it built in old houses, it made use of mud. These old houses are the deserted huts of settlers, who have abandoned their worn-out farms; and the nests are constructed on the wall-plates, as they are called, in the colony. Of the nests which have been brought to me, I have observed that the outside was made of mud, and the inside lined with feathers. Though I have seen swallows more or less throughout the year, yet it is my belief that they are migratory."

Among the swallows with elongated lateral tail feathers, some have these feathers acuminated, while others have them of equal breadth all the way. Of the latter, the *New Holland Swallow*, engraved from a drawing by Mr. Lewin, differs from the species as described in the Indian Ornithology, in having the crown of the head, nape, and all the back blue; the forehead and throat brownish-yellow, passing into yellowish-white on the belly and vent; the wings and tail alone are dusky-brown.

Although in our general introduction to birds we have noticed the nests of the *esculent swallow*, it will be necessary to add a few details on the subject in this place.

This bird, which inhabits banks, shores, &c., is designated in the Philippine Islands by the name of *Salangana*, and is celebrated by the very singular nests which it constructs. These nests have been compared to those productions called by the ancients the nests of the halcyon; but the comparison does not hold good, for those maritime productions so called by the ancients are not birds' nests, but the receptacles of the polype, which at present are termed *alcyonium*.

All authors are agreed on the estimation in which the Chinese, and other Asiatics, hold the nest of the Salangana as a delicacy of the table, and on the high price and value which they bear; but they differ much as to their nature, form, and the places in which they are to be found. According to some, the substance of these nests is a sort of froth of the sea, or of the spawn of fish, which is strongly aromatic, though others assert that it has no taste at all; some pretend that it is a kind of gum, collected by the birds on the tree called *Calambone;* others, a viscous humour, which they discharge through the bill at the season of reproduction. Many again declare, that these swallows compose their nests of the *débris* of *holothuria*, &c.

As to their form, some say that it is hemispherical; others

Lewin del.
New Holland.

Griffith sc.

NEW HOLLAND SWALLOW.

H. PACIFICA.

London. Published by G.B. Whittaker March 1828.

ESCULENT SWALLOW.

H. ESCULENTA.

London. Published by G.B. Whittaker, Nov. 1828.

represent it like the valve of a shell, with similar stria or rugosities.

With regard to the places where these birds construct their nests, some assert, that they attach them to rocks, very near the level of the sea. Others that they are in the hollows of these same rocks; and others, that they conceal them in holes in the earth.

According to Kœmpfer, these nests, as we are acquainted with them, are nothing more than a preparation made from the flesh of the polype. From such contrarieties many were inclined to believe that either these nests did not exist at all, or that nothing to be depended on was known concerning them.

Montbeillard, for the purpose of settling these doubts, addressed himself to M. Poivre, a very enlightened and accurate observer. This traveller, while he was employed in collecting shells and corals, in a little islet called *Petit Toque*, situated near Java, entered into a tolerably deep cavern, hollowed in the rocks which border on the sea, and he found its sides, &c. hung with small nests, closely adhering to the rock. When these nests were removed, they were recognized by persons who had made several voyages to China, to be the nests held in such high esteem among the Chinese. The birds which had constructed them, were distinguished by M. Poivre to be true swallows, but whose size was nearly that of the colibris. He adds, that in the months of March and April, the seas which extend from Java to Cochin China to the north, and from the western point of Sumatra to New Guinea on the east, are covered with the spawn of fish, which forms on the water like a strong demi-diluted glue, and that he learned from the Malays, the Cochin Chinese, the Indians of the Philippine Islands, and the Moluccas, that the Salangana makes its nest with this fish-spawn. They were all agreed on this point. The bird collects it either as it shaves the level surface of the sea, or when it places itself on the rocks where this spawn is coagulated and deposited. At the epoch of the construction of the nest, threads of this viscous matter have been observed to hang from the bill of the bird. The observer we have quoted, collected some of this fish-spawn, and having dried it, declares it to be similar to the substance of the nests of the Salangana.

Notwithstanding all this, Sir Everard Home, who has investigated the stomach of a bird of this description, has given his decided opinion that the edible nests are composed of a substance secreted by the glands of that organ.

In our introductory essay on birds, we have cited another opinion, which appears to be by no means void of foundation. But we must refer our readers to the part in question, and forbear all repetition here.

At the end of July and in the commencement of August the Cochin-Chinese collect these nests. The species does not suffer by this, as these swallows multiply in March and April.

The form of the nests, according to the most accurate observers, is that of a semi-ellipsoid, hollow, elongated, and cut at right angles through the centre of its grand axis. They are composed externally of very slender laminæ, nearly concentric, and inclining so as to cover each other. The interior exhibits many layers of irregular net-work, with very unequal meshes, superposed on each other, and formed by a multitude of threads of the same substance as the exterior, and which cross and re-cross each other in every direction. This composition, which has a slight flavour of salt, is of a yellowish demi-transparent white. It softens in hot water without dissolving, and swells as it softens. It constitutes a substantial nourishment, full of prolific juices; and its effect, says Mauduyt, is excellent on exhausted constitutions, and where the stomach performs its functions badly. Poivre assures us, that he never ate any thing more nourishing and restorative than the soups made of these nests and some good meat combined.

These nests are of two kinds, the white and black. Some inhabitants of Sumatra are of opinion that they are the work of swallows of two different species. But Marsden, who relates this opinion, presumes that the white nests are those of the same year, and the black the old ones. He cites a fact, which seems to set the matter at rest. The Sumatrans, who collect these nests, destroy the old ones which they cannot bring away, in great quantities, that they may have a fresh supply of white ones in their place the approaching season.

These birds employ nearly two months to construct their nests, lay two eggs in each, and hatch them for about fifteen days. When the young are feathered is the time to carry off the nests, which is done three times a year. It is certain that these birds have three broods every year. They do not appear, according to Kirker, on the coasts, except in the laying season: but M. Poivre tells us, that they live all the year round in the islets, and on the rocks where they were born.

In the opinion of Sir George Staunton, followed by Dr. Latham, there is more than one species of swallow engaged in the construction of these curious and celebrated nests. The Doctor gives a figure of what he names the esculent swallow (for the subject we are now on he calls the *edible*), with its nest, which is as large as the river swallow. This was sent from Sumatra, with its nest and young, to Sir Joseph Banks, who presented them to Dr. Latham.

We shall now make a few observations on the swallows of South Africa, from M. Levaillant.

All the swallows which nestle in this part of the globe remain there only during the summer season, when there is the greatest degree of heat; and all the species of hirundo seen there during the winter of that country, or the rainy season, called in these climates the bad monsoon, are birds which come from other countries of Africa, after having had their brood in them, and which do not make a second there. This is evident from these last always bringing their young ones with them; and as the swallows which return to pass the hot weather in these same countries, for the purpose of reproduction, do not bring young birds with them, it is a proof that they have not brought forth during their absence. This is an additional argument that the birds do not reproduce in the two seasons of the year, although they change countries. We may further remark, that the birds which are purely birds of passage in a country, and do not reproduce there, never arrive during the brooding season of that particular country; and it is also not to be expected that they should lay at a season when none of the indigenous birds of the country do so. Such observations must of necessity throw great light on the migrations of birds, if travellers attend particularly to discover, in the countries they visit, the species which reproduce there, and those which do not.

The observations of M. Levaillant on this subject are so interesting, that we shall make a further extract of their substance.

He remarks, and the observation is of equal interest and importance, that a certain species which makes its brood in one country, quits that country and goes elsewhere; while, frequently, other individuals of the same species which have reproduced elsewhere, come and replace the former. This would very naturally lead us to conclude that it is not always the want of food that induces birds to expatriate, but an absolute necessity of changing climate. In Europe, where during winter no insects are found, it is very natural that the species which have no other nutriment, should, all of them, or for the most part, migrate. Some few, indeed, as we have seen, remain, and pass this rigorous season among us, still finding wherewithal to subsist. But in the very warm climates, from which all the swallows which have nestled there depart on the entrance of the rainy monsoon, most assuredly it is not the want of nutriment which causes them to migrate, for in this very same season other swallows arrive, which remain there, and find wherewith to subsist. Among migrating birds, we must also make a difference between those that only traverse a country, and those which make a regular sojourn of many months.

" It would appear," says M. Levaillant, in conclusion, " that the birds of every country may be divided into three distinct Classes: the birds proper and peculiar to the country, that is, those which nestle there; the stationary birds, which sojourn there without reproducing; and, finally, the birds of passage, which only traverse the country without either stopping or leaving any offspring. But before such a division can be established with any thing like accuracy, how much labour, observation, and patience, will be necessary! One thing we may venture to predict, that the work will never be performed by the ornithologists of the closet."

The *Cape Swallow* (*Rousseline*, of Levaillant) passes the entire summer at the Cape. It is also the species found most frequently and abundantly over the entire southern point of Africa. It is met with every where, but more particularly in inhabited places. These birds are so familiar that they enter houses, and especially those belonging to the colonists of the interior, for in the town their visits are not much encouraged, an account of the dirt they occasion in apartments. The peasants, who are less scrupulous on this point, suffer them not only to establish themselves very quietly in their dirty halls, which bid defiance to contamination, but also behold them nestle with great pleasure in their chambers, because they regard them as birds of good omen. The nest of these birds, when constructed in a chamber, is attached to the ceiling against a beam, and built with clay, mixed up like those of our European swallows. But its form is altogether different. It is a hollow ball, to which a long tunnel is adapted, through which the female glides into the interior of the nest. Inside it is furnished with a profusion of the most downy materials. The eggs are four to six in number, white, sprinkled with small brown spots, and the incubation lasts from sixteen to eighteen days.

This species, in relation to its form, is probably that which some travellers, who have only seen it fly, have taken for our chimney swallow. Its gait, cry, and physiognomy are exactly similar.

The *Rufous-fronted Swallow* is exactly similar to the preceding in form, but characterized by a red band across the forehead. M. Levaillant only met this species during the rainy monsoon, and never in the breeding season. He therefore thinks it probable that it only passes the winter in the South of Africa, and never nestles there. The only difference he observed between the sexes was, that the female was not quite so large as the mail, nor had the tail so long.

Many of these swallows were found by this naturalist in a collection of birds made in Senegal. He thinks it probable that their true country is situated near the equinoctial, and that they quit this region to betake themselves southward to pass the rainy season, after having reared their young ones. The young which they bring with them, when they arrive at the Cape, incontestably proves that they nestle elsewhere.

The *Sharp-quilled Swallows* have a short, robust, and rounded tarsus; strong claws; the thumb articulated more highly in the tarsus than in the preceding species, and the croup is muscular. But they are more particularly distinguished by thick stalks to the caudal quills, which terminate in a sharp point, furnished with barbs, from which the name *Acutipennis* has been attached to them.

These swallows take the place of the martens in America, where the latter birds are not found; for the birds of this continent, on which this name has been imposed, are true swallows. They rise to a very great height in the air, and are of a distrustful disposition, which always makes them keep out of range of shot. Their flight is uncertain and rapid, and they execute every motion which pleases them with the most perfect facility. Sometimes they may be observed clapping their wings with precipitation, sometimes extending them at full length, like broad vans winnowing the air. They pursue every direction which suits them, either horizontally, perpendicularly, or obliquely. They pass with much dexterity among the dry branches, and they are so essentially organized and destined for flight, that they never stop, or repose for a moment in the day.

The South American species nestle in the hollows of trees, or unite together in families. Those of North America usually establish their domicile in chimneys, where they construct a nest very ingeniously formed. In their plumage, manners, and style of life, they all present so much analogy that they would seem to compose only distinct races of one and the same species.

Among the most remarkable of these swallows is that called *Bibombi* (*Hirundo oxyura Vieil.*), which we shall take leave to call the *Sharp-tailed Swallow*, to distinguish it from the *aculeated Swallow* of Latham. M. D'Azara has described this swallow under the name of *Petit Martinet*, because he thinks that it resembles the marten of Europe. Like the latter, it cannot remain on the ground; but its crooked claws, very strong and sharp, give it a facility of climbing. Some, says M. D'Azara, name it *Mbiyuimbopi*, which means *bat-swallow*, because it has some resemblance to the bat in colour, and in its mode of flying, which is more rapid and uncertain than that of any other species. It is called by abbreviation *Bibompi* or *Bibombi*.

This species is very common in the woods of Paraguay. It always flies above the highest trees, and if in the open plain it sometimes approaches within forty or fifty feet of the ground, it rapidly remounts to its ordinary elevation. It is a sedentary bird, and extremely wild. Like the other swallows, it drinks flying, and sometimes catches, as it passes, the spiders which weave their nets under the branches. It passes the night in the holes of trees. These swallows come to their retreats about sunset, in small flocks. Before they enter, they fly round about three or four times, and at a considerable distance from the tree. The cry which they utter in flying is something like the sound of a small castanet. There is no difference between the male and female, the young and adult.

The *Aculeated Swallow. Hirundo Pelasgia.* It is an inhabitant of Louisiana and Carolina. It is also found still farther north, even to Pennsylvania, and beyond. Everywhere it fixes its nest on the top of chimnies, or in the crevices of rocks, if it has no other choice of place. It constructs this nest with an industry peculiar to itself. First of all, it establishes a sort of platform, composed of dry branches, and cemented together with peach-tree gum, or that of liquid amber. These materials are in such abundance, that they sometimes stop up the aperture of the chimney. It is said, that the bird supports itself, during this labour, by applying the sharp points of its tail against the wall. On this sort of scaffolding it places the cradle of its young, which is composed only with small sticks, glued together with the same gum, and arranged like the osiers of a basket, such as is given to pigeons to hatch in. It is open at top, and forms about the third of a circle. The eggs are four or five, of an elongated form, very gross in proportion to the size of the bird, spotted and streaked with black and grayish brown towards the gross end, on a white ground.

The Goatsuckers (*Caprimulgi*) are so named from a most absurd notion, that they suck the mammæ of goats, a notion which may perhaps have originated in the enormous depth and aperture of the gape. This vulgarism is by no means modern, for it appears, by the Greek appellative, to have existed in the time of Aristotle, though it seems probable, that the first application of the name might have had rather a figurative than a literal meaning. Many of the insectiverous birds, it is true, are found frequently near the persons of cattle and sheep while grazing—for the purpose, doubtless, of preying on the numerous insects which feed on the excretions from these animals; but this habit is common to many genera of birds, and gives no reasonable support to the notion in question, which is incompatible with the organization of the whole class. They are known among the peasantry in France by the name of *Engoulevent*, or Wind-swallowers, and *Crapaud-volant*, or Flying-toads, names which seem also to have reference to the extraordinary capacity of the gape.

These birds are inhabitants of Europe, and, indeeed, are found in almost all parts of the world; but they are rare here, and more so in appearance than reality, from their crepusculous habits. It is in the new world, especially South America, that they most abound, and are divisible into many species. Asia, and New Holland, moreover, are not without them.

Unfitted, like the owls, for full day-light, the Goatsuckers hide themselves in some obscure retreat. Twilight is their short period of activity, but the rapidity of their flight, and the size of the mouth, enable them to make the most of this limited time in procuring food. They devote no time to nidification, but deposit their eggs in simple concavities on the ground, and thus the time necessary for the two great occupations of animal existence, self-support and propagation, are proportioned to the comparative short periods of their activity. In the day, they sometimes utter a plaintive cry, repeated rapidly three or four times, and indicative of the then negative character of their desires, for they seem to want nothing but retirement and repose.

The *European Goatsucker* (*Caprimulgus Europeus*) is the only species known here. This bird has received a variety of popular names, which have been, many of them, adopted by naturalists. Such as flying-toad, square-tailed swallow, night-raven, night-hawk, dorr-hawk, churn and fern owl, &c. Its food, mode of taking it, and style of flying caused it to receive the name of square-tailed swallow.

This bird is solitary; two of them are rarely seen together, and even then they preserve a tolerable distance from each other. It frequents mountains and plains, almost always conceals itself under a bush, or in the young coppices, and seems to give the preference to dry and stony soils, and those which are covered with briars. Its mode of perching differs from that of other birds; it fixes itself longitudinally on the branch, which it seems to tread like the cock, from which circumstance it is often called *chauche-branche* in many parts of France. It is difficult to be perceived, when in this position, from the dulness of its colours, and their approximation to that of the branches. It is when thus situated that it principally utters its peculiar cry, which having some resemblance to that of a toad, has gained for it, in France, the popular name of *crapaud-volant.* This cry is a plaintive sort of sound, repeated three or four times in succession. The noise which it makes in flying is dissimilar, and, according to some, is a cry of another kind. But, according to others, it is caused by the air, which it engulphs in its large throat, since it flies with its mouth open, which produces a humming noise, like to that of a spinning-wheel. This opinion seems probable, especially if it be true, as is asserted, that the sound varies according to the different degrees of velocity in the flight.

This demi-nocturnal bird never quits its retreat but towards twilight; or, if it ever do so, it is only in sombre and cloudy weather, for it is dazzled by a strong light. The feeblest degree of light suits it best. If it happen to be disturbed and roused of a fine day, its flight is low and uncertain. The reverse, however, is altogether the case after the setting of the sun. It is then quite lively, and active in its flight, which is necessarily irregular, like that of the winged insects which constitute its prey, and which it can only seize by short and rapid zigzags.

It feeds on insects, especially those of the nocturnal kind, as beetles, cockchafers, moths, &c. It will also eat wasps, drones, &c. It has been observed, that this bird has no occasion to close its bill to secure the insects, the interior being provided with a kind of glue, which appears to come from the upper part, and which is sufficient to retain them.

This bird has one habit peculiar to itself; it will make, about one hundred times, in succession, the circuit of a large leafless tree, with an irregular and very rapid flight. And from time to time, it will drop abruptly down, as if to fall upon its prey, and then suddenly rise again in the same manner. At such times it is exceedingly difficult to bring it within range of shot, for on the advance of the fowler, it disappears so rapidly, that it is impossible to discover the place of its retreat.

An elegant and artificially constructed nest, requires the assistance of day-light, and the love of labour. We must not then expect it from this bird, condemned by nature to remain during the day in a sombre and solitary state of inaction. A small hole at the foot of a tree, or of a rock, and sometimes even in the naked and beaten ground, is the place where the female deposits her eggs. These are rather more bulky than those of the blackbird, oblong, lightly shaded, and marbled with blackish points on a white ground. We are assured, that she hatches them with the greatest solicitude, and that when she discovers that they have been observed, she changes their place, by pushing them forward dexterously with her wings into some other spot, and sometimes even carrying them there with her bill.

This goatsucker is migratory. It arrives in our climates in spring, and departs in autumn. In the latter season the greater number of these birds are usually seen, but they are preparing to quit us for other regions, where their food is more abundant. These birds are found from the most northern parts of Europe even into Africa. They pass the month of April in Malta, whither they are carried by the south-west wind, and where they are also found in great numbers in the autumnal season. They do not arrive in this country until towards the end of May, and they leave us about the middle of August. They remain later in France, and may be seen even in November. It is said, that many birds of this species have been killed in the woods of Vosges, in the middle of winter and in the depth of snow. Still, however, this must be very rarely, as, at such periods, the insectivorous birds must experience the greatest difficulty in procuring an adequate supply of aliment.

The *Leona Goatsucker*, *Cap. longipennis* and *macrodipterus*, is remarkable for a long single feather issuing from the wing, the shaft being without web till near the end where the web is broad; this feather is much longer than the bird itself. The species, as to colour, is assimilated to the common goatsucker. Very little is known of its habits, though many specimens have been brought to England: our figure is from one in the late Riddell's museum.

The *Virginian Goatsucker* is found from Virginia as far as Hudson's Bay, after the middle of April. Unlike most of its congeners, this species does not prefer the tops of trees, but resorts to bushes, or the top of a post or rail, whence it will leap up to catch the insects, as they pass over it. For some time about sunrise, and again at sunset, it is constantly

LEONA GOAT-SUCKER.

C. MACRODIPTERUS.

London. Published by G.B. Whittaker. Nov. 1828.

repeating a cry in a very sharp tone, which has been expressed by the compound word *whip-poor-will,* and hence the bird has been so called. It is said to eat bees. Its eggs are greenish brown, with zigzag black stripes; these it deposits negligently, sometimes in the middle of a beaten path.

The *Cayenne,* or *White-necked Goatsucker,* of Latham, is said to be less nocturnal in its habits than the other species; it is also more sociable, frequenting the vicinity of highways, nor does it move until approached very closely, and then only to a short distance. It is said to utter two sorts of cry, one like that of the toad, and another which has been compared to the barking of a dog; while uttering the former, it is said to shake the wings.

The *Collared Goatsucker, C. Pectoralis,* probably the *C. Asiaticus,* or Bombay Goatsucker of Latham, was observed by Levaillant on the banks of the Gamtoo in Africa, where it is very common.

During the time of incubation the male begins its loud and singular song about an hour after sunset, and if the night be fine, continues till daylight. Levaillant states, that while he had the misfortune to be encamped in the vicinity of these birds, it was impossible for him to sleep on account of their ceaseless singing. The female lays two white eggs, with the shells so exceedingly thin and brittle, that it requires great caution to handle without breaking them. Notwithstanding this, the bird, like those before mentioned, deposits these eggs frequently in the middle of a path. The male sits as well as its female, and when so occupied, the bird will not move on the approach of any one until in actual danger of being trodden on. After the eggs have been touched, they are sure to be removed, and probably to a considerable distance; for Levaillant, under these circumstances, was never able to find them again. He had an opportunity of observing the mode in which they carried them off, by placing himself within sight of the eggs in a tree. The female first came to them after they had been handled by him, and flying to the ground a short distance from the eggs, approached them gently step by step, and having ascertained that they had been touched (by what faculty it does not appear), she walked several times round them, with her beak close to the eggs; she then uttered several cries, at the same time resting on her breast, and beating the ground with the wings; this seemed to bring the male bird, which immediately commenced the same cry, and the same operations; after which they both flew a few times round the eggs, and then suddenly each took one in the mouth, and disappeared on the wing.

M. Levaillant found another species of goatsucker in South Africa, remarkable for its size—this was the *Fork-tailed Goatsucker*. This character of tail is unique in this genus. Of all the different species mentioned by nomenclators, this one alone has the tail of this form. It is six-and-twenty inches long from the point of the bill to the extremity of the longest feather of the tail, which is the last lateral one on each side. The bill of this large goatsucker is of an enormous width, and terminates in a small hook, more resembling a talon, than the end of a bird's bill. A proof how little it was the intention of nature that these birds should engulph a quantity of wind, when in pursuit of insects, is that there is no bird whose mouth is so firmly closed. In fact, the construction of the bill is so managed, that the lower mandible covers the upper at the corner of the mouth by a small projecting edge, and the upper, by a sort of lid, covers again the lower, which last is firmly enclosed as far as a very marked notch about the middle of the upper. After this notch, the upper mandible grows suddenly very narrow, and emboxes itself in the lower, which is conformed to receive it with a proper edge, and is itself finally surmounted by the end of the upper one, which holds it firmly, and curves beyond it in the form of a hook. From this perfect union of the two mandibles, this bird appears, when the mouth is closed, to have a very small bill—" Those who imagine," says M. Levaillant, " that these birds always fly with their mouths open, are very grossly deceived. They frequently place themselves on the ground, for the purpose of collecting the insects there, and in taking them on the wing, it does not appear necessary that they should always keep their mouths gaping open. We find the bee-eaters, the martens, and all the swallows, take insects on the wing, and we never see them open the bill until the moment they are near enough to snap them up. It is probable that the goatsucker does the same. Nature, who is never deceived, nor does any thing in vain, would scarcely have constructed the bill of this bird with so much care, and sealed it so hermetically, if it was always to be wide open, that the bird might procure its nutriment. M. Levaillant has given an excėllent representation of the bill of this bird, both open and closed, of the natural size. This goatsucker, from the forked tail, has a still greater analogy with the swallows than the rest of its congeners. The wings are about forty inches from tip to tip, and when folded extend as far as the feathers of the tail.

The fork-tailed goatsucker was discovered by the naturalist we are citing, on the banks of the river of Lions, in the country of the great Namaquois, in the interior of the Cape. It was by a singular chance that he procured both male and female of this species. One day, hunting on the banks of this river, accompanied by his native attendant, they were assailed by a hurricane with tremendous rain, which forced them to retire under some very large mimosas for shelter. Looking around, they beheld a very thick tree quite dead, whose stalk, almost entirely hollowed, contained a vast hole, which communicated into the whole body of the worm-eaten trunk. Hoping to find some insects under the bark of this tree, they approached, but when they came there, they heard, in the interior of the tree, a low humming noise. Not knowing what this could be, they took some precaution to ascertain from what kind of animal the noise proceeded, fearing, with some reason, that a nest of serpents might be lurking there. They were, however, much surprized by finding two large birds, which they drew out of the hole, one after the other. M. Levaillant preserved them alive for two days. The light of the sun so affected them, that they did not attempt to escape during the day, but they made a desperate bustle at night in a basket in which they were confined.

These were the only two of the species which M. Levaillant ever saw. During the night alone they uttered a tremulous guttural sound, with the mouth so wide open that a large apple might have been introduced into it. The tongue of this bird is very small, and placed at the entrance of the throat. The species does not appear to be common.

The *Urutau* (*Caprimulgus cornutus* of Vieill.) is so named from the short erect feathers over each eye having a fancied resemblance to horns. D'Azara informs us, that they remain in Paraguay from October to February, and perch on high trees, clinging by the claws in the manner of the woodpecker, and very seldom come to the ground. The male and female answer each other during the whole night by long and melancholy cries. They lay two brown and spotted eggs in a small cleft of a dry tree, and without any nest; and, according to credible information received by D'Azara, the female sits on, or rather covers, these eggs by hanging vertically on the side of the tree, and bringing the breast on the small cleft which contains the eggs. The report in the country is, indeed, that these birds have the means of affixing their eggs to the tree by means of some agglutinous substance, but M. D'Azara was not able to verify this statement. It may derive some degree of probability, from the consideration that the mouth of all the species of this genus seems to be furnished with a sort of gluten which assists them in the capture of insects, and

SCISSARS-TAILED GOAT SUCKER.

C. PSALURUS, Tem.

of insects, and by means of which, in some species, at least, it seems they are enabled to retain many insects of small size at a time in their mouth, and thus to swallow a number of them together.

M. D'Azara, in his work on the Animals of Paraguay, has described several species of these birds, short specific characters of which are given in the table. His valuable observations, however, afford little on their habits, instincts, and manners, to interest, beyond what is above stated.

One species, however, is too remarkable to be confounded. It is the *Ibigau* of Paraguay, the Scissars-tailed Goatsucker, *Caprimulgus psalurus*. D'Azara saw but few of these birds, always alone, and in the middle of winter only frequenting rivers, and flying near the surface of the water. When they alter the direction of their flight they open wide the scissars-like feathers of the tail. These feathers, of which the outermost run out to a considerable distance from the rest, bear some analogy to the long feathers which proceed from the shoulder of the Leona goatsucker, except that in the latter species the feathers are barbed only at the end, whereas in this the barbs are longest near the insertion, and diminish by degrees toward the tip. This character is much more remarkable in the old males than in the females, or young birds. We insert a figure from M. Temminck's work.

Like the rest of the genus, the prevailing colour of this bird is dark brown, varied with spots of different forms and sizes, lighter brown, and white; the neck is yellow, and the ground colour of the throat yellowish white.

The strength of the bill is the principal character which distinguishes our author's genus, PODARGUS, from the common goatsuckers. If their bill were more compressed it would resemble that of the owls, to which, indeed, this group of the goatsuckers is much assimilated, by nocturnal habits, the nature of the plumage, and the stiff bristly hairs which surround the base of the upper mandible.

The opposite figure of one of these birds is from a specimen in the museum of the Zoological Society. Its characters are like those of Podargus Javanensis of Dr. Horsfield, having many long loose feathers proceeding horizontally from the root of the upper mandible, which gives the bird a singular and grotesque appearance. The general colour is dark brown, lighter on the lower part, with transverse bars on the wing-coverts and tail. Mr. Vigors, who is about to describe it in the "Zoological Journal," names it *P. Auritus*.

The *Trinidad Goatsucker* of Dr. Latham, which seems also to be the *Steatornis* of Humboldt; seems either to belong to this division, or else to be entitled to a generic separation. We subjoin Dr. Latham's account of it, as communicated to him by Mr. Thompson, except of the specific characters, for which we refer to the table; he names it the *Trinidad Goat-sucker*.

They inhabit coves of the islands forming the Bocate, an entrance into the Gulf of Paria, accessible only at the very lowest ebb tides, and in moderate weather; and as they are never observed on the wing in day-time, most probably, like the rest of the genus, they seek their food in the absence of the sun. Here they breed during the early part of the spring; and it is at the time of new and full moons in April and May that the people who are acquainted with these coves, resort thither to take the young, and such of the old birds as they can knock down with sticks.

They have a strong and disagreeable fishy smell, but some people resemble it to that of the cockroach; and when dressed they look like a round lump of fat, the little flesh there is tasting more like that of a sucking-pig than any other, but yet with a flavour and lusciousness peculiarly its own. But what is most extraordinary is, that in a family supposed to be wholly insectivorous, this should constitute a singular and solitary exception, and be found to subsist, at least during the breeding season, entirely on fruit; for on examining the

H. Kearsley del.[t]

Mus. Zool. Soc.

LONG-EARED PODARGUS.

PODARGUS AURITUS. Vigors & Horsfield.

London Pub.[d] by G.B. Whittaker Nov 1828

C. Hamilton Smith Esq.[e] del.

Rev. M.[r] Hennah's Mus.

GOATSUCKER FROM PERU.

CAPRIMULGUS.

London. Published by G.B. Whittaker, July, 1828.

stomach of a dozen of them, young and old, no other species of food whatever but the fruit of the palm appeared. The collector in ornithology will find a very troublesome task in preserving this bird, as the skin adheres with uncommon closeness and tenacity to the granular fat, which every where covers the body, and which liquifies under the touch.

This, or a species greatly similar, is mentioned by M. de Humboldt, as inhabiting a dark cavern, formed by rocks, thrown together by the hand of nature, in the Cordilleras. "Numberless flights of nocturnal birds," says he, "haunt the crevice, and which we were led at first to mistake for bats of a gigantic size; thousands of them are seen flying over the surface of the water. The Indians assured us that they are the size of a fowl, with a curved beak, and an owl's eye. They are called cacas, and the uniform colour of their plumage, which is brownish gray, leads me to think that they belong to the genus Caprimulgus, the species of which is so various in the Cordilleras. It is impossible to catch them, on account of the depth of the valley, and they can be examined only by throwing down rockets to illuminate the sides of the rock."

We have inserted a figure, from a specimen brought from South America by the Rev. Mr. Hennah, which we cannot with certainty refer to either of the described species. It may possibly be the female of Dr. Latham.

The Third Family of the Passeres, or the Conirostres,

Comprehends the genera with strong beak, more or less conical, and not toothed. They live more exclusively on grain, in proportion as the bill is stronger and thicker.

We distinguish at first among them, the genus of the

Larks. Alauda. *Lin.*

By the claw of their thumb, which is altogether straight, strong, and much longer than the others. These are granivorous birds, pulverators, which sojourn and nestle on the ground.

The greater number have the straight bill moderately bulky and pointed.

The Sky Lark. Al. arvensis. Enl. 368. 1.

Is universally known by its perpendicular flight, which it performs singing with force and variety, and by the abundance in which it is provided for our tables. Plumage, brown above, whitish beneath, spotted throughout with a deeper brown; the two external quills of the tail brown without.

The Crested Lark. Alauda cristata. Enl. 503. 1.

Very nearly of the same size, and of the same plumage; the feathers of the head capable of being erected into a tuft; less common than the preceding; approaches villages and coppices.

The Wood Lark. Cujelier. Lulu. Al. arborea. Al. nemorosa. Enl. 503. 2.

Also bears a small tuft, but less marked. Is smaller than the preceding, and otherwise distinguished by a white mark round the head. Particularly delights in the brambles in the interior of woods.

The Italian Lark. Al. Italica.

Is considered as a variety of the sky lark.

The Undated Lark. Al. Undata. Enl. 662.

This is said to be a variety of the crested lark.

Short-toed Lark. Al. brachydactyla. Naum. 98.—2. *Calandrelle.* Temm.

Upper wing-coverts longer than the quills; toes very short; red, with white spot under the eye. Southern Europe and Africa.

The Clapper Lark. Al. Apiata. Vieil. Vail. O. A. t. 194.

Above, varied chestnut, black, and white; throat, white; belly, orange-white. Cape of Good Hope.

The Red-backed Lark. Al. Pyrrhonotha. Vieil. Vail. O. A. 197.

Reddish above, whitish underneath, with brown lines on the chest. Cape.

Red-crowned Lark. Al. rufipilea. Vieil. Vail. O. A. 198.

Red above, whitish beneath, with black cross-streaks. Cape.

Al. Magna, is only the Sturnus Ludovicianus.

We sometimes see in Europe,

Al. Alpestris. Al. Flava, and *Al. Sibirica.* Gm. Enl. 652. 2 Naum. 99. 2. 3. Wilson, I. 5. 4.

A bird proper to Siberia and North America, with the forehead, cheeks, and throat yellow, and a large black patch across the upper part of the chest. The male has a small pointed tuft behind each ear.

Others have the bill so short and thick, that in this respect they approach the sparrows: such is

The Calandre. Al. Calandra. Enl. 363.

The largest species of Europe; brown above, whitish beneath; a large blackish spot on the chest of the male. Of the south of Europe and the deserts of Asia.

This is the Al. Sibirica of Pallas, figured by Edwards, p. 268.

But especially

The Black or mutable Lark. Al. Tatarica et *mutabilis,* et *Tanagra Sibirica.* Gm. Sparm. Mus. Carls. pl. xix. Vieil. Gal. 160.

The plumage of the above is black, waved above with grayish. It sometimes wanders into Europe.

The *Tracal.* Vaill. Afri. pl. exci.

Is considered as a variety of the last.

Great-billed Lark, Al. crassirostris. Vieil. *l'Alouette à gros-bec.* Levaillant. Afri. IV. 128. pl. 193.

Brown above, white beneath. South Africa.

Others have the bill long, slightly arched, and compressed, in which particulars they are allied to the Hoopoes and Promerops.

The Sirli. Al. Africana. Gm. Enl. 712. Vieil. Gal. 159.

A common bird in the sandy plains from one end of Africa to the other; its plumage is not unlike that of our common lark.

Temminck refers them to Anthus.

The Double-banded Lark. Al. bifasciata. Lich. Pl. Col. 293.

Ochraceous-yellow; throat and belly white; the former with black spot; quills and tail brown. Nubia, and rarely in Provence *

* The following species of Alauda have been named by other naturalists :—

Alauda rufescens. Vieil.
Azara's lark: Al. Cunicularia. Vieil. Azara.
Double-crested lark. Al. bilopha. Tem. Pl. Col. 244.
The Desert lark. Al. Deserti. Pl. Col. 244.
White-banded lark. Al. melanocephala. Licht.
Lapland finch. Fr. Laponica. Gm. *F. Calcarata.* Pall., &c.
Al. Mirafa. Tem. Pl. Col. 305. *Mirafa Javanica.* Horsf.
Al. crocea. Vieil.

THE TITMICE. PARUS. Lin.

Have the beak slender, short, conical, and straight; furnished with little hairs at its base, and the nostrils concealed in the feathers. These are very lively little birds, flitting and climbing incessantly through the branches, suspending themselves there in all manner of ways, tearing the grains on which they feed; eating also many insects, and not sparing even small birds, when they find them sick, and are able to despatch them. They are in the habit of collecting provisions of grain. They nestle in the hollows of old trees, and lay more eggs than any other of the passeres.

We have in France six titmice, properly so called.

The Great Titmouse. Parus Major. Enl. 3. 1.

Rather olive above; yellow beneath; the head, and a longitudinal band on the breast, black; a white triangle on each cheek. Very common in coppices and gardens.

The Cole Titmouse. Parus Ater. Lin. Frisch. I. pl. XIII. 2. A.

Smaller than the preceding, with some ashen instead

Malabar lark. Al. Malabarica. Gm. Ind. t. 116.
Gingi lark. Alauda Gingica. Lath. Ind. t. 113.
Alauda Yeltoniensis. Forster.
New Zealand lark. Al. Novæ Zelandiæ. Gm. Portlock's Voy. t. 37.
Cinereous Lark. Alauda Cinerea. Gm.
Senegal lark. Al. Senegalensis. Gm. Pl. Enl. t. 504.

of olive, and whitish instead of yellow. It inhabits, by preference, large woods of fir-trees.

The Black-cap Titmouse. Parus Palustris. L. Enl. 3. 3.

Ash-colour above, whitish beneath: a black coif.

The Blue Titmouse. Parus Ceruleus. Enl. t. 3. f. 2.

Rather olive above; yellowish beneath; the top of the head a fine blue; the cheek white, edged on all sides with black; the forehead white. A pretty little bird, common enough in coppices.

The Crested Titmouse. Parus Cristatus. Enl. t. 502. f. 2. Albin. t. 57.

Brown above; whitish beneath; the throat and circle of the cheek black; a small tuft, speckled with black and white.

The long-tailed Titmouse. Parus Caudatus. Enl. 502. 3. ♀ Frisch. t. 14. f. 2. ♂

Black above; wing-coverts brown; top of the head and all beneath, white; tail longer than the body. It makes its nest on the branches of small trees, and covers it above.

The Crested Titmouse. Parus Bicolor. Lin. Wils. A. O. 8. Cat. 1. 57.

Crested; lead-coloured; beneath whitish; flank tinged with dull reddish orange. North America.

Azure Titmouse. Parus Cyanus. Pal. Nov. Com. Acad. Pet. XIV. 23. *P. Cyaneus.* Falk. *P. Sælbyensis.* Sparm. *P. Kujæsuk.* Gm.

Azure; forehead, temples, spot on nape, and beneath, white; tail long. North Europe and Asia.

The Black-cap Titmouse. P. Palustris, var. Lath. *P. Atricapillus.* Wils. A. O. I. 8. Pl. Enl. 502.

Lead-coloured; crown and nape black. North America.

Siberian Titmouse. P. Sibiricus. Gm. Pl. Enl. 708.

Reddish ash above; head and neck brown; throat and upper part of chest black, with a white band extending up the sides of neck. North Europe and Asia.

Javan Titmouse. P. Atriceps. Lin. Trans. 13. p. 160. Horsfield. Pl. Col. 287.—2.

Above, bluish gray; beneath, whitish; head, blue-black. Length, five and a half inches. Java.

The Parus Malabaricus. (Sonnerat. 2. Voy. Pl. cx. 1.), and *coccineus* (Sparm. Mus. Carls. 48. 49.) *P. furcatus.* Col. 287.—1., are traquets or fly-catchers, bordering on *Oranor.* Vail; on *Mot. ruticilla.* L; and *turdus speciosus.* Lath.

We may remark, that whenever the characters of a bird are not very clearly marked, authors have been in the habit of shuffling it from one genus to another without end.*

These have been also named:—

Parus lugubris. Lath. Tem.
Muscicapa fuliginosa. Sparm.

THE BEARDED TITMICE

Differ from the titmice proper, in the upper mandible of the bill, the end of which is a little recurved over the other.

They form the genus *Mystacinus* of Bore.

The Bearded Titmouse. P. biarmicus. Lin. Enl. 618. Vieil. 69. Naum. 96.

Fawn-coloured; the male with ash-coloured head, with a black band, which surrounds the eye, and terminates in a point behind. This bird nestles in the thickest reeds. It is found in all the ancient continent, though but rarely.*

THE PENDULINE TITMICE. REMIZ. CUV.

Have the bill more slender (more straight) and pointed than the common titmice. They employ more art

* To this add from other authors :—

Great headed Titmouse. P. Macrocephalus. Lath.
New Zealand Titmouse. P. Novæ Zelandiæ. Gm.
Norway Titmouse. P. Stornei. Lath. *P. ignotus.* Gm.
White-cheek Titmouse. P. Cinereus. Vieil. Vail. O. A. 139.
Grayish Titmouse. P. Cinerascens. Vieil. Vail. O. A. 190.
Black Titmouse. P. Niger. Vieil. Vail. O. A. 137.
Fork-tailed Titmouse. P. Indicus. Lin. *P. furcatus.* Pl. Col. 287.
P. Griseus. Muller, appears to be a regulus. *P. Virginianus and Americanus,* a sylvia.

Also have been named :
P. Afer. Gm.
Guiana Titmouse. Lath. *P. Celer.* Lin.
Amorous Titmouse. Lath. *P. Amatorius.* Lin.
Alpine Titmouse. P. Alpinus.
Chinese Titmouse. P. Chinensis. Gm.

in the construction of their nests. We have but one in Europe.

The Penduline Titmouse. Parus pendulinus.
Enl. 618. 3.

Ash-colour; wings and tail brown; a black band on the forehead, extending even behind the eyes in the male. This little bird, inhabiting the South and East of Europe, is famous for the pretty nest, tissued with the down of the willow and poplar, furnished within with feathers, and shaped like a purse, which it suspends to the flexible branches of aquatic trees.

The young is the Languedoc Titmouse.

Parus Narbonensis. Gmel. Pl. Enl. t. 708.

Appears to be the female of *pendulinus.*

Cape Titmouse. Lath. *Parus Capensis.* Gmel. Sonn. Ind. t. 115.

Gray-ash; quills black, white edged; tail black, beneath white. Cape of Good Hope.

The nest of this bird, made of cotton, in the form of a bottle, has a stand for the male bird near the neck.

THE BUNTINGS. EMBERIZA. Lin.

Have a character extremely distinct, in their conical beak, short and straight, the upper mandible of which, more narrow, and re-entering into the lower, has in the palate a hard and projecting tubercle. They are granivorous birds, of little sagacity, blindly running into all the snares which are laid for them.

The Yellow Bunting. Emberiza citrinella. L.
Enl. t. 30. f. 1. Alb. 1. t. 66.

Fawn-coloured; back, black-spotted; head, and all the body beneath, yellow; the two external quills of the tail, on the internal edge, white. Nestles in the hedges; approaches habitations in winter, in very numerous flocks, with the sparrows, &c., when the snow covers the ground.

The Foolish Bunting. Emb. cia. Lin. En. t. 511. f. 1.

The young male differs in being reddish gray beneath, the sides of the head whitish, surrounded with black triangular lines. Of mountainous countries.

The *Emb. Lotharingica.* Pl. Enl. t. 511. f. 1. does not differ.

The Cirl Bunting. Emb. cirlus. Lin. En. 653. f. 1.

Male, or young male, black throat; sides of the head yellow; nestles in thickets on the borders of fields.

The Passerine Bunting. Lath. *Emb. passerina.* Gm. Lin. 1. 871. Ind. Orn. 1. 403. Tem. Man. 182.

Head ferrugineous ash-colour; white line on each side of chin; back, gray brown; fore-neck, black; underneath, cinereous white. Russia.

Mustachio Bunting. Lath. *Emb. provincialis.* Gm. Lin. 1. 881. Pl. Enl. 656. 1.

White streak through the eye, beneath it a black patch; breast and sides, pale brown; under part white. Provence.

Lesbian Bunting. Lath. *Emb. Lesbia.* Gmel. Lin. 1. 882. Enl. 656. 2.

Like the last, but without the black spot; instead of which are three narrow black bands. Provence.

These two are, perhaps, but accidental varieties.

Reed Bunting. *Emb. schœniclus.* Lin. Enl. t. 247. f. 2. 5.—t. 477. f. 2.

Has on the head a black coif, and spots of the same colour on the chest. Nestles at the foot of bushes by the water-side.

M. Wolf believes that *Emb. chlorocephala,* and *Emb. badensis,* should be joined to this.

The largest species of this country, is

The Common Bunting. *Emb. Miliaria.* Enl. t. 233. Gray brown, spotted throughout with deep brown. It nestles in grass and corn.

The most celebrated, from the flavour of its flesh, is

The Ortolan. *Emb. hortulana.* Enl. t. 247. f. 1. Back olive-brown; yellowish neck; two external tail-feathers white within. Nestles in the hedges; is common and very fat in autumn.

The *Emb. Melbensis.* Sparm. Mus. Carls. 1. 21. is only a young ortolan.

After all the repetitions of synonyms which we have stigmatized, it will be still necessary to remove from this genus *Emb. brumalis*, which is the same bird as *Fring. citrinella.* Enl. 658. 2. *E. rubra*, the same as *Fring erythrocephala.* Enl. 665. 1. 2.—All the widow-birds, as I shall remark by and by:—*Emb. quadricolor.* Enl. 101. 2. *Emb. cyanopis.* Briss. III. pl. viii. fig. 4. *Emb. cœrulea.* id. ib. xiv. 2., the same as *cyanella.* Sparm. Carls. II. 42, 43, which are three loxiæ;—*Emb. quelea.* Enl. 223. 1. *Emb. capensis.* Enl. 158 and 564;—*Emb. borbonica.* Enl. 321. 2;—*Emb. braziliensis*, ib. 1, which are four sparrows;—*Emb. ciris*, Enl. 158, which is a linnet;—and in fine, *Emb. oryzivora.* Enl. 388, which has the bill of the linnets; not to reckon the species which I have not been able to examine.

But the following must be placed in this genus:—

The Commanding Bunting. Emb. Gubernatrix. Tem. Pl. Col. 63. *male*, 64. *female.*

Small crest; a band of pure yellow from the nostrils to beyond the eyes; top of head, crest, throat, and part of the fore-neck, black; sides of head and neck, and underneath, yellowish; length, six inches 3 lines. Buenos Ayres.

Also, *Emb. striolata.* Ruppel. A. O. Pl. 10, a;—*Emb. cæsia*, id. ib. b.; the *tanagra cristatella, graminea, ruficollis.* Spix. 53, are also buntings.

The Emberizoides. Tem. Col. 114. appear to be buntings with long and wedged tail, and whose bill approximates a little to that of the sparrows.

In the South of Europe is also sometimes found

The Black-headed Bunting. (*Emb. melanocephala.*) Scop. Naum. 101. 2. *Fring. crocea.* Vieil. Ois. tab. 27.

Fawn-colour above, yellow underneath, with a black head.

The Pine Bunting. (*Emb. pithyornis.* Pall.) Naum. 104.—3.

The throat, and a mark on the side of the head, of a moronne-red.

M. Meyer distinguishes, under the name of PLECTROPHANES, the buntings which have the thumb-claw elongated like the larks. Such is

The Snow Bunting. (*Emb nivalis.* L.) Enl. 511. Naum. 106 and 107.

Which is recognized by a wide longitudinal white band on the wing. It is a northern bird, and becomes almost all white in winter.

Emberiza montana, and *Emb. mustelina*, are only different states of the snow-bunting.

We must add to this genus,

The Lapland Bunting, Grand Montain of Buffon. (*Fring. Lapponica,* Gm. or *Calcarata,* Pall. See Trad. fr. III. pl. 1. 1.) Naum. 108.

Spotted with black on a fawn-coloured ground; throat and upper part of the chest black in the male. It inhabits the same countries as the preceding, and in like manner, comes to us only in winter, and much more rarely.*

* The following have been also attributed to Emberiza, by other naturalists:—

Chinese bunting. Emberiza Sinensis. Gm.

Yellow-winged bunting. Emb. Chrysoptera. Portl. Voy. t. at p. 35.

Louisiana bunting. Emb. Ludoviciana. Lin. Pl. Enl. t. 158. f. 1. *Passerina ruficapilla.* Vieil. *Emberiza rutila.* Pallas.

Rusty bunting. Emberiza ferruginea.

Black-throated bunting. Emberiza Americana. Gm.

Mexican bunting. Emberiza Mexicana. Pl. Enl. t. 386. f. 1.

Military bunting. Emberiza Militaris. Hasselt.

Angola bunting. Emb. Angolensis. Gm.

Barred-tailed bunting. Emberiza fusca. Gm.

Emberiza coccinea. Sander. Naturf. XIII. 194.

Familiar bunting. Emb. familiaris. Lin. *Mot. familiaris.* Osb.

Amazon bunting. Emb. Amazona. Lin.

Olive bunting. Emb. olivacea. Lin. Brisson. t. 13. f. 5.

Oonalashka bunting. Emb. Oonalashkensis. Gm.

Black-crowned bunting. Emb. Atricapilla. Gm. *Emberiza mixta.* Lin.

Green-bunting. Emb. viridis. Gm.

Plata bunting. Emb. Platensis. Gm.

Red-eyed bunting. Emb. Calfat. Gm.

Gray bunting. Emb. grisea. Gm.

Towhe bunting. Emb. Erythropthalma. Fringilla Erythropthalma. Lin.

Emberiza rustica. Pall.

Emberiza pusilla. Pall.

The Sparrows. Fringilla.

Have the beak conical, and more or less gross at its base, but its commissure is not angular. They generally subsist on grain, and are, for the most part, voracious and mischievous.

We subdivide them as follows:

The Weavers. Ploceus. Cuv.

With beak sufficiently large to have caused them to be partly classed with the Cassiques; but its straight commissure distinguishes them. They have, moreover, the upper mandible slightly convex.

They are found in both continents. Most of those in the Old World construct their nest with much art, interlacing it with blades of grass, whence the name of weavers *(Tisserins)* is given to them.

Sandwich Bunting. Emb. Sandwichensis. Gm.

The following are very doubtful:—

Surinam bunting. Emb. Surinamensis.
Gaur bunting. Emb. Asiatica.
Emberiza fucata. Pall.
Emberiza Chrysophrys. Pall.
Emberiza Superciliosa. Vieil.
Emberiza melanodera. Quoy and Gaim. t. 109.

The following are *Emberizoides* of Temminck, and *Tardivola* of Swainson:—

Emberizoides Melanotes. Temm. Pl. Col. 114. 1. *Chipui.* Azara. No. 140.

Emberizoides Marginalis. Tem. Pl. col. 114. f. 2. *Fringilla Macroura.* Lath.

Such is the *Philippine grosbeak. The Toucnam-courvi* of the Philippines. *Loxia Philippina.* Lin. En. 135. f. 2. ♂ Brisson. III. t. 12. f. 1. t. 18. En. f. 1. 2.

Yellow, spotted with brown, and black throat. Its nest, suspended, is in the form of a ball, with a vertical canal, and open beneath, which communicates through the side into the cavity where the young are placed.

Weaver Oriole. Oriolus textor. Gm. pl. Enl. 375. and 376. *Agelaius.* Vieil.

Fulvous-yellow; head gold-colour; quills and tail black, fulvous-edged. Senegal.

Crimson Bunting. Fringilla rubra. Kuhl. *Fring. Erythrocephala.* Cuv. *not.* Gm. pl. Enl. 665. 1. 2. *Emberiza rubra.* Gm.

Scarlet; nape and back, olive, varied with black, and scattered scarlet streaks; belly, ash; quills and tail, brown. Isle of France. *Female,* olive green; beneath, paler.

Malimbic Tanager. Lath. *Tangara de Malimbe.* Daud. An. Mus. I. p. 148. pl. X. *Textor Malimbicus.* Temm. *Fringilla textrix.* Licht. *Ploceus cristatus.* Vieil.

A scarlet crest; this colour extending over cheeks, throat, and upper part of breast; rest of the plumage lustrous-black. Western Africa.

Ploceus Aurantius. Vieil. pl. 44. des. Ois. chant.

Lores-black; head, throat, and body beneath, orange; above, olive: middle wing-coverts yellow. Congo.—Length, five inches.

Ploceus Nigricollis. Vieil. t. 45. Ois. Chant.

Throat and nape black; body above, greenish brown; beneath, and head, yellow; bill black. Congo.

Ploceus Aurifrons. Tem. Pl. Col. 175.—176. *male and female.*

The forehead, and top of the head, a fine gold colour; the cheeks and throat, a yellow tint less brilliant; sides of neck, and beneath, citron yellow; nape, above, tail, and edges of wings, yellowish green; quills blackish, bordered with yellowish green. Southern point of Africa.

Loxia Abyssinica. Gm. *Ploceus Baglafecht.* Vieil.

Body yellow; crown, cheeks, throat, and chest, black; wings and tail, brown. Abyssinia.

The Nelicourvi. Loxia pensilis. Gm. *Pensile grosbeak.* Lath. Sonnerat, 2d. voy. Pl. CIX. Willoughby, t. 77.

Green; gray beneath; vent red; lower part of neck yellow; quills and tail black. Madagascar.

The Worabée. Fringilla Abyssinica. Gm. *Black-collared Finch.* Lath. Vieil. Ois. Chant. 28.

Black collar on the neck; sides of head, throat, fore-neck, and upper belly, black; rest of body, and lower belly, yellow. Abyssinia.

Red-headed Finch. Lath. *Fring. Erythrocephala.* Gm. Lin. 1. 903..

Head and neck, rich scarlet; back, breast, and belly, olive; wings and tail, black; two bars of white on the coverts. Mauritius.

Textor Alecto. Temm. *Tisserin Alecto.* Pl. Col. 446.

Corneous protuberances at the base of the bill; externally, the plumage lustrous-black; white for half the length of the feathers within. Length, nine inches, six lines. Galam. Western Africa.

Some of these birds approximate their nests together in great quantities, so as to form, in one mass, several compartments.

Such is,

The Sociable grosbeak. (*Loxia Socia.* Lath.) Paters. Voy. pl. XIX.

Olive brown; yellow underneath; head and quills brown or blackish.

The Surinam Crow. Lath. *Rice Oriole* id. *Petit Choucas de Surinam, de la Jamaique, Cassique Noir, &c.* (*Oriolus niger, Or. oryzivorus, Corvus Surinamensis. Gm.*) Enl. 534. Brown. Illustr. X. Wilson. Am. III. xxi. 4.

Which devastates, in innumerable flocks, the fields of many of the warmest regions of America. Its general colour is black, changing into magnificent reflexions of all the tints of burnished steel.*

* Nomenclators have not yet been able to put in order the black birds of America, which approximate more or less to the Cassiques, because the descriptions given by travellers are insufficient for the purpose.

We think it proper in this place, to point out the principal ones, and also what is most clear in their synonymy.

1. *The black and mantled Cassique*, mentioned below among the Cassiques.

2. The bird (above mentioned), well drawn, but painted without reflexions, Enl. 534., and cited under *Oriolus Niger. Oriolus Ludovicianus.* Enl. 646, is only an albine variety of this. It is evidently *Corvus*

The Sparrows Proper. Pyrgita. Cuv.

Have the bill a little shorter than the preceding, conical, and only slightly convex towards the point.

Surinamensis of Brown, III. pl. X. *The little Jackdaw* of Jamaica, Sloane, *Jam.* II. 299. pl. 257, 1, cited by Pennant under *gracula barita*, and *quiscala*, is again the same bird. It is also impossible to doubt that Latham had it under his eyes when he was describing his *Oriolus Oryzivorus.*

3. The *true black carouge*, changing to violet, with bill rather short but very straight, given as a tanager, Enl. 710, and of which the *tanagra bonariensis* has been made; but this figure really represents *Oriolus minor (le petit troupiale noir).* The bird, figured Enl. 606. fig. 2. is erroneously attributed as a female to this species, but is, in fact, altogether different.

4. A true *troupiale*, of a deep black, with violet reflexions, with a sharp bill a little arched, and which hollows the upper part of its tail like a boat. This is the *boat-tailed grakle* of Pennant and Latham, which these two writers regard as synonimous with *gracula barita;* and yet it is certainly the bird of Catesby, pl. 12. of which Linnæus has made his *gracula quiscala;* but Catesby has represented the bill badly.

5. A black bird, with violet and green reflexions; with tail somewhat wedged, having the bill of a troupiale, but more arched towards the end, &c. Cuvier.

The following species, not noticed by our author, have been named elsewhere:—

Gambia Grossbeak. Loxia Melanocephala. Albin. t. 62.

Fringilla Velata, under which name Lichtenstein unites *Lox. Melanocephala, Lox. Abyssini‹a,* and *Oriolus textor.* Gm.

Tanagra capitalis. Lath. *Fring. Capitalis.* Licht.

Fringilla Vitellina. Licht.

Fringilla Luteola. Licht.

Loxia Macroura. Gm. *Fring. Macrocerca.* Licht.

Fringilla laticauda. Licht.

Bengal grosbeak. Loxia Bengalensis. Lin. Pl. Enl. t. 393. f. 2.

Weaver Bunting. Emberiza Textor. Gm.

Lichtenstein observes that *Fring. Phalerata, Oryx ignicolor, &c.* should be referred to this genus.

Ploceus nigerrimus. Vieil.

The House Sparrow. (Fring. domestica.) En. C. 1. Naum. 115.

Nestles in the holes of walls, and infests inhabited places with its audaciousness and voracity. Brown, spotted with blackish above; gray underneath; a whitish band on the wing; cap of the male, red on the sides, throat black.

There is in Italy a species, or variety, the male of which has the head entirely moronne. *(Fr. Cisal pina.* Tem. *Fr. Italiæ.* Vieil. Gal. 63.) The black of the throat extends sometimes over the chest. It is then *Fr. Hispaniolensis.*

The Tree Finch. (Fring. Montana). Enl. 267. 1. Naum. 116. 1. 2.

Keeps more remote from habitations. It has two white bands on the wing, a red cap, and the side of the head white with a black spot.

Loxia Hamburgia. Gm. is only the tree-finch disfigured by Albin. Ois. III. Pl. 24.

We must join to the common sparrows the birds scattered as follows by naturalists:—

Crescent Finch. Lath. *Fringilla arcuata.* Pl. Enl. 230. fr. 1.—*Much too red.*

Bill black; head and fore-neck to breast, black; a streak of white on each side of neck, surrounding

Ploceus jonquillaceus. Vieil.
Ploceus Velatus. Vieil.
Ploceus bicolor. Vieil.
Ploceus flavocapillus. Vieil.
Ploceus collaris. Vieil

fore part, like a crescent; back scapulars and lesser wing-coverts, chestnut. Cape of Good Hope.

Emberiza Capensis. C. Enl. 389. 2. and g. Enl. 664, 2. *Yellow-bellied Bunting of Latham.* H. 28.

Top and sides of head, yellowish white; back, brown; rump, gray; underneath, yellow, inclining to red on the breast, and white on the chin and vent. Cape of Good Hope.

Tanagra Silens. Enl. 742. The genus ARREMON of Vieil. Gal. 78., belongs to this genus. It is inserted at the end of the Rhamphoceline Tanagers in "The Animal Kingdom," Vol. VI., p. 318.

Beautiful Finch. Lath. *Fringilla Elegans.* Enl. 205.—1.

Forehead and throat red; breast yellow; back of head and neck dusky ash; back and wings green; belly, breast, and sides, crossed with irregular lines of white spots; abdomen, thighs, and vent, white; rump and tail, red, inclining to chestnut. Mosambique.

Painted Bunting. Lath. *Emberiza ciris.* Gm. Pl. Enl. 159. Genus PASSERINA. Vieil. Gal. 66.

Head and neck, violet; upper part of the back and scapulars, yellowgreen; lower part, rump, and beneath, dull red. Warmer parts of North America, to Brazil and Guiana.

Grenadier Grosbeak. Lath. *Loxia Oryx.* Enl. 6. 2.

Forehead, sides of head, chin, breast and belly, black; wings and tail, brown, with paler edges; rest of body, fine red. St. Helena and Cape of Good Hope.

Fire-coloured Sparrow. Loxia ignicolor. Vieil. Ois. Chant. 59. *Fringilla ignicolor.* Id. Dict. Hist. Nat. Tom. 12. Art. Fringille. *Moineau ignicolor*, ib.

Velvet black on the head to below the eyes, and on most part of breast and belly; entire throat, fiery orange-red; fiery red on the neck, back, upper wing, and tail-coverts. Senegal.

Dominican Grosbeak. Lath. *Loxia Dominicana.* Gm. pl. Enl. 55-2., and the other species. 103. Var. A. Lath. and *Crested Dominican Grosbeak.* Lath.

Head and throat, deep red. Above, generally blackish, or dusky grey; beneath, whitish. Of this Latham makes two species and one variety; the differences of plumage very trifling. Brazil.

Black-faced Finch. Lath. *Fringilla cristata.* Enl. 181. The *Dioch. Emberiza quelea.* Vieil. Ois. Chant. 23. *Fringilla Quelea.* Id. in Dict. H. N. Tom. 12.

Plumage above, wings and tail, reddish brown; beneath, and rump, crimson; cheeks and chin, black. Cayenne and Paraguay. Found also in Senegal.

Le Dioch Rose. Vieil. Ois. Chant. 24. et Dict. H. N. Tom. 12.

Head, neck, breast, and middle of belly, a lively rose-colour; bill, deep crimson. Probably a variety of the last.

Cape Grosbeak. Lath. *Loxia Capensis.* Lin.

Head, neck, upper part of back, body beneath, and tail, deep black; shoulders, lower part of back, and rump, deep yellow. Coromandel, and Cape of Good Hope. This bird commences a little approximation to the grosbeaks.

Fringilla Cruciger. Temm. pl. col. 269, f. 1.

All above, wing coverts and sides, grayish brown, except the cheek, forehead, top of head, and sides of chest, which are whitish; a broad longitudinal black band from base of bill to abdomen, crossed by a similar band at the thoracic region. Bengal. *

THE FINCHES. FRINGILLA. Cuv.

Have the beak a little less arched than the sparrows, a little stronger and longer than the linnets. The habits are more gay, and their song more varied, than those of the sparrows.

We have three species:

The Chaffinch. Fringilla Cælebs. Enl. 54. 1. Naum. 118.

Brown above; vinous red, in the male, underneath; grayish in the female; two white bands on the wing; some white at the sides of the tail. Eats all sorts of grains, and nestles on all sorts of trees. It is one of those birds which renders the country particularly cheerful.

The Mountain Finch. Fring. Mantifringilla. Enl. 54. 2. Naum. 119.

Black, speckled with fawn, above; breast, fawn; under

* The following species have been attached to this subdivision by others:

Fox-colored Sparrow. Fring. illiaca, Merrem.

Fringilla rufa. Wils. Amer. Orn.

Fringilla erythrophthalma. Lin. Wils. O. A.

P. Macronyx. Swainson.

P. Maculata. Swain. Phl. Mag

P. Fusca. Swainson.

part of wing, fine citron. This bird, which varies much, nestles in the thickest forests, and comes into the plains only in winter.

The Snow Finch, Fring. Nivalis. Brisson III. xv. 1. Naum. 117.

Brown, speckled; with a clearer shade above; white underneath; head, ashen; the wing-coverts, and almost all the secondary quills, white; throat, black in the male. It nestles in the rocks of the high Alps, from which it descends only in the depth of winter to the inferior mountains.*

The LINNETS and GOLD-FINCHES. CARDUELIS. Cuv.

have the beak exactly conical, without being swollen out in any point. They live on grains. Those are particularly named gold-finches which have the bill a little more long and sharp.

The Common Gold-Finch. Fring. Carduelis. L. Enl. 4.

One of our handsomest European birds; brown above, whitish underneath. The masque of a fine red; a beautiful yellow spot upon the wing, &c.

* To these have been added the following:—

Fringilla cinerea. Swa n.
Emberiza borbonica, pl. Enl. 321. 2.
Emberiza Braziliensis, pl. Enl. 321 1.

It is also one of the most docile of birds, learns to sing extremely well, and play all sorts of tricks. It derives its French name (*Chardonneret*) from the grain of thistle, &c. on which it feeds by preference.

Parrot Finch, Lath. *Fringilla Psittacea*, Gm.

Bill black; face, throat, rump, and vent, deep scarlet; rest of the body, parrot-green. New Caledonia.

Green Gold-Finch, Lath. *Fringilla Melba*, Gm. Edw. pl. 272 male, 128 female.

Fore part of head and throat, bright red; upper parts of body, yellowish-green; breast, olive-green; rest underneath, white, variegated with dusky. China. Edwards's bird from Brazil.

Scarlet Finch. *Fringilla Coccinea*, Gm. Vieil. Ois. Chant. pl. xxxi.

General colour of the plumage, brilliant deep orange, verging to scarlet. Sandwich Islands.

Fringilla Leucocephala. Lath. Vieil. Ois. Chant. 26, pl. G. 20. f. 2. Dict. H. N.

Head, neck, throat, middle of belly, and under parts, white; small black crescent between bill and eye; breast, alar, and caudal quills, black. Australasia.

Fringilla Magellanica. Vieil. Ois. Chant. 30. *Fringilla Spinus*, Var. Lath.

Head, throat, and half the alar and caudal quills, black; anterior part of wing, middle of coverts, top and front of neck, chest, and lower parts, yellow. South America to Straits of Magellan.*

The LINNETS. LINARIA. Bechst.

Have also the beak exactly conical, but shorter and more obtuse than the gold-finches. They also live on the grains of plants, particularly those of hemp and flax, and bear the confinement of a cage extremely well.

We have here two brown species, with some red tints, more particularly named LINNETS. The young and the females vary in the quantity of red, or want it altogether. The first has the beak almost as much pointed as the gold-finch. It is,

The Lesser Red-headed Linnet. Fr. Linaria. Linn. Enl. 485. 2. Vieil. Gal. 68. Naum. 126.

Brown, spotted with blackish above; two white bands across the wing; the throat black; the upper part of the head red, as well as the breast of the adult male, and sometimes even the croup. It is a northern bird, of which a small and greater race have been lately supposed to exist. See M. Vieillot's Mem. Acad. de Turin, tom. xxiii. p. 193, &c.

* Others add the following here:
Carduelis Mexicanus. Swain.
Fringilla tristis. Lin. pl. Enl. 202. f. 2.
Arkansaw. Siskin. Fring. Psaltria. Say. Bonap. Amer. Orn. t. 6. f. 3.
Pine Thrush. Fringilla Pinus. Wils. A. O. t. 6. f. 3.

The Great Red-headed Linnet. Fring. Cannabina. Lin. Enl. 485. t. 151, f. 2. Old Male.

Back, fawn-coloured brown; quills of the wing and tail, black, edged with white; whitish underneath; of a fine red on the head and chest of the old male. Often nestles in the vines, otherwise in coppices and bushes.

An intermediate species, more approximating to the second (*Fring. Montium.* Gm. Naum. 122.) comes to us sometimes from the north. Its bill is yellow, and the crupper of the male a little red.

Other species, more or less greenish, bear the names of Serins or Siskins.

The Common Siskin. Fring. Spinus. Enl. t. 485. f. 3. Male, Alb. t. 76.

Has also the bill more approximating to the goldfinch, and resembles in many points the small linnet. It is olive above, yellow beneath; a coif, the wing, and tail, black; two yellow bands on the wing. It nestles only in the summits of fir-trees.

Citril Finch. Fringilla Citrinella. L. Enl. 658, 2.

Rather olive above; yellow underneath; the hind part of the head and neck, ash.

The Serin Finch. Fring. Serinus. Lin. (Enl. 658. 1. Naum. 123.)

Olive above, yellowish beneath; spotted with brown;

a yellow band on the wing. Two birds of the mountains of the South of Europe, nearly the size of the Siskin.

Canary Finch. (*Fring. Canaria.* Lin.) Enl. 202. 1.

Is larger, and its facility of propagation in captivity, as well as the charms of its song, have extended it every where, and caused it so to vary in colour, that it is difficult to assign its primitive hue. It mixes with the majority of the other species of this genus, and often produces with them mules, more or less fruitful.

Among the foreign birds, which cannot be distinguished from the linnets by any generic character, we place

Lepid Finch. Lath. *Fringilla lepida.* Gm.

Bill and eyes, black; general colour of the plumage, greenish brown; chin, fulvous; breast, black. Havannah.

American Yellow Finch. Fring. tristis. Enl. 202. 2.

Fore part of the head, black; rest of the body, yellow; wing-coverts, black, crossed with a white band. North America.

Mozambique Serin. Fringilla ictera. Vieil. Enl. 364. 1. 2. *Fringilla canaria.* Var. Lath.

Larger than canary; beneath, yellow; and on croup, upper wing, and tail coverts, bordered with yellowish;

brown above; brown and yellow band alternately on the head. Mosambique.

Glossy Finch. Lath. *Fringilla nitens.* Enl. 291. Var. A. Lath. *Fringilla Æthiops.* Gm. Ind. Orn. 1. 442. B.

Plumage, wholly blue-black, with a polished steel gloss. Brazil,—Cayenne.

Senegal Finch. Lath. *Fring. Senegala.* Gm. pl. Enl. 157. 1. Vieil. Ois. Chant. 9.

Greater part of plumage, vinaceous red; hind part of head and neck, back, scapulars, and wing-coverts, brown; abdomen, thighs, and under tail-coverts, greenish brown. Senegal.

Amaduvade Finch. *Fring. Amandava.* Gm. Enl. 115. 2 and 3.

Above, brown, with a mixture of red; beneath, same, but paler; tail, black. Bengal, Java, Malacca, &c.

Brazilian Finch. Lath. *Fring. granatina.* Gm. Enl. 109. 3.

Sides of head, about the eye, blossom-coloured, inclining to violet; throat, lower part of belly, and thighs, black; rest of head and body, chestnut; back and scapulars, brownish; tail, black. Brazil and Guiana.

Blue-bellied Finch. Lath. *Fring. Bengalus.* Briss. II. 203. t. 10. f. 1.

Head, and upper parts of body, gray; lower part of back, rump, and all beneath, blue. Africa.

Angola Finch. Lath. *Frin. Angolensis.* Gm. Pl. Enl. 115. 1.

Upper parts, brownish ash-colour; under parts, plain dull orange.

Carduelis. Cucullata. Swain. Zool. III.

Head, crested red, and to the middle of the breast.

Many more species may be found under the names of *Astrils*, *Bengalis*, and *Senegalis*, in the work of M. Vieillot, intitled "*Oïseaux Chanteurs de la Zone Torride*; such as,

Bahama Finch. Lath. *Fringilla Bicolor.* Gm. Vieil. Ois. Chant. pl. 9.

Bill, head, throat, and breast, black; rest, dusky green. Bahama Islands.

Fringilla Tricolor. Vieil. Ois. Chant. pl. 20.

Top of head, throat, and all the posterior parts, azure blue; croup, red; rest of plumage, green olive. Timor.

Fringilla Cinerea. Vieil. pl. 6. Ois. Chanteurs.

All the upper parts, ashen gray; upper tail-coverts and quills, black; lower, white; breast, and upper belly, flesh-colour, succeeded by a lively rose, as far as anus. Africa.

Fringilla Cœrulescens. Vieil. Ois. Chant. 8.

For the most part bluish-gray, clearer on throat, and deeper on the hinder parts. Intertropical countries.

Fring. Melpoda. Vieil. pl. 7. Ois. Chant.

A band of deep orange across the eye, and extending over the cheek; feathers of lower belly more yellow. India and Western Africa.

Fringilla Viridis. Vieil. pl. 16. Ois. Chant.

Head, iron-gray; upper part of neck and body, wings, and tail, olive-green. Western Africa.

Fringilla Erythronotos. Vieil. pl. 14. Ois. Chant.

Above, brown transverse stripes on a gray ground; flanks, back, croup, and upper tail coverts, fine red. India.

Fring. Quinticolor. Vieil. 15. Ois. Chant.

Head and all beneath bluish-gray; fine red on croup and eye-brows; olive green on neck, back, and extremity of alar quills; dull brown on their inner barbs; tail, black. N. Holland.

Fringilla Rubriventris. Vieil. pl. 13. Ois. Chant.

Head, neck, and body above, gray-brown; middle of the belly, fine red. Senegal.

Fringilla Frontalis. Vieil. pl. 16. Ois Chant.
Loxia Frontalis. Lath.

The forehead is black, punctated with white; top of head and nape, orange; upper part of body, and neck, alar, and caudal quills, ferrugineous gray.

Fringilla guttata. Vieil. pl. 3. Ois. Chant.

All the upper parts ash; sides of the chest and belly reddish, and spangled with white spots. Moluccas.

Fringilla Melanotis. Tem. pl. col. 221. 1.

Head, hinder part, and sides of neck, lead-colour; back and wings, olive green; black round the eyes and ears. South Africa.

Fringilla Sanguinolenta. Tem. pl. col. 221. 2.

Above, wings, and two middle tail quills, earth-brown; a large red streak, like an eye-brow, above the orbits; blood-red on chest, belly, and abdomen. Senegal and the coast of Guinea.

Fringilla Polyzona. Tem. pl. col. 221. 3.

Forehead, cheeks, and throat, black; above, wings and tail, ashen-brown; reddish on the middle of belly, assuming a whitish cast on the abdomen. Gambia.

Fringilla Otoleucus. Tem. pl. col. 269. 2. 3.

Pure white on the region of the ears; head, black; streak of white on the nape; back and scapulars, brick-red; black below. Senegal.

Fringilla Simplex. Licht. Tem. pl. col. 358. 1. 2.

Throat, fore-part of neck, and lores, perfect black; Cheeks, sides of neck, and under parts, whitish; head and back, clear ashen. Nubia.

Fringilla Lutea. Licht. Tem. pl. col. 365. 1. 2.

Fine citron-yellow on all parts of head, neck, croup, and underneath; mantle, back, and scapulars, moronne. Dongola, in Nubia.

Fringilla Ornata. Pr. Max. Temm. pl. col. 208.

Top of head, lore, throat, breast, and middle line of belly, black; sides of breast and flanks, reddish-yellow;

body above, ashen-gray; wings, black; tuft on head. Brazil.

The pretended *Emberiza Oryzivora.* Enl. 388, has also the same bill; but the quills of its stiff and sharp tail distinguish it.

The reader may also consult the numerous fringillæ characterized by M. Ch. Bonaparte. Lyceum of New York, II., Dec. 1826, pag. 106, &c.*

* Also have been named—

Fringilla flammea. Bechst. Allem. 33. 2.

Lazuli Finch. Fringilla Anæna. Bonap. Amer. Orn. t. 6. f. 4. *Emberiza.* Say.

Indigo bird. Fringilla Cyanea. Wils. Am. Orn. 6. 5. Bonap. A. O. 11. 3.

Black-throated Bunting. Emberiza Americana. Gm. Wils. A. O. 1. t. 3. f. 2. *Fringilla flavicollis.* Gm. *Fringilla Americana.* Bonap.

White-crowned Bunting. Embertza leucophrys. Wilson. Am. Orn. iv. t. 31. f. 4. *Fringilla.* Tem.

Lark Finch. Fringilla graminea. Say. Bonap. Am. Orn. t. 5. f. 2.

White-throated Sparrow. Fringilla Pennsylvanica. Lath. *F. Albicollis.* Wils. A. O. t. 22 f. 2.

Bay-wing Bunting. Fringilla graminea. Gm. *Emberiza.* Wils. Amer. Orn. iv. t. 31. f. 5.

Song Sparrow. Fring. fasciata. Gm. *F. Melodia.* Wils. A. O. ij. t. 16. f. 4.

Savannah Finch. Fringilla Savanna. Wils. Amer. Orn. iv. t. 34. f. 4. iij. t. 22. f. 3.

Snow bird. Fringilla nivalis. (Hudsonia.) Wils. Am. Or. ij. t. 16. f. 6. *Fringilla Hyemalis.* Lin. *not* Cuvier.

Yellow-winged Sparrow. Fring. Passerina. Wils. A. O. iij. t. 24. f. 5. *Fringilla Savannarum.* Gm. § *F. Caudacuta.* Lath. §§

Tree Sparrow. Fringilla Arborea. Wils. A. O. ij. t. 16. f. 3. *Fr. Monticola.* Gm. *Fr. Canadensis.* Lath.

Chipping Sparrow. Fringilla Socialis. Wils. A. O. ij. t. 16. f. 5.

Field Sparrow Fring. Pusilla. Wils. A. O. ij. t. 16. f. 2.

Swamp Sparrow. Fringilla Palustris Wils. A. O. iij. t. 22. f. 1.

The Widow Birds. Vidua. Cuv.

Are African and Indian birds, with the bill of the linnet, sometimes a little more swollen at its base. They are distinguished by having some of the tail quills, or upper tail-coverts, excessively long in the males.

One species (*Vidua longicauda*) has only the coverts elongated; the others have the quills so.

Linnæus and Gmelin have unaccountably united them to the buntings, under the following names: —

Shaft-tailed Bunting. Lath. *Emberiza Regia.* Gm. Enl. 8. 1.

Sides of head, beneath, and round the neck, rufous; back of neck, spotted black; four middle tail-feathers, nine or ten inches long. Africa.

Dominican Bunting. Lath. *Emberiza Serena.* Gm. Pl. Enl. 8. 2.

Upper part of head black; crown, back of neck, and all beneath, rufous white; back, black; feathers, edged with dirty white. Habitat. ?

Sharp-tailed Finch. Oriolus caudacutus. Gm. *Fringilla caudacuta.* Wil A. O. ij. t. 34. f. 3. not Lath.

Sea-side Finch. Fringilla maritima. Wils. A. O. t. 34. f. 2.

Prince Musignano observes, that the three last, and especially the two latter, have a peculiar form and general habit, which might entitle them to a separate sub-genus or section. Mr. Swainson names the group *Ammadovamus.*

Ammadovamus bimaculatus Swain.

Fringilla Georgica. Lath. Licht.

Whidah Bunting. Lath. *Emberiza Paradisea*. Gm. Pl. Enl. 1941.

Head, chin, fore-neck, back, wings, and tail, black; neck behind, pale orange; breast, and half belly, full orange; rest white. Africa.

Panayan Bunting. Lath. *Emberiza Panayensis*. Gm. pl. Enl. 647.

Plumage, wholly black, except one large bright red spot on the breast. Panay.

Caffrarian Grosbeak. Lath. *Emberiza longicauda*. Gm. *La veuve à épaulettes*. Buff. pl. Enl. 535.

Plumage in general, black; shoulders, crimson; wing-coverts, white; tail, sometimes twice the length of body. Cape of Good Hope.

Fringilla Superciliosa. Vieil. Gal. 61.

A white band above the eyes, which is prolonged over the sides of the nape; another from the base of the bill extending over the vertex; top of head, and sides of head and neck, black; also, mantle, coverts, wing-quills, and tail above; white beneath. Africa.

Variegated Bunting. Lath. *Emberiza principalis*. Gm. Edw. 270. *Moineau du Brésil*. Pl. Enl. 291. 292.

Above, black and rufous, mixed; sides of head and below, white. Angola.

This and *Emberiza vidua*. (Aldrov. Ornit. II. 565.) *Long-tailed Bunting*. Lath. appear to be the same kind, in different states of plumage. *Emb. psittacea*, Seb. i. pl. lxvi. fig. 5. isnot very authentic. *Angolensis*. Salern. Orn. 277, *La veuve chry-*

soptère, Vieil. Ois. Ch. pl. xli., and *lox macroura.* Enl. 283. 1., which perhaps do not differ from each other, are not widow-birds, but common grosbeaks.

There is a gradual passage, and without any assignable interval, from the linnets to the

GROSBEAKS (COCCOTHRAUSTES.) Cuv.

Whose beak, exactly conical, is distinguished only by its excessive bulk.

This passage takes place, in the species which I have had an opportunity of examining, pretty nearly in the following order, the bill still increasing in bulk:—

Red-rumped Bunting. Lath. *Loxia quadricolor.* Cuv. *Emberiza quadricolor.* Linn. Pl. Enl. 101. 2. *Gros-bec longicone.* Temm.

Head and neck, blue; back, wings, and end of tail, green; upper part of tail, tail-coverts, and middle belly, red; breast and lower belly, pale brown. Java.

Red-billed Grosbeak. Lath. *Loxia Sanguinirostris.* Gm. Pl. Enl. 183, 2.

Forehead above the eye, and round to the chin, black; rest of head, and above, rufous gray; underneath, pale rufous. Africa.

Molucca Grosbeak. Lath. *Loxia Molucca.* Gm. Enl. 139, 2.

Fore part of head, sides, and fore part of neck, black; hind head and above, brown; rump, and under breast, cross-barred, black and white. Moluccas.

Loxia Variegata. Vieil. Ois. Chant. pl. 51.

Head, cheeks, and throat, black; croup and beneath, white, with vermiculated stripes; above, gray-brown, shaded with yellow.

Loxia punctularia. Vieil. pl. D. 14. f. 1. Dict. d'Hist. Nat.

Head, top of neck, back and wings, brown moronne; front of neck, chest, and flanks, varied with white marks, surrounded by a blackish border. India.

White-headed Grosbeak. Lath. *Loxia Maia.* Gm. Pl. Enl. 109. 1.

Head and neck, dirty white; body above, wings, and tail, chestnut brown; belly and vent, blackish. Malacca, China, Java.

Striated Grosbeak. Lath. *Loxia Striata.* Gm. pl. Enl. 153. 1.

Head, neck, and above, brown; throat and neck before, blackish; beneath, from breast, white. Bourbon, Java.

Nitid Grosbeak. Lath. *Loxia nitida.* Ind. Orn. Vieil. Ois. Ch. 60.

Bill and irides, crimson; plumage in general, pale olive-brown above; dusky white beneath. Australasia.

Malacca Grosbeak. Lath. *Loxia Malacca.* Gm. Pl. Enl. 139. 3.

Head, neck, middle of belly, and under tail-coverts, black; breast and sides of belly, white; back, wings, and tail, chestnut. Java.

Wax-billed Grosbeak. Lath. *Loxia Astrild.* Gm. Pl. Enl. 157. 2.

Above, brown; beneath, reddish-gray, crossed every where with blackish lines. Canaries, Madeira, Senegal, Angola, Cape, India, &c.

Black-lined Grosbeak. Lath. *Loxia bella.* Ind. Orn. Vieil. 55. Ois. Chant.

General colour, gray; paler beneath; crossed every where with slender black lines; bill, crimson. New South Wales.

Warbling Grosbeak. Lath. *Loxia cantans.* Gm. Vieil. Ois. Chant. 57.

Above, brown, marked with narrow dusky lines; belly, white. Senegal.

Java Grosbeak. Lath. *Loxia oryzivora.* Gm. Pl. Enl. 153. 1.

Head and throat, black; rest above, pale ash-colour; belly and thighs, pale rose-colour; tail, black. Java.

Brown Grosbeak. Lath. *Loxia fusca.* Gm. *L. fuscata.* Vieil. pl. 62. Ois. Chant.

Bill, lead-colour; head and above, brown; beneath, pale ash-colour.

Azure Grosbeak. Lath. *Loxia Cyanea.* Gm. Vieil. Ois. Chant. pl. 64.

Bill, lead-colour; plumage in general, deep blue; quills and tail, black. Cayenne, Brazil, and Paraguay.

Loxia Atricapilla. Vieil. Ois. Chant. 53.

Black capote on the head; hinder parts above, black; belly, moronne. India.

Black Grosbeak. Lath. *Loxia nigra.* Gm. Catesb. 1. 68. Vieil. Gal. 57.

Bill and plumage, black; a little white on the fore-part of wing, and base of two first quills. Mexico.

Brazilian Grosbeak. Lath. *Loxia Braziliana.* Ind. Orn. Pl. Enl. 309. 1.

Bill, flesh-coloured; head and chin, red; back and wing-coverts, brown; quills and tail black; breast, belly, and sides, reddish white; middle of belly, red. Brazil.

Red-breasted Grosbeak. Lath. *Loxia Ludoviciana.* Gm. Pl. Enl. 153. 2. and Vieil. Gal. 58.

Head, upper part of the body, wings, and tail, black; breast and under coverts, rose; belly, thighs, and vent, white. North America.

We have, besides the common grosbeak, two species in Europe, with bills less gross.

Ring Finch. Lath. (*Fringilla Petronia.* Lin.) Enl. 225. Naum. 116. 3. 4.

It has been usual to join this to the sparrows, with which it assimilates in colour; but, independently of the gross bill, it is easily distinguished from them by a whitish line round the head, and a yellowish spot on the breast. It is as evidently a grosbeak as the following.

Green Grosbeak. Lath. *Loxia Chloris.* Linn. 267. 2. Naum. 120.

Greenish above, yellowish underneath; the external edge of the tail, yellow. Inhabits coppices, and eats all kinds of seeds.

Loxia hæmatina. Vieil. pl. 67. Ois Chanteurs. *Bill too slender.*

Above, wings, tail, and middle of belly, black; rest red; bill, lead-colour. Africa.

Loxia guttata. Vieil. 68. Ois. Chant., is a variety of the last.

Loxia quinticolor. Vieil. Ois. Ch. pl. 54.

Wings and tail, brown, growing reddish on cheek and back; top of head and neck, gray; rump, upper tail-coverts, and outer edge of quills, orange; dull black on lower tail-coverts; legs and top of throat, front of neck, breast and belly, pure white. Moluccas.

Fasciated Grosbeak. Lath. *Loxia fasciata.* Gm. Vieil. Ois. Chant. 58. Brown. Ill. III. xxvii.

Top of head, gray, brown-barred; rest above, cinnamon brown; across the throat, a crimson band. Africa.

Madagascar Grosbeak. Lath. *Loxia Madagascariensis.* Gm. Pl. Enl. 134. 2.

General colour, red; middle of each feather on the back, red. Madagascar.

Blue Grosbeak. Lath. *Loxia cærulea.* Gm. D'Azara. Voy. III. No. 192. Catesb. 1. 39.

Black feathers at base of bill; rest of plumage, deep blue; quills and tail, brown, with green mixed. Georgia, Cayenne, &c.

Cardinal Grosbeak. Lath. *Loxia cardinalis.* Gm. Pl. Enl. 37.

Head, greatly crested, red, and body above, same; beneath, chestnut. North America.

Gray-necked Grosbeak. Lath. *Loxia Melanura.* Gm. Saun. Voy. Ind. II. 199.

Head, black; neck gray, before; belly, pale rufous; tail, black. China.

Common Grosbeak. Haw Grosbeak. Lath. (*Loxia Coccothraustes.* Lin.) Enl. 99. and 100. Naum. 104,

Is one of those which best deserve this name. Its enormous bill is yellowish. The back and cap are brown, the rest of the plumage grayish; the throat and wing-quills black; a white band on the wing. It lives in the mountain woods; nestles in beech-trees and fruit-trees, and eats all kinds of fruit and almonds. Europe.

Loxia Ostrina. Vieil. Ois. Chant. 48. gal. 60.

Head, throat, neck, breast, thighs, and tail, poppy red; rest of plumage, bill, and feet, black; bill higher at base than at the forehead. India and Africa.

Loxia rosea. Vieil. pl. 65. Ois. Chanteurs.

Plumage, fine rose-colour, varied with grayish brown, on the occiput, top of neck, back, and wing-coverts, and gradually growing white beneath. India.

Some foreign species must be distinguished from the grosbeaks.

PITYLUS, CUV.

With a bill equally thick, a little compressed, arched above, and sometimes with a salient angle at the middle of the edge of the upper jaw.

Such are,

White-throated Grosbeak. (*Loxia grossa.* L.) Enl. 154.

Deep blue; base of bill surrounded with black, passing downwards, and covering fore part of neck; in the midst of which, on chin and throat, is a patch of white; bill, red. Length, seven and a half inches. America.

Canada Grosbeak. (*L. Canadensis.* L.) Enl. 152. 2.

Upper parts, olive green; underneath, paler, inclining to yellow; chin, and base of bill, black. Length, seven inches. Canada.

* The following have been also attached to this genus:

Evening Grosbeak. Fringilla Vespertina. Cooper. Bonap Amer. Orn.

Brisson Grosbeak. Lox cærulea. P. Gm. *Lox Brissonii.* Gm. *Fring. Brissonii.* Licht. *Gros-bec bleu de ciel.* Azara. 118.

Purple Finch. Fring. purpurea. Gm. Wilson. Amer. Orn. 1. t. 7. f. 4. v. t. 42. f. 3.

Guiraca melanocephala. Swainson.

Black-headed Grosbeak. (*L. erythromelas.* L.) Lath. pl. 47. and Vieil. Gal. 59.

Deep crimson, inclining to chestnut above, and to pink beneath; quills and tail, dusky red; bill, black. Length, nine inches. Cayenne.

Porto-Rico Grosbeak. (*L. Portoricensis.* Daud. II. 411. *Pyrrhula auranticollis.*) Vieil. Gal. Ois. t. 55.

Deep black: on the top of the head, a small ferugineous crescent prolonged to sides of neck; throat, neck, and vent, same. Length, seven inches. Porto Rico.*

From these have been long distinguished

The BULFINCHES (PYRRHULA,)

whose beak is rounded and gibbous.

We have one,

The Bulfinch. (*Loxia Pyrrhula.* Lin.) Enl. 145. Vieil. Gal. 56. Naum. III.

Ashy above; red underneath, with black cap; the female has reddish gray, instead of red; builds in trees, copses, and hedges. Its natural warbling is

* And, according to others,
Fringilla gnatho. Licht. *Tanagra psittacina.* Spix.
Pitylus atrochalybeus. Vigors. Jard. Ill. Orn t. 3.
Fringilla Martinicensis. Lath.
Fringilla rufobarbata. Jacq. t. 11. f. 8.
Fringilla noctis. Lath. pl. Enl. t. 201. f. 1.
Loxia violacea. Lath. *Pyrrhula superciliosa.* Vieil.

soft. It is easily tamed, and learns both to sing and speak.

A variety is known one third larger than the common race.

Lineated Grosbeak. (L. lineola.) Enl. 319. 1.

Glossy blue-black, above; beneath, white; quills black; base of primaries, white; base of beak, a white spot; white streak from forehead to crown. Size of Titmouse. Asia.

Minute Grosbeak. (L. minuta. L.*)* Enl. 2.

Gray-brown above; beneath, and rump, chestnut; quills, white at base. Size of a wren. Surinam.

Nun Grosbeak. (L. collaria. L.) Enl. 393. 3.

Top of head, and body above, greenish-blue; rump, and under part, rufous white; wings, rufous yellow and black, mixed. India.

Siberian Grosbeak. (L. Sibirica. L.) Falk. Voy. iij, t. 28.

Head and back, deep vermilion, marked with brown; tips of feathers on the head, white; wing-coverts white, with black tips. Siberia.

Pyrrhula cinereola. Tem. pl. col. II. 1.

Head, cheeks, back, and scapulars, bluish-ash; wings and tail, more blackish; underneath, white; thighs, ashy. Brazil.

Pyrrhula falcirostris. Temm. pl. col. 11. 2.

Above, ashen-brown; gray, tinted with clear rose-colour, beneath; bill, red. Egypt and Nubia.

Pyrrhula mysia. Vieil. Ois. Chant. pl. 46. Spix. 59 and 60.

Above, lustrous black ; cheeks, partly, and, underneath, all white, except rump and thighs, which are bluish-gray ; bill, black. Guiana.*

The Crosbills. (Loxia. Briss.)

Have the bill compressed, and the two mandibles curved in such a manner, that their points cross each other, sometimes on one side, sometimes on the other, according to the individuals. This extraordinary bill enables them to pluck the seeds from under the scales of the pine-apple.

Loxia, from the Greek word λοξος, (curved), is a name invented for this bird by Conrad Gesner, and extended by Linnæus to all the Grosbeaks.

The European species is frequent wherever there are large woods of green trees ; it is

The Crosbill. (Loxia Curvirostra.) Lin. Enl. 218.

The plumage of the young male is bright red, with brown wings ; that of the adult, and female, greenish above, yellowish underneath. Two races, different

* Others have named,

Pyrrhula frontalis Bonap. Amer. Orn. I. t. 6: 1. 2. *Fringilla.* Say.
Loxia noctis.
Loxia Torrida.
Loxia Angolensis.
Pyrrhula longicauda. Vieil.
Pyrrhula crispa. Vieil.
Social Pyrrhula. Pyrrhula synoïca. Temm. Pl. Col. 375. 1. 2.

in size, and even varying, as is said, in the form of the bill, and in the voice, are known.

They are,

Loxia Curvirostra. Naum. 110. and *Loxia Pytiopsittacus*. Becht. Naum. 109.

Add,

White-winged Crossbill. Lath. *Loxia Leucoptera*. Gm. Vieil. Gal. 53. and Wils. Orn. Amer. IV. t. 41 .f. 1.

Head, neck, back, and beneath, whitish, deeply margined with crimson; wings black, with a bar of white. N. America.*

We must not remove far from the bulfinches and crossbills,

The HARDBILLS. (CORYTHUS. Cuv.)

Whose bill, altogether gibbous, has its point bent above the lower mandible.

Corythus is the Greek name of an unknown bird. M. Vieillot has changed this name into that of STROBILIPHAGA.

The species most known, is

The Pine Grosbeak. *Loxia Enucleator*. Lin. Enl. 135. or better, Edw. 123. 124. Vieil. Gal. 53. Naum. 112.

Inhabits equally the north of the two continents, and

* To these is attached, elsewhere,

Loxia falcirostris. Lath.

lives in the same manner as the crossbill. It is red or reddish; the feathers of the wings and tail are black, bordered with white.

Loxia Flamengo (Sparm. Mus. Carl. pl. 17,) appears to me to be but an albino variety of the *enucleator*.

Loxia Psittacea of the Sandwich Islands, Lath. Syn. II. pl. XLII. or PSITTACIROSTRA *Icterocephala*. Temm. col. 457, does not appear to differ from the hardbills, but by a little greater elongation of the recurved point of its bill.

The north of the globe also possesses species approximating to these, of equally fine colours, some individuals of which occasionally arrive as far as Germany.

Crimson-crowned Finch. Lath. *Loxia Erythrina*. Pall. *Fringilla Flammea*. Linn. Naum. 113. 1. 2.

Plumage above, wings, and tail, brown; rump, upper and lower tail-coverts, and beneath, rose-colour. Sweden.

Rosy Finch. Lath. *Loxia Rosea*. Pall. Naum. 113. 3. *Fringilla Rosea*. Gm.

Head, neck, and throat, red; nape and back, reddish-ash; beneath, white; breast and sides, red-tinged. Siberia, &c.

Purple Finch. Lath. *Fringilla Purpurea*. Gm. Wilson. Am. Orn. i. pl. 7. f. 4.

General colour of the plumage, crimson, with a tinge

of purple; middle of belly, thighs, and vent, dusky white. N. America.

The Colies. Colius. Gm.

Also approximate considerably to the preceding. Their bill is short, thick, conical, slightly compressed, and its two mandibles are arched, without one passing the other. The quills of the tail are wedged, and very long; the thumb, as in the martens, is capable of being turned forward with the other toes. Their feathers, fine and silky, have generally cinereous tints. They are African and Indian birds, which climb nearly in the manner of the parrots, live in flocks, approximate their nests in great numbers on the same boughs, and sleep suspended to the branches, with the head downwards, and crowded against each other. . They subsist on fruits.

The word *Colius* comes from the Greek Κολοιος, which is the name of a small species of crow.

Cape Coly. Colius Capensis. Gm. Pl. Enl. 282. 1. Vaill. 258, and the young, 256. This last is *C. Striatus* and *Panayensis.*

Outer side of the outer tail-feathers, white; body above, ash, beneath white. Cape of Good Hope.

Radiated Coly. Colius Striatus. Gm. and *Panay Coly. Col. Panayensis.* Gm. Son. Voy. t. 74.

Crested; ash above; black beneath, and cross-banded; chest, grayish-red; belly, red; tail, green. Cape of Good Hope.

White-backed Coly. Lath. *Colius Erythropus.* Gm.
Col. Leuconotus. Lath. Vaill. 217.

Plumage, in general, bluish ash-colour; crested; back, rump, and under tail-coverts, purple, with a middle stripe of pure white. South Africa.

Black-throated Coly. Lath. *Colius Gularis.* Cuv.
Vaill. 259.

Throat, black; on hind head, a pendulous crest; beneath, light rufous brown; above, light vinous brown, cross-barred. Angola and Malimba.

I approximate to the Colies the birds named *Merion Natté* (*Malurus Textilis.* Less.) and *Merion Leucopteré.* (*M. Leucopterus*, id.) Voy. de Freycinet, Zool. pl. 23.*

It is also here that we must place

THE BEEF-EATERS. (BUPHAGA Briss.)

A small genus, whose bill, of moderate length, at first cylindrical, swells at the two mandibles, before its extremity, which terminates in rather a blunt point. It enables the birds to compress the skin of oxen, to force out the larvæ of the *œstrus*, which lodge there, and on which these birds feed.

Attached to this genus by other writers are—

Senegal Coly. Lath. *Colius Senegalensis.* Gm. *Lanius macrourus.* Linn. pl. Enl. 282. 2. Briss. iii. 16. 3. *Colius Indicus.* Lath.

Colius Coromandelensis. Licht. 1793.

African Coly. Lath.? *Colius Erythromelon.* Vieil.

Green Coly. Colius viridis. Lath.

But one species is known; it is of Africa; brownish, with a moderately sized wedged tail, and is almost the size of a thrush. *(Buphaga Africana).* Enl. 293. Vail. Af. pl. 97. Vieil. Gal.

THE CASSIQUES. (CASSICUS. CUV.)

Have a large conical bill, thick at the base, but very sharp at the point; small round nostrils on its sides; the aperture of the mandibles in a broken line, forming an angle, as in the starlings. These are American birds, whose habits are similar to those of our starlings, living, like them, gregariously, building their nest near each other, and using much art in their construction. They live on insects and grains, and their numerous flocks do great damage to the cultivated fields. Their flesh is not good.

We subdivide them as follows:

THE CASSIQUES, properly so called. (CASSICUS.)

In which the base of the bill mounts upward on the forehead, and encroaches on the plumage by a large semicircular notch. It is among these that the largest species are found.

Cassicus Bifasciatus. Spix. Spix. Braz. t. 61. a.

Bill black, with two reddish bands; crest of linear feathers, deep chestnut; head and neck, black; quills, black, brown-edged. Brazil.

Cassicus Angustifrons. Spix. 62. *Young. Oriolus Cristatus.* B. Var. Lath. *Cassicus Viridis.* Vieil. pl. Enl. 328.

Bill, orange; crest, of long feathers, olive green beneath paler; wing-coverts, brown tipped. Brazil.

Cassicus Nigerrimus. Spix. Braz. t. 63. f 1.

Bill and body, all black. Brazil.

Crested Oriole. Oriolus Cristatus. Enl. 344. *L. Yapu.* Azara, n. 57.

Head, neck, and body, to the middle, black; rump and vent, deep chestnut; wings, black; size of a magpie. Cayenne.

Red-rumped Oriole. Lath. *Cassicus Hæmorrhous.* Enl. 482. Brisson, II. t. 8. f. 2.

Black, with a greenish gloss; lower part of back, rump, and tail-coverts, crimson.

Black and yellow Oriole. Lath. *Oriolus Persicus.* Enl. 184. Edw. 819. Bris. 9. 1.

Black patch on wing; rump, belly, and vent, yellow; size of a black-bird. Brazil. Not found in Persia, but in America, like the other species.

Cassicus Ater. Vieil. *Grand Troupiale* d'Azara. Voy. III. p. 167.

Black, with metallic reflexions; the feathers of the neck forming a sort of mantle.

THE TROUPIALES. (ICTERUS.)

Whose bill enters on the plumage only by a small notch, but is arched lengthwise.

M. Vieillot has changed this name of *troupiale* for that of *carouge*, which I have given to the following division. He translates *carouge* by *pendulinus*. Galer. pl. 186.

Chestnut Oriole. Lath. *Oriolus Varius*. Gm. *O. Castaneus*. Lath. *O. Mutatus*. Wils. A. O. 4. 4. pl. Enl. 607. 1.

Shining black; sides of chest and body beneath, edge of wings, and rump, shining chestnut.

Yellow-winged Oriole. Lath. *Oriolus Cayanus*. Pl. Enl. 535. 2. Azara, n. 67. *Agelaius Chrysopterus*. Vieil.

Bill, straight, slender, and very acute; tail, wedge-shaped, black; smaller wing-coverts, yellow.

Olive Oriole. Lath. *Oriolus Capensis*. Enl. 607. 2.

Olive-brown above; yellow beneath; length, 7 inches. Louisiana, and not the Cape.

St. Domingo Oriole. *Oriolus Dominicensis*. Gm *Pendulinus Flavigaster*. Vieil. pl. Enl. 5. 1.

Black; lesser wing-coverts, and lower part of belly and vent, yellow; length, 8 inches. St. Domingo.

Oriolus Chrysocephalus. Gm. *Pendulinus*. Vieillot. Spix. Braz. 67. 1. Merrem, Beytr. I. pl. 3.

Black; forehead, crown, side of nape, tibia, and smaller wing-coverts, yellow; bill, slender and arched.

Quiscalus Versicolor. Vieil. Gal. 108. Wils. III. xxi. 3. *Gracula Quiscala.* Linn. Catesb. pl. 12. *Gracula Barita.* Lath. I. pl. 18.

A black species, with metallic reflexions; the tail assuming all kinds of forms by the direction of its lateral plumes, sometimes on a level with the others, and sometimes raised, and boat-shaped. Antilles, Carolina, &c.

From these must be separated *Icterus Sulcirostris.* Spix. 64, whose bill, much more gross, has the lower mandible furrowed obliquely at the base.*

THE CAROUGES. (XANTHORNUS.)

Differ from the troupiales only in having the bill altogether straight.

M. Vieillot gives to my *Carouges* the names of *Baltimore* and of *Yphantes.* Galer. pl. 87. He separates some of them, which he particularly names *Troupiales*, or AGELAIUS. Pl. 88.

Among the number, we must distinguish one species, with the bill a little shorter, which in that respect approaches to the finches:—

Icterus Pecoris. Temm. *Emberiza Pecoris.* Wils. Am. II. xviii. 1. 2., and Enl. 606. 1.

Of a violet black; head and neck, gray-brown. This bird lives in flocks, near cattle; but the most remark-

* To these have been added—

Oriolus Mexicanus. Leach.-Zool. Misc. I. t. 2. *Psarocolius leucopteryx*—*Wagner.*—*Oriolus olivaceus.* Gm. Pl. Enl. 606. *Pendulinus rufigaster.* Vieil.

able trait in its habits is, that it deposits its eggs in strange nests, like the cuckoo.

Oriolus Icterus. Linn. Pl. Enl. t. 532. *Peudulinus Longirostris.* Vieil. Brisson. II. t. 8. f. 1.

Head, neck, middle of back, wings, and tail, black; rest, golden yellow; size of a blackbird. Carolina.

Oriolus Minor et *Tanagra Bonariensis.* Enl. 710.

Glossy black, tinged with blue about the head; length, 7 inches. Dr. Latham makes it a variety of the Cowpen.

Oriolus Citrinus. Spix. 76.

Face and streak on each side of the chin, black; head and body beneath, yellow; back, wings, and tail, black. Brazil.

Xanthornus Gasquet. Quoy et Gaim. Voy. de Freycinet. pl. xxiv. *Agelaius Guirahuro.* Vieil.

Head, and neck in front, blackish; above, brown, with slight tint of yellow; rest, yellow. Paraguay.

Oriolus Phœnicus. Gm. *Sturnus Predatorius.* Wils. Pl. Enl. 402.

Black, rather shining; smaller wing-coverts, orange red, edged behind by a yellow band. North America.

Red-breasted Oriole. Oriolus Americanus. Gm. Pl. Enl. 236. 2.

Dusky brown or black; chin and breast, red; 7 inches

White-winged Oriole. Oriolus Leucopterus. Lath. Syn. frontisp.

Glossy black; patch of white on wing-coverts; size of a lark. Cayenne.

Bonana Oriole. Oriolus Bonana. Enl. 535. 1.

Black above; beneath, deep orange-red; 7 inches. West India islands.

Oriolus Icterocephalus. Gm. Edw. 323. Pl. Enl. 343.

Black; head and throat, yellow; spot from base of bill to the eyes, black.

Oriolus Xanthocephalus. Ch. Bonaparte, I. IV. 1. 2.

Head, neck, and breast, yellow. American Continent.

Oriolus Mexicanus. Gm. pl. Enl. 533.

Head, neck, belly, sides, &c., yellow; crown, back, rump, &c., brown. Mexico.

Oriolus Xanthornus. Gm. Pl. Enl. 5. 1.

Yellow; face, throat, and wing-coverts, black; 7 inches long. Jamaica.

Oriolus Baltimore. Gm. Pl.. Enl. 506. 1. Vieil. Gal. 87. and Wils. I. 1. 3.

Black above, orange beneath; white bar on the wing; bill, lead-colour. North America.

Oriolus Spurius. Pl. Enl. 506. 2. Wils. I. iv. 1. 4. Var. A. of *Baltimore.* Lath.

Forehead and cheeks, black, mixed with yellow; hind head, nape, and upper part of back, olive grey;

lower part, fore neck, to vent, and beneath wings, orange. North America.

Oriolus Melancholicus. Gm. Pl. Enl. 448.

Throat, white ; each feather elsewhere black ; brown in the middle, bordered with orange on wings, tail, and lower part of body, and with yellowish above. Cayenne. The adult is *Or. Guyanensis*. Enl. 536. Vieil. Gal. pl. 88. *

THE OXYRHINCI. Temm.

Have a pointed and conical bill like the Carouge, but shorter than the head.

The species known, *Oxyr. Flammiceps*. T. *O. Cristatus*. Swain. III. Col. 125 , bears a tuft, mixed with red, like many of the tyrants.

THE PIT-PITS. Buff. DACNIS. Cuv.

Represent the Carouges in miniature, in their conical and sharp bill.

They connect that genus with the fig-eaters. The

* Also by others—

Le Dragon, Azara, 66. *Agelaius virescens*, Vieil. *Icterus Anticus*, Licht. *Agelaius frontalis*, and *ruficapillus*, Vieil. *Le troupiale à calotte rousse*, Azara, 72.

Oriolus Jamaicaii, Gm. *Agelaius longirostris*, Vieil.

Agelaius Cyanopus, Vieil. *Le Troupiale noir et varié*, Azar, 71. *Icterus-Tanagrinus*. Spix. Braz. 64. 1.

Emberiza Oryzivora. Wils.

Agelaius badius. Vieil.

Agelaius pyrrhopterus. Vieil.

known species (*Mot. Cayana.* L.) Enl. 669. Vieil. Gal. 165. is a small bird, blue and black.*

THE STARES. STURNUS. Lin.

Differ from the Carouges only in having the bill depressed, especially towards the point.

The Common Stare. Sturnus Vulgaris. Lin. Enl. 75. Naum. 62.

Black, with violet and green reflexions, spotted throughout with white or fawn-colour. The young male is brownish-gray.

This bird, much extended through the old Continent, feeds on insects of all kinds, and renders a service to the cattle in ridding them of such encumbrances. It flies in numerous and crowded flocks, is easily tamed, and learns to sing, and even to speak. It quits us in winter. Its flesh is disagreeable.

* Other Pit-pits have been named:

Blue-striped warbler. Mot. Lineata. Gm.

Blue-headed warbler. Mot. Cyanocephala. Gm. Briss. iii. 28. 4.

Blue winged yellow warbler. Sylv. Solitaria. Wils. A. O. ii. 15. 14.

Worm-eating warbler Motacilla vermivora. Gm. Edw. 305. Wils. A. O. iii. 24. 4.

Protonothary warbler. Mot. protonotharius Gm. pl. Enl. 704. 2. Wils. A. O. iii. 24. 3.

Golden-winged warbler. Mot. Chrysoptera. Lin. and Mot. *flavifrons.* Gm. Wils. A. O. ii. 15. 5. Bonap. A. O. i. 1. 39. ♀

Tenessee warbler. Sylvia peregrina. Wils. A. O. iii. 25. 2.

Nashville warbler. Sylvia ruficapilla. Wils. A. O. iii. 27. 3.

Orange crowned warbler. Sylvia celata. Say. Bonap. A. O. i. 5. 3.

Sturnus Unicolor. Temm. Pl. Col. III. Viel. Gal. Pl. xci.

All black, with slight purple reflexions, not very brilliant. South of Europe, and Egypt.

Cape Stare. Sturnus Capensis, Enl.' 280.

Black; white beneath; round the eye, bare, and orange-coloured; patch of white on side of face. Cape of Good Hope. *St. Contra*, of Albin, does not differ, but is of the East Indies, not the Cape.

Magellanic Stare. St. Militaris. Enl. 113.

Brown above; chin, breast, and upper part of belly, crimson; white about the head. Straits of Magellan.

Louisiana Stare. St. Ludovicianus. Enl. 256.

Brown and rufous-gray above, underneath yellow; white about the head; size of a lark. The same as *Alauda Magna.* Gm. Catesb. 1. 83., *Stournelle à collier.* Vieil. Gal. pl. xc. and Wils. III. xix. 2.

Red Oriole. Oriolus Ruber. Gm. *L'Etourneau à Camail rouge.* Sonn. Nouv. 9. pl. lxviii. *Amblyramphus Tricolor.* Leach. Zool. Misc. pl. xxxvi.

Head, neck, back, and thighs, vermilion red. A fine species of the Steppes of Buenos Ayres, not of India, as Sonnerat affirms.

The *Sturnus Cinelus* forms a genus, as the reader has seen above, near the thrushes: the *Sturnus Sericeus.* Brown, ·III. 21. is rather a marten.

The *Sturnus Collaris* is the same as the Alpine warbler (*Accentor*). *The Sturnus Caruncalatus* should, in my opinion, be placed with philedon.

The species mentioned by Osbec, by Hernandez, &c., are but ill authenticated. As for those of Pallas, it is to be regretted that there are no figures. The stares of Daudin should go back to the thrushes and philedons, and his quiscala partly to the martens and partly the cassiques. In general, Daudin may be said to have completed the confusion of this genus, already very badly defined by his predecessors.*

We see no sufficient character whereby to distinguish accurately from the conirostres the genera of the corvine family, which have all the same internal structure, the same external organs, and are characterized only by a generally larger size, which enables them sometimes to hunt the smaller birds. Their powerful bill is most frequently compressed at the sides.

These genera are three in number, the crows, birds of paradise, and rollers.

The Crows. (Corvus. Lin.)

Have a strong bill, more or less flatted at the sides; the nostrils are covered with stiff hairs, directed forward. They are subtle birds, whose scent is very fine, and who have, in general, a habit of taking and hiding things which are even useless to them, as pieces of money, &c.

* To these, others have added—

§ *Sturnus Dauricus.* Lath. *Gracula Sturnina.* Pallas.

§ *Sturnus Olivaceus.* Lath. *Osbec.*

§ *Sturnus Sericeus.* Lath. pl. 21. Illust. Brown.

§ *Sturnus Viridis.* Lath. *Osbec.*

The larger species, with a stronger bill, compared with the others, and the ridge of the upper mandible more arched, are more especially called Crows. Their tail is round or square.

The Raven (Corvus Corax. Lin.) Vaill. Ap. pl. 51.

Is the largest bird of the passerine order found in Europe, being as big as the cock. Its plumage is entirely black, the tail round, the back of the upper mandible arched forward. It lives more retired than the other species, flies well and high, smells carcasses at a league distance, and feeds moreover on all sorts of fruit and small animals, and will even carry off the tenants of the poultry-yard; builds singly on high trees or sharp rocks; is easily tamed, and will learn to speak tolerably well. Its flight is elevated and easy. It seems to be found in all parts of the world.

In the north its plumage is often mixed with white (Ascan. Ic. Nat. pl. viii.) It is then the *Corvus Leucophæus.* Temm. Vieil. Gal. 100.

Enl. 495 appears to be simply a crow, and 483 a young rook. M. Temminck thinks that the figure cited above from Levaillant is of a distinct species, peculiar to Africa, and which he names *C. Montanus.*

The Carrion Crow. (C. Corone. L.) Enl. 495.
Naum. 55.

One-fourth smaller than the raven, the tail more square, and the bill less arched above.

M. Temminck makes a distinct species of the Cape Crow, and names it *C. Segetum.*

The Rook. (*C. Frugilegus.* Lin.) Enl. 484. Naum. 55.

Still a little smaller, and with the bill more straight and pointed than the carrion crow. Around the base of the beak is without feathers, except during the nonage of the bird, probably owing to its raking the ground in search of food. These two species live in large flocks, and assemble together to build: they devour as much grain as insects. They are found in all Europe, but they remain in winter only in the warmer countries.

Hooded Crow. (*C. Cornix.* L.) Enl. 76. Naum. 54.

Ashy, with the head, wings, and tail, black; less frugivorous than the other species; frequents the sea-shore, and lives on shell-fish, &c. Naumann tells us, that it often pairs with the black crow, and produces fertile mules.

Jackdaw. (*C. Monedula.* L.) Enl. 523. Naum. 56. 1.

Still a fourth smaller than the last, being nearly the size of a pigeon, of a less deep black, which becomes ashy round the neck, and under the belly; sometimes also entirely black. Builds in towers; lives in flocks; has the same food and habits as the crows, and often associates with them. The birds of prey have no more vigilant enemy.

The jackdaw terminates the tribe of true crows, because its upper mandible is not much more sensibly arched than the lower. But to this tribe we must add,

Chattering Crow. (*Corvus Jamaicensis.*) Gm.

Entirely black; size of a common crow. *Corneille à duvet blanc*, *Downy crow* of Latham, *Corvus Leucognaphalus*, Daud. is the same, furnished with fine white down at the base of the feathers.

White-breasted Crow. *Corvus Dauricus*. Gm. Enl. 327.

Black; head and throat, glossed with blue; neck, breast, and belly, white. Twelve inches. Senegal, China, and Persia.

Corvus Scapulatus. Daud. Vaill. 53.

Heretofore classed with the preceding; but M. Temminck thinks them distinct. Scapulars, white. Africa.

White-necked Raven. Lath. *Corvus Albicollis*. Vail. 50.

Black, with white patch on nape and sides of neck. eighteen or nineteen inches. S. Africa. It might constitute a separate subgenus, from its compressed and elevated beak, with trenchant back.

Corvus Splendens. Vieil. Temm. pl. col. 425.

Forehead, mask, and throat, lustrous black; head, cheeks, nape, and breast, ashen-gray, reddish-tinted; beneath, slate-colour; wings, back, and tail, lustrous black, with violet and purple reflexions. East Indies.

Clark's Crow. *Corvus Columbianus.* Wils. Amer. Orn. iii. pl. 20. f. 2.

General colour above, light silky drab; breast and belly, dove-colour; vent, white. Banks of Columbia river.

Corvus Nasicus. Temm. Pl. Col. 413.

Beak, very bulky, much dilated at the edges of the upper mandible, and strongly curved; nostrils, naked; all the plumage black, but not lustrous. Island of Cuba.

Fish Crow. *Corvus Ossifragus.* Wils. Amer. Orn. v. 37. 2.

Plumage, wholly black, with reflexions of steel-blue and purple. N. America. M. Cuvier seems to doubt its difference from our common crow.*

THE PIES. (PICA. CUV.)

Less than the crows; have also the upper mandible more arched than theirs, and the tail long and cuneiform.

* M. Temminck has the following additional species in his genus *Corbeau,* Pl. Col. 70ieme. Livraison.—

Corvus Montanus. Vail. 51. as above noticed.
Corvus Leucophæus. Vieil. Gal. 100. *Borealis* of Brisson.
Corvus Segetum. Vail. O. A. 52., above mentioned.
Corvus Australis. Lath. Ind. Sp. 2.
Corvus Enca. Horsf. Cat. Java.

Also from others,—

Corvus Mexicanus. Lath. Vieil.
Corvus Clericus. Lath. Vieil.
¿ *Corvus Versicolor.* Lath. Vieil.
Corvus Spermologus. Frisch. Vieil. Var. of *Corvus monedula.* Lath.

The Magpie. (C. Pica. L.*)* Enl. 488. Naum. 56. 2.

Is a fine bird, of a silky black, reflecting purple, blue, and gold; belly, white, and a large spot of the same colour over the eye. Its perpetual chattering has rendered it famous. It prefers inhabited places, and feeds on all sorts of matter, and will even attack the small domesticated birds.

Senegal Crow *Corvus Senegalensis.* Enl. 538.

Violet black above; dusky beneath; size of a magpie. Tail longest in the males. Senegal.

Corvus Ventralis. Shaw. Vaill. Afr. 55.

Above, glossy black; belly, thighs, and vent, flesh-colour. Said to be from South Sea Islands.

Red-billed Jay. *Corvus Erythrorhyncos.* Enl. 622. better Vail. 57.

Body, brown above; white beneath; head, neck, and breast, black. China.

Cayenne Jay. *Corvus Cayanus.* Enl. 373.

Black; wings and tail, violet; last tipped with white; head, black, with some white spots. Cayenne.

Peruvian Jay. *C. Peruvianus.* Enl. 625.

Forehead, blue; hind-head, whitish; green above; yellow beneath. Peru.

Blue Crow. *Corvus Cyaneus.* Pall. Vail. Afr. 58. 2.

Top of head to nape, glossy deep-black; body, ash-colour; paler beneath; wings and tail, fine blue. Mongolia.

Rufous Crow. C. Rufus. Vail. 59.

Body, and tail-coverts, red-brown; wings, black, with broad gray stripe; tail long, cuneiform.

Paraguan Jay. Lath. *Corvus Pileatus.* Illig. *Acahé.* d'Azara. pl. col. 58. *Pica Chrysops.* Vieil. Gal. 101.

Top, and sides of head, black, soft, and velvety; above, deep blue; beneath, yellow; tail, white tipped. Paraguay.

Garrulus Gubernatrix. Temm. Pl. Col. 436.

Head-tuft like a military plume. M. Temminck makes this a jay.

Corvus Azureus. Azara. Temm. Pl. Col. 108.

Head and neck, full black; other parts of body, wings and tail, sky-blue. Paraguay and Brazil.

Corvus Cyanopogon. P. Max. Temminck. Pl. Col. 109.

A long coronal tuft; forehead, neck, and breast, full black; belly, thighs, and under wing-coverts, white, or whitish. Brazil.*

* M. Vieillot admits into this genus—

Pica Albicollis. Vieil. *Corvus Caledonicus.* Lath.
La Pie Houpette.
Pica Olivacea. Vieil. *Corvus Olivaceu* Lath.
Corvus Africanus. Lath.
Pica Vagabonda. Vieil. *Coracias Vagabonda.* Lath.
Corvus Zanöe. Lath.

THE JAYS. (GARRULUS. CUV.)

Have the two mandibles but little elongated, and terminated by a sudden, and nearly equal bend. When their tail is cuneiform, it is not long, and the feathers of the forehead, generally pliant and slender, are more or less elevated when the bird is angry.

The Jay of Europe. C. Glandularius. Lin. Enl. 481. Naum. 58. 1.

Is a fine bird, of a vinous gray, with black moustachios, and quills particularly remarkable for a large spot of brilliant blue, striped with deeper blue, which forms a portion of the wing-coverts. The acorn constitutes its principal food. It is one of those birds which evince the greatest disposition to the imitation of all kinds of sounds. It nestles in our woods, and lives in pairs, or in small flocks.

Add

Blue Jay. Corvus Cristatus. Enl. 529. Viel. Gal. 102.

Blue, above; white, underneath; black streaks on side of face, and crescent on the breast. Eleven inches. America.

Steller's Crow. C. Stelleri. Vail. Ois. de Par., and c. I. 44.

Purple-greenish-black; blue on quill feathers; 15 inches. North America.

Siberian Crow. C. Sibiricus. Enl. 608.

Cinereous above; ferrugineous-orange beneath; ten inches. Siberia.

C. Infaustus et *Russicus*. Var. Lath. *Cinereous Crow*. *C. Canadensis*. Enl. 530, and a variety. Vail. 48.

Brown, above; pale ash-colour, underneath; forehead and throat, dirty yellow; 10½ inches. Canada.

Corvus Cristatellus. Vel. *Corvus Cyanoleucus*. Pr. Max. Temm. Pl. Col. 193.

Tuft of fine curved feathers, and plumes of full deep black; occiput, nape, neck, and breast, bláckish-brown; back and scapulars, bluish-brown; underneath, light yellowish white. Brazil.

Corvus Ultramarinus. Cuv. *Garrulus Ultramarinus*. C. Bonap. Temm. Pl. Col. 439.

Long square tail; head, nape, and cheeks, ultramarine; mantle and middle of back, ashy, shaded with ultramarine; wings and tail, last colour; throat, abdomen, and tail-coverts, whitish. Mexico.

Corvus Floridanus. Ch. Bonap. I. 13. 1.

Approximating to the last, but the blue of the back is more lively, and all the azure colours have more brilliancy. Wings, in proportion to tail, longer. North America.

Corvus Torquatus. *Garrule Torquéole*. Temm. Pl. Col. 444.

An extensive white space on all the neck, breast, and part of belly; black, with reflexions of polished steel, on all other parts of body, wings, and tail. New Caledonia, Celebes, and perhaps Borneo. *

* M. Vieillot has

Garrulus Melanogaster. Ois. de Par. Vail.

Garrulus Cærulescens.

Garrulus Galericulatus.

Garrulus Auritus.

Corvus Purpurascens. Lath.

The Nutcrackers (Caryocatactes.) Cuv.

Have both mandibles equally pointed and straight. Only one is known.

The *Nutcracker*. (*C. Caryocatactes*. L.) Enl. 50. Naum. 58. 2. Vieil. Gal. 105.

Brown, spotted with white all over the body. Builds in the clefts of trees in elevated thick woods; climbs the trees and pierces the bark, like the woodpecker; eats all sorts of fruit, insects, and small birds; and comes occasionally, but without regularity, in large flocks into the open country. It is famous for its confidence.

The *C. Hottentottus*, Enl. 226., appears to me to be allied to the Tyrants. *C. Balicassius*, Enl. 603., is a Drongo. *C. Calvus*, Enl. 251, is a Gymnocephale. *C. Novæ Guineæ*, Enl. 629, and *C. Papuensis*, Enl. 630, are Choucaris. The *C. Speciosus* of Shaw is the Chinese Roller, Enl. 620. *C. Flaviventris*, Enl. 249, is a Tyrant. *C. Mexicanus* is probably a Cassique, or a Weaver, and *C. Argyrophtalmus*, Brown. Ill. 10, is certainly so. *C. Rufipennis*, Enl. 199, is a blackbird, the same as *Turdus Morio*. *C. Cyanurus*, Enl. 355, *C. Brachyurus*, Enl. 257 and 8, *C. Grallarius*, Enl. 702, of Shaw, are Pittæ and Anteaters. *C. Carunculatus*, Daud. is a Philedon.

I have approximated to the Blackbirds *C. Pyrrhocorax*, Enl. 531, and to the Upupæ, the *C. Graculus*, Enl. 155. I believe the *C. Eremita* not to exist. The *C. Caribæus*, Aldrov. I. 788,is a Bee-eater, whose description has been used by Duterte to describe an object he had nearly forgotten. *C. Gymnocephalus*. T. Col. 327, appears to me to be of the Dentirostral family.

The Temia. Vaill.

Have, with the carriage and tail of the pies, an elevated bill, with the upper mandible gibbous, and the base furnished with velvety feathers almost like the Birds of Paradise.

The most anciently known, (*Corvus Varians.* Lath.) Vail. Afr. 56. Vieil. Gal. 106, is of a bronze-green. It is found in India and Africa.

M. Vieillot has called this genus Crypsirina. Gal. 106. Dr. Horsfield, Phrenotrix. M. Temminck unites *Temia* to *Glaucopis.*

Add,

Glaucopis Leucopterus. Temm. pl. col. 265.

Size of our European jay; all the plumage perfect black, except a white band on the wing parallel with the body, and a white stripe on the external barb of the first two secondaries of the wing. Length, fourteen inches. Sumatra.

Glaucopis Temnura. Tem. Pl. Col. 337.

Plumage, dusky black, shaded with dark gray; tail curiously scolloped. India. ?

The Glaucopis. Forster.

Have the same bill, and the same carriage; but under the base of the bill, hang two fleshy caruncles.

The species known (*Glaucopis Cinerea. Wattle-bird.* Lath. Syn. I. pl. 14.) is of New Holland; as large as a pie; blackish, and with wedged tail. It lives on insects and berries; perches little. Its flesh is considered excellent.

Bechstein substitutes for *Glaucopis* the name *Callæas.*

THE ROLLERS. CORACIAS. Lin.

Have the bill strong, compressed towards the end, and the point a little crooked; the nostrils, oblong, placed at the edge of the feathers, and not covered by them; the feet, short and strong. They are birds of the Old Continent, very like the jays, in their manners, and the loose plumes of the forehead. Their colours are lively, but seldom harmonious. Their anatomy presents certain peculiarities which approximate them to the king-fishers and pies; such as two notches on the sternum, a single pair of muscles at the lower larynx, and a membranous stomach.

The name *Coracias,* consecrated by the authority of Linnæus, has been changed by M. Vieillot, into that of GALGULUS, which, among the ancient Latins, belonged to the Oriole.

THE ROLLERS, properly so called,

Have the bill straight, and more high than broad in all the species.

We have one in Europe,

Garrulous Roller. (*Coracias Garrula.* Lin.) Enl. 480.

Sea-green, with back and scapulars, fawn-colour;

pure blue on the tip of wing; pretty nearly the size of the jay. It is a very wild bird, though sociable enough with its consimilars; it is clamorous, nestles in the hollows of trees, and quits us in winter. It lives on worms, insects, and small frogs.

Some foreign rollers have, like ours, the tail square; but still the external tail-quills in our roller are a little elongated in the male, which is the first indication of their great elongation in many other species.

Bengal Roller. C. Bengalensis. Enl. 285.

Top of the head, green; hind part of neck and back, fulvous; lower part of back, blue; belly, blue-green; throat, reddish white. The same as *Indica*, Edw. 326, and Albin's figure, 1. 17, cited under *caudata*.

Coracias Viridis Nob. Vail. I. 36. Vieil. Gal. 110.

Forehead and throat, reddish white; top of head and neck, of back, scapulars, wing-coverts, and underneath, aquamarine. East Indies.

Coracias Temminckii. Vail. pl. G.

Tuft of feathers on head, aquamarine; neck, throat, underneath, cropper, and tail, indigo-blue; back, scapulars, and wings, full green.

Abyssinian Roller. C. Abyssinica, Enl. 626.

Head, neck, and wing-coverts, green; shoulders, quills, and rump, blue; back, brown; 18 inches. Abyssinia. *C. Senegala.* Enl. 326. Edw. 327, is a variety. *C. Caudata* is an individual disfigured by the head of *Bengalensis*. (Vail. loc. cit. p. 105.)

Coracias Cyanogaster. Cuv. Vail. Ois. de Par. pl. 26.

Head, neck, and, breast, red, shaded with green; beneath, crupper, and wing-coverts, blue; tail, green. Africa.

Cor. Caffra, where Shaw cites, Edw. 320, is a *blackbird*. (*Turdus Nitens*.) *C. Sinensis*, Enl. 630, from its notched bill, must also be approximated either to the blackbirds or shrikes. M. Temmnick makes it a *Pyroll*. Shaw thinks that *C. Viridis*. Lath. is a *King-fisher*. *C. Strepera* and *C. Varia*. Lath. are Cassicans. *C. Militaris* and *C. Scutata*, Shaw, are Piauhaus. *C. Mexicana*. Seb. 1, pl. 64. f. 5. is the Jay of Canada. *C. Cayana*. Enl. 616. is a Tanager.*

The Rolles. (Colaris. Cuv.)

Differ from the rollers by a shorter bill, more arched, and especially enlarged at its base, so as to be less high there than broad.

* The following have been named elsewhere—

Coracias Docilis. Lath.
Coracias Cœrulea. Lath.
Coracias Pilosa. Lath.
Coracias Cyanea Lath.
Coracias Puella. Lath.
Galgulus Melanops. Vieil. Vail. 30.
Coracias Nigra. Lath.
Coracias Striata. Lath.
Coracias Pacifica. Lath.

Oriental Roller. (*C. Orientalis.* 619.)

Head and neck, brown; green-brown above; blue-green underneath; quills, blue and black; size of a jay. East Indies.

Madagascar Roller. (*C. Madagascariensis.* Enl. 501.)

Plumage in general, rusty purplish-brown; rump and vent, blue-green; tail, blue-green; tip, blackish.

African Roller. (*C. Afra.* Lath.)

Pale cinnamon above; paler beneath; quills, blue; tail, tipped black. Eight and a half inches. Africa.*

Birds of Paradise. (Paradisea. L.)

Have, like the corvi, a straight, compressed, strong, toothless bill, and the nostrils covered; but the influence of the climate they inhabit, and which extends itself to the birds of many genera, has given the feathers which cover the nostrils a sort of velvet texture, and frequently a metallic brilliancy, while it has, at the same time, greatly developed the feathers of many parts of the body. These birds are originally from New Guinea and the neighbouring islands. At first they were only to be obtained from the barbarians of those countries, who prepared them for making fans, and cut off the feet and wings, so that

* M. Vieillot, who calls this genus Eurystomus, has placed here:

Eurystomus Cyanocollis. Vail. O. de Par. pl. 96.
Eurystomus Gularis. Vieil.
Eurystomus Purpurascens. Vieil. Vail. Ois. de Par. 35.

it was thought, for some time, in Europe, that the first species was really without these members, and lived altogether in the air, supported by the long feathers of the sides. Some travellers, however, having procured perfect specimens of some species, it is now known that their wings and feet indicate the station to which we assign them. They are said to feed on fruits, and especially to seek spices. Some have the feathers of the sides silky, and of great length, forming a fan larger than the body, which gives such a resistance to the wind that these birds are often carried away by it, in spite of their efforts; and have, moreover, two barbed threads adhering to the croup, and running out as far, or farther, than the feathers of the flanks.

The Great Bird of Paradise. (*P. Apoda.* L.) Enl. 254. Vaill. Ois. de Par. Pl. 1. Vail. pl. 1.

As large as a thrush; moronne colour, with the upper part of the head and neck, yellow; round the bill and throat, emerald-green; the male of this species has the long bundles of yellow feathers used by the ladies in dress. There is a race rather less.

The Red Bird of Paradise. Par. Rubra. Vail. pl. 6. Vieil. pl. 3.

With the lateral plumes of a bright red, and the threads wider, and concave on one side.

In the other birds of paradise we still find threads;

but the feathers of the sides, though a little elongated, do not exceed the tail in length.

King Paradise Bird. (*Paradisea Regia.*) Enl. 496. Vail. 7. Vieil. 5.

As large as a sparrow; moronne-purple; a band across the breast; the extremity of the feathers of the sides, and the barbs which widen the end of the two long threads, emerald-green.

This M. Vieillot makes his genus CINCINNURUS.

Magnificent Paradise Bird. (*Par. Magnifica.*) Sonnerat. 98. Enl. 631. Vail. 9. Vieil. 4.

Moronne above; green underneath and on the flanks. The quills of the wings, yellow; a bundle of straw-coloured feathers on each side of the neck; another more yellow opposite the fold of the wing.

Others again have slender but short feathers on the flanks, and are destitute of threads at the crupper.

Gold-breasted Paradise Bird. (*Par. Aurea.* Gm. *Sex-setacea.* Shaw.) Sonnerat. pl. 97. Enl. 633. Vail. 12. Vieil. 6. et Galer. 97.

As large as a blackbird; black; a golden-green breast-plate on the throat, terminated by a small disk of webs golden-green.

M. Vieillot makes of this species his genus PAROTIA. Gal. 67.

Others, in fine, have neither thread, nor elongated feathers on the flanks.

Superb Paradise Bird. (*Par. Superba.*) Sonnerat. 96. Enl. 632. Vail. 14. Vieil. 7. Galer. 98.

The feathers of the scapulars are nevertheless prolonged into a sort of mantlet, which can cover the wings, and those of the breast, into a sort of coat of arms, pendant, and furcated. All the plumage is black, except the pectoral coat, which is of a brilliant burnished-steel-green.

Golden Paradise Bird, (*Par. Aurea.* Sh. *Oriolus aureus.* Gm.) Edw. 112. Vail. 18. Vieil. 11.

Has no extraordinary development of plumage, and is only recognized by the velvet of the feathers which cover the nostrils. The male is of a most lively orange, the throat and primaries of the wings being black. The female is brown, instead of orange. M. Vieillot's genus, LOPHORINA.

I refer to the black-birds, *Paradisæa Gularis,* Lath., or *Nigra,* Gm. Vail. 20. 21. Vieil. 8. and 9., and *Leucoptera,* Lath. to the Cassicans. *Par. Chalybœa.* Enl. 663. Sonn. 97. Vail. 23. Vieil. 10. *Cirrhata.* Aldrov. is too much mutilated to be characterized; and *Furcata,* Lath., appears to be an imperfect individual of *Superba.*

* M. Vieillot distinguishes—

Oiseau de Paradis noir. Valentyn.

Paradisea Alba. Lath.

Paradisea Minor Papuana. Lath.

Dr. Latham enumerates twenty species and varieties of the Birds of Paradise, lamenting the imperfection of their descriptions, from the little knowledge to be obtained concerning them. He adds a hope that future naturalists or travellers visiting New Guinea, may be capable of discriminating the subjects found there, and thus supplying the present desiderata in this department of ornithology. He has, besides those we have mentioned—

Doubtful Paradise Bird.
Hackled Paradise Bird.
Emerald-breasted Paradise Bird.
Frosted Paradise Bird.
Crisped Paradise Bird.
Twelve-wired Paradise Bird.

Generic Characters of Birds

ORDER PASSERES.

Fam. Fissirostres

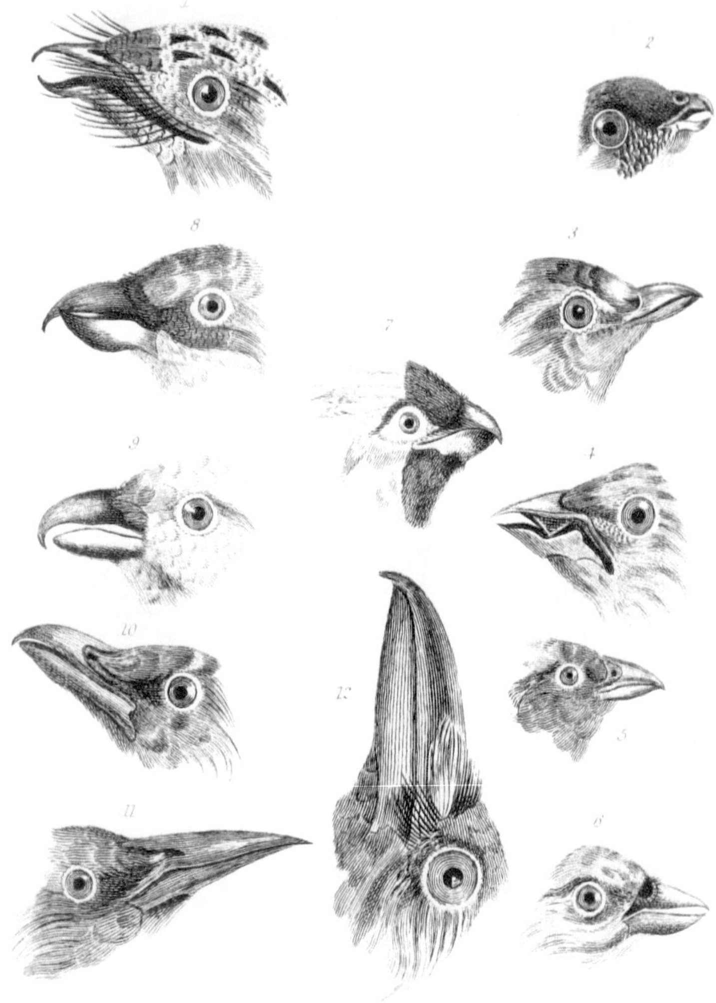

Fam. Conirostres.

1. *Caprimulgus*	2. *Hirundo*

Fam. Conirostres.

3. *Alauda*	8. *Loxia*
4. *Emberiza*	9. *Corythus*
5. *Ploceus*	10. *Buphaga*
6. *Fringilla Cuv.*	11. *Cassicus*
7. *Pyrrhula*	12. *Corvus*

London. Published by Whittaker & Co. Ave Maria Lane, June 1828.

SUPPLEMENT ON THE CONIROSTRES.

THE first genus to be considered in this family, is that of the LARKS. To the very brief indication of their generic characters in the text, we may add, that they are distinguished by a cylindrical and subulated bill, furnished at the base with small feathers, directed forwards, and completely covering the nostrils. This bill is straight in some, and more or less arched in others; the nostrils are rounded, and half closed by a membrane; the tongue is cartilaginous, and cleft at the point; the hinder claw is straight, or nearly so, acuminated, and usually longer than the thumb. The first remex is sometimes shorter than the fourth; but more generally they are of equal length; the second and third are the longest of all. There are two secondaries, nearly as long as the primaries, notched on the end, as well as the intermediate feathers. There are four toes, three before and one behind.

All the larks nestle on the ground; and the majority rise so high that they are often lost sight of, and sing during their flight. Some of them perch, but not often; they are seminivorous, insectivorous, and herbivorous, and they swallow grains entire; they are to be found in all parts of the world.

Some naturalists have divided the larks into three sections, according to certain differences of conformation observable in the bill. The first of these differences is that the bill is conical, straight, or nearly so, more high than wide at the base, and, on the whole, rather slender; the second is when the bill is gross, more high than wide, and gently curving into an arch; the third is when the bill is long and very arched. As these divisions, however, are not recognised in the "Regne Animal," we shall pursue, in noticing the species, the order adopted there, adding such accounts of others, not mentioned by our author, as may prove interesting.

The first to be noticed is the *sky*, or *field* lark, also sometimes called the *common* lark, and to which naturalists have, as usual, added a crowd of epithets. Notwithstanding that it is so common in our climates, it has not unfrequently been confounded with other species of the same genus. One cause of this is, that its plumage is at once extremely varied, and yet it presents no prominent colour, or even a very decided tint of any kind; another reason of this uncertainty is, that, in many cases, authors of works on natural history have thought that they might dispense with the necessity of describing a bird so very generally known. This mode of avoiding the difficulties and the tediousness of a minute description, is convenient enough to a writer, in the way of saving trouble, but it is calculated to abridge the utility of his work. The object of the naturalist is to instruct, as well as to amuse; and, therefore, accurate descriptions, however tiresome, are yet necessary. A foreigner has just right to complain of an author of this kind, who does not extend his views beyond the circle of his native country, and endeavour to render his book one of general instruction. Now, in the instance in question, the species of which we are speaking, is a stranger to America; and a native of that country would have some reason to find fault with an European naturalist who should neglect to describe it. In this respect the work of Brisson is valuable, from the extent and accuracy of its descriptions; and though not adapted to be read through with pleasure, is extremely useful for the purposes of consultation. Actuated by similar views, we have endeavoured to render our book worthy of general approbation, by combining close description with matter more popular and entertaining; and, by separating the two departments, we trust to have succeeded better in our object. In the latter part we have, in general, rather avoided minute description; but as in the "Regne Animal" itself the descriptions, in consequence of the original plan of that work, are often purposely imperfect, it is necessary, occasionally, to attempt to supply such deficiencies in the present portion of our own. To our author's account, therefore, of this lark, we beg to add the following particulars.

The length of the sky-lark is usually about six inches ten lines from the extremity of the bill to that of the tail; the extended wings, two inches and a half, and in a state of repose, they reach to two-thirds of the tail; the twelve quills of the latter are two inches nine lines, with the exception of the middle ones, which being a little shorter than the lateral, render the tail slightly forked. The claw of the posterior toe is sometimes nearly two inches long; it appears longer in proportion to the age of the bird; for in a bird of eighteen months old, it is usually no more than six lines; of the three anterior toes the lateral are slightly arched, while the middle one is straight.

A mixture of blackish and of gray, tinted with red and dirty white, constitute, properly speaking, the colour of the upper plumage of the sky-lark. A narrow band of reddish white passes above the eyes, on each side of the head; the throat is white; the fore part of the neck and all the under parts of the body, reddish white, with longitudinal blackish spots. The small upper wing-coverts are gray, tinted with reddish, and bordered with white; the large coverts, the most remote from the body, have a fawn-coloured border, around a brown ground, and those which are nearest have a gray brown ground, the extremity being fawn-colour on the border. The quills of the wing are brown, their external edge fawn, just tipped with white, and the last three or four nearest the body, have a deeper shade, and their internal edge is fawn. The upper half of the bill is blackish, and the lower rather whitish.

The male is a little browner than the female; he has a sort of black collar, and his hinder claw is longer. He is also something more bulky, although the heaviest of the larks rarely weighs above two ounces; the stomach of these birds is fleshy, and tolerably ample in proportion to the volume of the body.

Some individuals have more or less of reddish, and more or less of the wing-quills edged with this colour. There are also in this species certain individual varieties more strongly decided: first, the *white lark*; secondly, the *black lark*; and, thirdly, the *Isabella lark*, which last is not found so frequently as the first. This tint may possibly be produced by the age of the bird; for all the species of Alauda exhibit variations in the plumage as they grow older, and before they become entirely white pass through this colour. M. Picot la Peyrouse has observed many that were variegated with isabella on their ordinary plumage, and others with isabella and white.

The common lark is the musician of the fields; its delightful song, like a hymn of joy, announces the approaching spring, and accompanies the earliest blush of morning. It is heard from the very commencement of those fine days which succeed the cold and gloom of winter; and its notes are the first which strike upon the ear of the vigilant cultivator of the ground. The matin song of the lark was in ancient Greece the signal for the reaper to commence his labours, which he suspended during that portion of the day in which the ardours of a summer's noon imposed silence on the bird. The lark is always silent during the noon ; but when the sun begins to verge towards the west, it fills the air anew with its varied and sonorous modulations. It is silent when the sky is overcast and the weather rainy; but, generally speaking, it sings during the entire of the fine season.

Like all other species of the feathered class, the song in the lark is the peculiar attribute of the male; he rises almost perpendicularly, and by starts, in the air, and describes, in rising, a sort of screw-like curve. He often mounts very high, always singing, and forcing his voice in proportion as he recedes from the earth, so that he is easily heard, even when he is scarcely visible; he sustains himself a long time in the air, descends slowly as far as ten or twelve feet above the ground, and then precipitates himself downward like an arrow. His voice grows feeble in proportion as he approaches the earth, and he becomes completely mute as soon as he alights upon it.

From a certain elevation in the air the male is observed, in the season of reproduction, to look out for, and attract some female. When he has succeeded in drawing the attention of one, the latter from below regards him earnestly for some little time, and then flutters lightly towards the place where she observes him about to alight. Constancy, however, is no characteristic of this species of birds, and their unions are but transitory. We must not search amongst them for those models of tender affection and fidelity, so often to be found in some other divisions of the feathered race.

The female, when fecundated, very soon makes her nest: she conceals it carefully between two lumps of earth. It is flat, not very concave, and nearly devoid of firmness or consistence. It is composed of grass, small dried roots, and the hair of cattle. The eggs, to the number of four or five, have brown spots on a greyish ground. The female hatches only about fourteen or fifteen days, and in somewhat less time than that the young ones are in a state to quit the nest, and dispense with her farther care. After having fed them for a few days, she instructs them how to procure their food, and makes them leave the nest before they are entirely covered with feathers. The fowler is often deceived by this, not finding in the nest the young ones, which but a few days before he had viewed recently broken from the shell, and almost entirely naked.

The vernal amours of the larks leave them sufficient time to have several broods in the summer. With us, and in France and Germany, they have usually but two: but in more southern countries, as Italy, for instance, they have three; the first in the commencement of May, the second in July, and the third in August.

The first nourishment taken by the young larks is composed of chrysalides, of worms, of caterpillars, and even of the eggs of locusts. This last sort of food makes them in high consideration in the countries which are exposed to the ravages of those insects. They were on this account considered as sacred birds in the island of Lemnos, where the locusts still create, as they do in many other countries of the Levant, incalculable ravages. The services which the same birds render to us, in destroying the germs of the generations of many species of insects which devastate our crops, should induce us to spare them a little more.

When they are adult, the larks feed principally on different grains, herbs, and vegetable substances in general. They seldom go to water, but quench their thirst most generally by inhaling the dew-drops.

When people are desirous of bringing up the young of this species, they feed them with paste, made of meat and crumb of bread, bruised hemp seed, and with crumb of bread and beef's heart hashed. This paste is improved by an admixture of poppy seed grated. In Flanders they feed the young larks with this seed moistened. As soon as they begin to sing they are fed with sheep's heart, or boiled meat hashed with hard eggs, to which Olina recommends the addition of corn, of spelt and oats properly cleaned, millet, flax seed, poppy, and hemp seed bruised, and steeped in milk. They are in course of time accustomed to live on grain of every kind; but Frisch tells us, that when they receive nothing but hemp seed, they are liable to become black. It is said that mustard seed produces a contrary effect. We are assured by Frisch, that they have this peculiar instinct, to taste their food with the tongue before they eat it.

The males are brought up in aviaries, or spacious cages, for the enjoyment of their song throughout the whole year. They have sufficient memory and flexibility of throat to retain and imitate foreign sounds, and to repeat them more agreeably than almost any other bird can do. At Paris, a lark has been seen which could whistle distinctly seven airs after the bird organ.

The males intended to be preserved for singing, should be taken in October or November; but their voice does not acquire its full developement until after two years. If they are taken large, their wings must be tied, lest by springing violently they should break their heads against the roof of their cage; but it is better to cover it with stuff of some kind, to prevent the danger of killing themselves, by following their natural habit of rising perpendicularly. The cage should be without any cross stick, and furnished at bottom with fresh turf, often renewed. Another indispensable precaution is to place fine sand within their reach, in which the larks are fond of rolling, to rid themselves of the little insects which incommode them. They soon become so familiar as to eat out of the hand, on the table, &c.

The lark usually lives nine or ten years in a state of captivity, and sometimes longer. Instances have been known of their having attained even to four-and-twenty years. Albertus Magnus was of opinion, that towards its ninth year the lark lost its sight. Like all other birds in a state of captivity, it is subject to epilepsy; from which circumstance the ancient physicians imagined that its flesh was injurious to those persons who were attacked by this terrible malady. They regarded it, however, as a specific against gravel, stone, and colic. In our times, on the contrary, it has been said to produce the latter complaint; and probably there is as much truth in one statement as in the other.

The lark, however, is generally considered as an wholesome, delicate, and light food. It is prepared in a variety of ways, on which, as we are not writing a book of cookery, we shall forbear to dwell. We must therefore return to our proper province, which is nature.

The extreme elongation of the hinder claw of the lark in a right line, gives the bird a facility in walking, but renders it incapable of seizing the branches of trees, and perching. Its walk and attitude are neat and graceful. It sometimes forms a little tuft by elevating the feathers of its head.

The larks are dispersed through the fields during the whole of the fine season, but assemble in large flocks in autumn and winter. They then become very fat; for the period of their amours, their song, and their breeding, being gone by, they keep continually on the ground, and have no other occupation than that of taking nutriment.

These numerous assemblages are only preparatory to an

approaching departure with a part of the birds which compose them. The majority of naturalists have denied, without reason, that the larks are birds of passage. They have been met at sea in crossing the Mediterranean, and several of them have dropped upon the decks of vessels. The island of Malta, and other eastern islands of the Mediterranean, serve them as resting places; and they terminate their voyage on the coasts of Syria and Egypt, from where they spread even into Nubia, and over the shores of the Red Sea into Abyssinia.

These migrations of the larks have been noticed by several scientific observers. M. Thevenot has seen them arrive in Egypt. The Chevalier Desmazis, quoted by Montbeillard, was an ocular witness of their passage in the island of Malta. M. Lottinger, an ornithologist of some eminence, has observed a considerable passage of them every year in Lorraine, terminating precisely at the time in which they arrived in Malta, at which time very few are seen in Lorraine, and the emigrants detach with them many that are born in the country. In fine, all the fowlers who are capable of observation make the same remark.

But though there can be no sort of doubt respecting the emigration of the larks, yet it is equally certain that this emigration is but partial, and that a great number of them remain in the countries in which they have been born. Similar, however, is the case of many other species of birds, as well as the larks, part of which migrate, while part remain sedentary. We are wholly ignorant of the motive and cause which determine this separation in the same family, and produce effects so very different in the same animals. To discover these would be an object worthy of the researches of a true philosopher, of an observer of expanded mind, who studies nature in her own proper and immense domain, and does not shut himself up in his cabinet, with the inanimate relics of her productions, founding his pretensions to science on the arrangement of synonimes, and sheltering his own dulness and ignorance under "words of learned length and thundering sound."

The larks which remain during the whole season in our climates retire, during considerable cold, into sheltered places, to the edge of waters which do not freeze, where they find small worms and insects, on which they feed, when they can no longer find grain for their sustenance. When the weather grows mild, they again spread themselves in the plains. They often disappear suddenly in the spring; when, after a few mild days, which have induced them to quit their retreats, there comes on some sharp weather, which obliges them to retire again, until the temperature becomes less rigorous.

The common lark is found in almost all the inhabited countries of the ancient continent, but is supposed not to exist in America. Dr. Latham, indeed, notices that Sloane mentions having met with one some leagues distant from the American coast, out at sea; but this will hardly be admitted as a proof.

Though very fruitful, this species is less numerous in our days than it was formerly. It has been remarked in France, that the quantity of larks has sensibly diminished within forty or fifty years back. Whether the same observation will hold good in this country, we have not heard. Many causes may concur to this diminution. Increased cold, and abundant snow, remaining a long time on the ground, have destroyed a prodigious quantity of larks. They have been seen, at such times, to unite in bands, approach the villages, and even take refuge in houses; and being totally exhausted, and without the power to fly, have suffered themselves to be killed with poles.

The birds of prey also destroy many of them in summer; but man (in this case as in others) is the most voracious, the most determined, and, we may add, the most improvident of destroyers. Four thousand dozen of them are annually taken in the neighbourhood of Dunstable, the greater part of which are sent to London. But they are far more plentiful in Germany than in this country. They are there subject to an excise, which Keysler affirms produces six thousand dollars yearly to Leipsic. It is also said that their flavour in this neighbourhood is superior to what it is elsewhere. The duty at Leipsic is about two and a-half sterling for every sixty birds; and it is sometimes known to produce twelve thousand crowns. The fields are literally covered with them from Michaelmas to the beginning of November.

The most convenient season for hunting larks, is from the month of September to the end of winter, particularly after the white frosts and snow.

There are a variety of modes of taking or destroying this bird. The least advantageous is that of the gun. The game scarcely compensates for the loss of time and trouble, powder and shot. There is a method, however, sometimes used, which renders the shooting of larks more productive to the fowler, and which, as it is curious, we may as well describe.

A piece of wood is taken, nine or ten inches in length, flat, and about two inches wide on the under side. On the upper, it rather shelves a little on both sides, is not exactly rounded, but divided into several narrow planes, and the extremities of it are cut into slopes, or very much inclined planes. Each of these planes is encrusted with little bits of looking-glass, cemented in certain notches with a kind of stucco, composed of three-parts of black pitch, and four of red cement—the whole melted together. A hole, about an inch deep, is made underneath this mirror, about the centre, into which an iron spit is passed, somewhat thicker than the little finger. This spit is hefted into a bobbin: a stake, about a foot long, sunk in the earth, and pierced above with a vertical hole, about two inches deep, receives in this hole the other extremity of the spit; and by means of a packthread wound around the bobbin, a man, seated on the ground, at a certain distance, in a box, or some hollow place which partly conceals him, turns the mirror round, as children turn the little hand-mills which they fabricate with a large apple, placed at the end of a little piece of wood, which crosses an empty nut. The mirror thus put in motion, attracts a sufficient number of larks to render it worth the fowler's while to fire at them.

This method, however, is more successful when the mirror is placed between cloth nets, and when a bird-call is used, or a living lark attached by a packthread to a stake, and forced to flutter about.

The mirror just described, may be used by the same person who manages the cloths of the net. But if it is employed with the gun, it is necessary that another person should put it in motion. Another mirror, however, has been invented which the fowler himself may use.

This is a machine of wood, something in the shape of a platform, furnished internally with a sort of pallet, on which are fixed steel buttons, or some bits of looking-glass. This machine, supported diametrically by two pivots, on a semicircle of iron, preserves an equilibrium, which does not require, like the other mirror, the assiduity and attention of the turner. The semi-circle which sustains the platform, is steel-like, and susceptible of a little elasticity. From the half of this semi-circle, proceeds a handle, at the extremity of which is a round or square hole, which serves to heft it into a stake which supports the platform at an elevation from the ground sufficient to afford it play. This platform must be horizontal, so as to receive vertically the rays of the sun. Motion is communicated to this machine by means of a pack-thread, which, attaching to the platform opposite the

DOUBLE-CRESTED LARK.

AL. CILOPHA TEM.

London Published by Whittaker & C° Ave Maria Lane, June 1829.

handle of the semi-circle, passes through a small slate, placed below, and extends to the hand of the fowler. This motion, though limited, becomes regular, and is multiplied by means of a very flexible little spring, attached to the platform, and the two extremities of which touch at intervals the semi-circle above and beneath. Between the two extremities of this little spring, there should be a distance of two or three inches, so that the platform may be balanced up and down, which puts into play the buttons or pieces of looking-glass.

Mirrors are made with springs, the mechanism of which is the same as that of a jack or turnspit; but the necessity of re-mounting them, renders them inconvenient. This may be avoided by supplying the place of the spring with two strings of cat-gut, wound in contrary directions round the same bobbin. To each of these strings, a packthread is attached, the extremity of which is in the hands of the fowler. One of the strings is unwound, while the other is wound, and thus the mirror is continually in motion without the fowler being obliged to draw the packthread so often. This machine is by far the most convenient of any.

The *Crested Lark (Alauda Cristata)*, is called in French *Cochevis*, which is an abbreviation of *visage de coq*, a name derived from the tuft with which its head is surmounted, and which gives it some resemblance to a little cock. The number of feathers composing this tuft is not the same in all individuals. It varies from seven to twelve, and the bird can raise and lower them at pleasure.

This lark is more bulky than the common lark. The bill is longer, and the wings and tail shorter. The wings, when folded, come to about half the length of the tail. Feathers of a deep grey, with an edgeing of a lighter tint, cover the head, and upper part of the neck and body. On each side of the head is a band of reddish gray interrupted by the eye. The lower parts are of an obscure white, slightly tinted with reddish.

The head is more thick, and the bill stronger, in the male than in the female, and it has more black on the breast. Both have the tongue wide, and a little forked.

Without being so common as the sky lark, the crested lark is pretty well spread throughout Europe, from Russia to Greece. It has been seen in Egypt. It seems very doubtful, according to Dr. Latham, whether it is ever found in this country. It does not seem to be migratory; at least, it does not quit France in winter. During this season it sojourns frequently on the borders of streams, and is often seen on roads, and sometimes in the midst of a flight of sparrows, seeking, like them, the undigested grains in horse-dung. It is usually found in fields and meadows, at the back of ditches, on the ridges of ploughed land, and sometimes at the entrance of woods. It is frequently seen at the entrance of villages, and sometimes will come in and plant itself on dunghills, on the walls of inclosures, and the roofs of houses. It neither flies in flocks like the common lark, nor rises so high; and it continues in flight a longer time without alighting. It is by no means wild, nor does it dread the appearance of man, but commences to sing at his approach. The males sing infinitely better than the females, and their voice is very sweet and agreeable. During fine weather there is no cessation to their strains; but they become silent when the sky is overcast, and rain descends; they forget their gaiety and their music until the re-appearance of a brilliant sun reanimates their vivacity. They usually sing until the month of September. In captivity they also sing, and retain more readily the airs which are taught them from the bird-organ, than almost any other bird. But they seldom survive the loss of their liberty, and it requires much care and difficulty to preserve them any time in cages.

The female places her nest on the ground, like the common species. She lays twice a year, about four or five eggs of a clear ash-colour, thick set with brown and blackish spots.

The young of this species can seldom be artificially brought up, and still more seldom kept alive for many years. Hashed beef's or sheep's hearts, the eggs of ants, millet, and bruised hemp seed, constitute their most appropriate food. They must be fed with very small morsels, and care must be observed not to wound their tongue in feeding them. The cage should be furnished with sand at bottom, and covered with a cloth, to prevent them from hurting their heads. The best season for catching them is autumn; they are then very numerous, and better in flesh.

The *Wood Lark* (*Lulu* of the text, *Al. Arborea*), has been confounded by ornithologists with the last, on account of the similar tuft with which its head is surmounted. Some separate it from the *cujelier*, which M. Cuvier gives as a synonime. It is smaller than the crested lark, and the tuft can hardly be considered as a genuine one, being only a little greater elongation of the feathers of the head than in the common lark. The male is more frequently observed to elevate these than the female.

This lark is found in Germany, Holland, Siberia, Poland, and Italy: it has also been observed in the Pyrenees, in the neighbourhood of Paris, Bourdeaux, Rouen, and is not unfrequent in the province of Lorraine. When these birds perch they sing agreeably. They are heard to warble in great numbers together, in the commencement of spring; but when these assemblages disperse in amorous couples, the male then displays all his vocal powers, and produces very melodious sounds, especially after sunset. Thus he soothes and charms his mate, engaged in her maternal cares. From the time the young family bursts the shell, the sire takes his share in their education; but his songs are over, for the love which created his melody is at an end.

The eggs are four or five in number, of a dirty white, tinted with brown, and picked out with reddish. The nest is usually concealed near the borders of woods, in some furrow covered with grass or brambles, and in the midst of a thick moss. Some stalks of dry grass constitute the external envelopement, and the inside is carpeted with soft grass and cattle hair. Spring is the season to look for these birds, and they are usually found on half-barren hillocks, where briars, &c. grow, but invariably on the edge of woods. During winter they occupy stony fields. In this season many families unite and form serried flocks, of from thirty to fifty in number, never mingling with any other species. They then utter a sort of plaintive cry, resembling the sound of the syllables *lulu*, from which they have been thus named. On the ground they always remain close together; and when flushed, to use the sportsman's term, they do not fly to a distance, but rise by degrees, always whirling, passing and repassing over the spot they have quitted, uttering from time to time, certain rallying cries, and frequently concluding by dropping down in the same place anew. They may, however, be occasionally met in this season in isolated couples. Some of them are also known to emigrate, while others remain in their native habitat.

The social disposition of these birds, and the uneasiness which they so frequently manifest, by their rallying cries after any of their strayed companions, present the means of catching them with greater facility. In hunting them, one of their own species is used as a decoy. They are frequently caught in nets with smaller meshes than those used for the sky larks. While their migration is in progress, which is during the months of October and November, is the most favourable time for taking them in abundance.

It is hardly to be supposed that birds which have so strong an attachment to their species, can live very comfortably in a state of isolated captivity. When taken adult, the lulus do not appear very sensible to the loss of their companions. They are tolerably tranquil, but they eat little; and on the return of spring, when new and more lively affections take possession of the bosoms of these little beings, they speedily perish of chagrin, unless restored to liberty, to friendship, and to love.

The *Short-toed Lark (Alauda Brachydactyla)*, is met with in the Canaries, in the southern provinces of France, and especially in Champagne, where the species is remarkably numerous. These larks arrive in the last mentioned country about the end of April, and are universally found in dry and sandy situations. They have several broods, and the first takes place soon after their arrival. The nest is constructed on the ground, of few materials, principally the blades of dog's-grass, and is usually found in a wheel-rut, or track of a horse's hoof. The eggs are three or four, gray in colour, and spotted with a brownish gray, which spots are more confluent towards the gross end. As soon as the young can manage for themselves, they quit the untilled lands of Champagne, unite in numerous bodies, and seek fresher abodes and oaten fields. They leave this province at the end of August, and do not return until the following spring.

Morning and evening, all the males of the plain assemble, and, at a very elevated height in the air, produce a concert, which is heard very distinctly, even though the birds are out of sight. This song is more agreeable and melodious than that of the common lark. They seldom sing in the middle of the day, and never when on the ground, but utter then a peculiar sort of cry.

This lark can run with the rapidity of a field mouse, especially when disturbed, and on the point of taking to flight. All the larks are pulverating birds; but this one is so particularly attached to powdering itself with dust, that, on being supplied with some in a state of captivity, it will immediately testify its joy by a little soft cry, frequently repeated, and by precipitate movements of the wings, and bristling of all the feathers. It will plunge instantly into sand or ashes, as other birds do into water, remains there a long time, wallowing in all sorts of ways, and does not come out of it until it is so covered with it, that its plumage is scarcely to be distinguished.

The *Clapper Lark (Alauda Apiata)* is of South Africa. It usually makes its nest in some small grass, and lays from four to five eggs, of a greenish gray. It seldom rises more than from fifteen to twenty feet above the ground, and makes a particular noise, occasioned by the precipitate motion of its wings, which being heard at a great distance, has caused the Dutch colonists to call the bird *Clapert-Liwerk*, which Levaillant has translated *Alouette Bateleuse*. When in the season of its amours it rises to the height above-mentioned, it utters a cry resembling the syllables *pi-ouit*, the last syllable of which is elongated during its descent. It descends with the wings closed, and in an oblique line to the earth, where it rests scarcely half a minute, and then rises again. It sings in the morning, in the evening at sun-set, and for most part of the night.

The *Red-backed Lark* chiefly delights in plains abounding with bushes. It perches readily on these, and even on the trees which are at the edges of woods. Its song is agreeable.

The *Alpine Lark (Al. Alpestris*) inhabits the most northern portions of the two continents. In the last days of summer, however, these birds quit the Frozen Zone, and advance in numerous flights towards the south. In America they do not pass the Carolinas; and in Europe, Russia appears to be the usual limit of their voyage, though some have been caught in the neighbourhood of Dantzic, in Germany, and even in Lorraine, but very few in number. In both quarters of the globe these larks, whose flesh is wholesome food, though without flavour, like that of most American birds, quit their winter retreat in the early days of spring, to withdraw into the countries which are nearest to the Pole, where, in perfect security from the aggressions of man, they may deliver themselves without disturbance to the education of their young families.

The *Calandre* is larger than the common lark, but yet has many points of resemblance to it, not only in conformation and colour, but also in habits and manners. Its voice is equally agreeable, but stronger; it possesses a similar levity of motion and disposition; it nestles in the same manner on the ground, under a clump of tufted grass, and lays four or five eggs. It has a similar facility of counterfeiting perfectly the song of many birds, and the cries of some quadrupeds, but its species is less numerous. It is found in the south of France, particularly in Provence, where it is common, and generally reared on account of its song; it is also found in Italy and the island of Sardinia, where it passes the entire year.

The calandres are not observed to congregate in flocks, but usually remain single; in autumn they grow very fat, and are then good eating; they are taken in nets, laid near the waters where they are accustomed to drink.

For the purpose of rearing them in captivity, they must be taken young, either on leaving the nest, or before the first moulting. Paste, partly composed of sheep's heart, must be first given to them; then they may have grain, or crumbs of bread. There must be kept in their cage some rubbish or plaster, on which they can whet their bill, and fine sand, in which they may powder themselves at their ease. Their wings should be tied, in the first instance, or their cage covered with cloth; for they are very wild, and may kill themselves in attempting to soar upwards. When these birds are accustomed to captivity, they will continually repeat their own strains, and those which they have learned to imitate, and which they easily retain.

There is nothing worthy of remark in the *great-billed lark*, except, as M. Levaillant assures us, that it never sings, or soars upwards: its eggs are fewer in number, of a greenish gray picked with red.

The *Sirli* is remarkable for its long and arched beak, and therefore, as we have seen in the text, constitutes a separate subdivision; in all other points it is a perfect lark. It is found in the southern parts of Africa, and even in Barbary, usually inhabiting the sandy downs; from its peculiar song, which it generally puts forth from some little eminence, its name is derived.

The *double-crested Lark (Al. Bilopha)* of Temminck, of which we have inserted the figure, is distinguished chiefly by the double crest, from which its name is derived. Of its habits we know nothing.

The majority of the titmice, particularly those which frequent woods, thickets, and orchards, are courageous, and even ferocious; they will attack the owl with greater boldness than any other bird, being always foremost in darting on him, and trying to pick out his eyes. They express their little rage and fury by the swelling of their plumes, by violent attitudes, and precipitate motions; they peck sharply the hand which holds them, strike it repeatedly with the bill, and seem by their cries to call others to their assistance, which usually attracts them in crowds, and produces abundant sport to the fowler, for a single individual can take them all. There are many traits of conformity in their manners and disposition with those of the crows, shrikes, and pies; they have the same appetite for flesh, and the same custom of tearing their food in pieces to eat it.

These birds being of a lively and active character, are incessantly in motion; they are continually fluttering from tree to tree, hopping from branch to branch, climbing up the trunk, crooking themselves to walls, and suspending themselves in all fashions, sometimes with the head downwards. Though fierce, they are social, seek out the company of their own species, and form little flocks, more or less numerous; and if any accident should separate them, they recal each other mutually, and are soon reunited. They then seek their food in common, visit the clefts of rocks and walls, and tear with their bills the lichens and the moss of trees, to find insects or their eggs. They also feed on seeds; but though in many species the bill is strong enough, they do not break them, like the bullfinches and linnets; they place them under their claws, and pierce them with their bills, like the nuthatches, with which they are sometimes seen to associate during the winter. If a nut be suspended at the end of a string, they will hook themselves to it, and follow all its oscillations without letting go, and keep incessantly picking at it. Such manœuvres indicate much strength in the muscles; it has accordingly been observed that the bill is moved by very robust and vigorous muscles and ligaments, as well as the neck, and that the cranium is remarkably thick. They will eat not only grains, but insects, as above hinted, and butterfly-eggs, and peck the growing buds. The largest species (the great titmouse) joins to its other aliments bees, and even little birds, if it finds them enfeebled by illness, or entangled in snares, but it usually eats only the head.

Almost all the species of titmice are very productive, even

more so than any other birds, in proportion to their size; their brood is said sometimes to consist of eighteen or twenty eggs. Some make their nests in the trunks of trees, others on shrubs, and give it the form of a ball, of a volume greatly disproportioned to their size; some suspend it at the end of a branch, in reeds or rushes. The materials which they employ are small plants, little roots, moss, flax, cattle hair, wool, the down of plants, cotton, and feathers; they tend their numerous family with the most indefatigable zeal and activity, are very much attached to it, and defend it with courage against the birds which attack it. They rush on the enemy with such intrepidity as to force him to respect their weakness.

The titmice are extended over the old continent, from the north to the south of Europe, through Africa, India, and China: they are also found in North America, but are as yet unknown in the southern part of that continent. Within a few years, several have been discovered in New Holland.

Among the titmice, those which are most easily caught in snares, &c. are the great, the black, and blue-headed species; the crested, the long-tailed, the bearded, and the penduline are not so easily managed. There are plenty of modes employed, with success, for the destruction of these little birds, the details of which would involve but little interest for our readers. Those who keep bees are very sufficiently justified, however, in destroying the titmice, as the latter wage a very cruel war upon these useful insects, particularly when they have young ones.

The Great Titmouse, the largest of the European species, is spread throughout the old continent, from Denmark and Sweden, to Africa. In France it is seen in all seasons, but is far most numerous in autumn; as at that season those which, during the summer, inhabited the lofty mountains, descend into the plains to seek a more abundant food. At this period, too, the majority of those from the north withdraw into more temperate climates.

The great titmouse, like its congeners, is lively, petulant, and continually in motion; it clears the buds, &c. of the little worms which harbour in them; destroys the eggs of butterflies, and feeds on caterpillars; it searches out, under mosses and lichens, the larvæ and small insects concealed there. Such are the services which this bird renders us; but on the other hand, it destroys an immense number of bees, more especially when it is rearing its young ones. From this circumstance it has acquired, in some provinces of France, the name of *croque abeilles*.

This species delights in large woods and thickets, in coppices and orchards; it is also found on lofty mountains, in extensive plains, on arid soils, and in verdant meadows—in short, wherever it can procure a suitable aliment. Besides insects, it lives on various grains, on hemp-seed, and even nuts and almonds; the latter it places between its little claws, pierces them with its bill, and very dexterously extracts the substance. It also attacks small birds, when sick, or entangled, not sparing even its own species under such circumstances; and, opening the cranium with its bill, it picks out the brain. It must not, therefore, when kept in captivity, be placed in an aviary with other birds, for it will incessantly pursue and kill them; even among its own species combats will take place, and some individuals be devoured by others. If a great titmouse be for some time alone, it will not suffer others to partake of its domicile; it will precipitate itself on the new-comers, and employ all the resources of its address and courage to give them the law. If they do not submit, it either falls itself or kills them, and feeds on the brain: these birds, however, do not become so cruel and voracious except when in want of food.

Notwithstanding all this, the titmouse soon grows familiar with its prison, and may be tamed to such a degree, as to come and eat out of the hand. It learns willingly all those little exercises that are taught to the goldfinch, such as rowing, drawing water, &c. and does not exhibit less address and docility. If it be desired to preserve these birds, they should not receive their seed without being bruised, for the labour of crushing it always makes them grow thin, and often brings on death, or, at least, blindness. They are not nice in their food, and will easily accommodate themselves to any thing. A paste, composed of hashed meat, crumbs of bread, and pounded hempseed, suits them very well. Suet may be added to this, of which they are so very fond, that it is used as a bait in the various snares which are laid for them.

Though so fierce in disposition, the great titmice are attached to the society of their consimilars. In autumn they migrate in little flocks, more or less numerous. A flock is usually composed of the individuals of the same family. They begin to pair from the month of January, and as soon as each has chosen his companion, the couples isolate themselves. This union appears indissoluble, for the male and female do not quit each other during life. The male is heard to sing during the fine days in autumn, but never puts forth the full compass of his voice but in spring. Independently of his song, he has two peculiar cries; one, from some fancied resemblance to the grinding of a file or bolt, has gained him the name of *serrurier*, in some parts of France.

From the earliest days of March, this bird fixes its nest in the hollow of some tree, but seldom in a wall. The male and female work together at its construction, and compose it of soft and pliant materials. They particularly employ plenty of feathers. The eggs are from nine to fourteen in number, white, and spotted with clear reddish, especially towards the thick end. The male partakes the incubation, which lasts twelve days. The little ones, when disclosed, remain longer with the eyes shut than other birds, opening them when the feathers begin to point, and about fifteen days after birth they quit the nest. All do not abandon their cradle, however, at the same time, for when the broods are numerous, it is not rare to see some only covered with down, while others are ready to fly away. This depends on the number of eggs. Buffon says, that, once departed from the nest, they never re-enter it; but this remark will not admit of generalization, for this species, like some others of the genus, usually sleep in the hollows of trees, and thus shelter themselves from cold during the long nights of winter. This habit is so natural to them, that when caged, something like a dove-cote is placed for them to sleep in, totally closed, where they will all go, if it be sufficiently spacious. They seem to fear that their retreat should be discovered, for previously to entering, they look round on all sides, and then pop in quickly. When they have made choice of a hole, they return there every evening. Once entered, it is difficult to force them out, even by the introduction of a small stick; and they can hardly be seized, but with the aid of a little harpoon. Still, they may be made to fly out quickly, by striking against the trunk of the tree, and this is often a certain means of discovering the nest.

If they are disturbed with a little stick, these titmice send forth a hissing sound, not very unlike that of a serpent. The young which issue first from the nest, remain under the neighbouring trees, calling to each other incessantly, a habit which they never lose at any age. This exposes them to the hunter; for, with one titmouse, as a decoy, numbers may be taken. It is not certain that they have more than two broods in the year, though young ones are sometimes found in the nest even at the end of June. It would appear, that if they have more, some disturbance has occurred to the first: but then the eggs are always less numerous.

This titmouse arrives to perfection in a very short time. In less than six months, it has arrived to its full growth, and can reproduce. This rapid development argues a short life, and the duration of that of the great titmouse is only five or six years. Gout, and defluxions from the eyes, are the infirmities which mark its decline.

The flesh of this bird is eatable, but not particularly good, and is seldom if ever fat. Empirics have attributed to it certain medicinal properties, such as anti-epileptic, diuretic, and remedial in calculous disorders. It is dried for these purposes, reduced to powder, and administered in doses of a scruple to a dram, in white wine, or some diuretic medium. But this appears to be without foundation.

The *Cole Titmouse (Parus Ater)* nestles, according to Meyer, under the roots of some elevated tree, or in a hole, which some mole has abandoned, and sometimes in a hollow tree or hole in the wall. It lays six or eight white eggs, spotted with points of a dusky hue, like the colour of musk. The male and female are alike. It is common on the continent of Europe, even as far north as Russia, and migrates southward in autumn. Appellants of its own species are necessary to draw it into snares. Forests of fir-trees and woods of ever-greens are its favourite abodes. In the after-season, it frequents orchards and gardens, especially the latter, when turnsols are found there, in the seed of which it much delights. It climbs and runs on trees like the rest of its congeners, and will suffer itself to be approached very near. It is equally courageous with the other titmice; but either less subtle or more bold, for it gives into all kinds of snares; and even those which have been caught, and escaped several times, may be taken again in the same way.

The *Marsh Titmouse (P. Palustris)*, delights not only in woods, but frequents orchards. It is common in England, and found in moist situations, chiefly where old willows abound. It makes its nest in a decayed tree, and composes it of moss and feathers, thistle-down, and sometimes a little wool. The eggs are five or six, white, marked with red.

Some individuals migrate, while others are sedentary. The migration takes place in September or October; and then these birds are observed in considerable numbers. They readily approach habitations, and frequent gardens. They live on hempseed occasionally, and even make a provision of it; but not being able to break it, they pierce it with the bill. They make war on wasps, bees, caterpillars, and other insects; easily give into snares, but do not live long in a state of captivity.

This species, extended throughout Europe, is more common in the north.

The *Blue Titmouse* is, of all the species, the most known and the most common. It is spread throughout all Europe; and is met with on the coast of Africa, and in the Canary Islands, with some little variation of plumage.

Like all the rest of its tribe, this titmouse has a bright and dark side in its character. It is useful in destroying an immense number of caterpillars and insects' eggs, especially of those insects which attack fruits. It is, however, injurious in our orchards and gardens, by biting the tender buds. It will even detach the fruit already formed, and carry it off. It has the same relish for flesh as its other congeners; and it bites so exactly that of the little birds it can seize, that Klein proposes to employ it in the preparation of skeletons. It also likes hempseed, which it cracks, like the great titmouse. It has been observed to exhibit more audacity than the latter, and to attack the owl with greater bitterness. But it is also more easily taken.

The usual dormitory of the blue titmouse is a hollow tree, or a hole in the wall: but it appears more careful than the other titmice in the choice of its abode. It is generally in a warmer situation in winter, and more elevated and difficult of access in the summer. It has a very singular habit when encaged; if deprived of a place where it can conceal itself.

it will pass the night hooked by the claws to the ceiling of its prison. It even delights to perform this manœuvre during the day-time. It makes its nest in the trunk of a tree, and uses abundance of feathers in the composition of it. It there deposits a great number of eggs, some say, from ten to two-and-twenty, while others make them only from six to ten. These eggs are white, according to some naturalists; but M. Meyer says they are of a reddish white, spotted and marked with red and brown. The great quantity of eggs, indicates that the bird has but one brood during the year, unless disturbed, and the second is then always less numerous. It is very easy to make the female renounce the care of her eggs, even when the young are actually formed. It is sufficient to touch or break a single one. But from the moment the young are disclosed, she manifests the greatest attachment to them, and the greatest courage in their defence. When these titmice are disturbed in their hole, they make an unpleasant grinding noise. They have also several other cries, either of appeal or terror, and a simple sort of song, without much variety, which is never heard but in spring. As soon as the young family can fly, they join with the parents, and quit the woods, where they have sojourned during the summer, and spread themselves through orchards and gardens, and frequently voyage in company with the great titmouse. The blue titmice, however, remain a longer time assembled together than the others. But from the month of January they are never seen but in couples. Those which are taken adult do not refuse the food which is offered them, and even grow familiar with their prison, if it be sufficiently large, and little holes be left where they can conceal themselves at pleasure, and pass the night. But they almost always perish at the end of winter.

The *Long-tailed Titmouse (Parus Caudatus)* rarely quits the woods during summer, but in winter approaches habitations, gardens, and orchards. These titmice also sojourn at times in marshy places, from which some have called them *reed titmice,* a name little suitable to them, as they retire to woods, and even to those situated on mountains, for the purpose of reproduction. These little birds are very lively and mobile, and great enemies to repose. They hop from bush to bush, and from tree to tree, traverse all the branches with an astonishing promptness, and fasten by the claws to the extremity of the weakest boughs. They seldom quit each other, and have a rallying cry, like *ti, ti, ti, ti,* at the sound of which they fly to their companions, and disappear with a *guickey,* sent forth by the chief when the band is troubled. These birds live in families from the time they leave the nest until spring; then each makes choice of a companion, retires into the thickest woods, and both are immediately occupied in constructing a cradle for the new and numerous progeny. Some of them suspend their nest to the branches; but they usually attach it, solidly, to the boughs of shrubs about three or four feet from the ground; give it an oval and almost cylindrical form, close it above, place an entrance, an inch in diameter, on the side, and sometimes two issues, which correspond. This nest is almost eight inches high, and four inches broad. Its texture is close, its exterior composed of blades of grass, moss, and lichens; and its interior furnished with a great quantity of feathers. The eggs, from ten to twenty, are hardly perceptible at the first look, so well are they concealed in the heap of feathers at the bottom of the nest. Their colour is gray, more clear towards the gross end, which is surrounded with a reddish zone. The father and mother feed the young with those aliments on which they live themselves, as caterpillars, gnats, insects of various kinds; sometimes small seeds, or morsels of the buds of trees, which they cut off adroitly and merrily. As soon as the young can quit the nest, the whole family, with the parents, form those flocks of from ten to twenty which are observed from the end of summer, and after the winter, flying, and uttering their shrill calls of appeal. Each family lives isolated, and do not unite with others. They emigrate, but only to short distances; where, during the inclement season, they may find more food, and better shelter. This titmouse has, besides its habitual cry, a tolerably pleasant song. They are very seldom caught, whether from their distrustful character, or our ignorance of the proper bait to employ—but the flesh is by no means a "*bonne bouche*."

This is an European species, and is even found in Siberia.

The *Bearded Titmouse (P. Biarmicus)* is found in considerable numbers in Holland, and is frequent enough here, in marshes where reeds abound, the seeds of which serve it as food. It also lives on small insects, and, in default of them, on water snails, which it swallows entire with their shells.

The biarmicus appears to have few or none of the habits of its congeners. It runs through the reeds at the edge of the water like a wagtail. Its crop has been found filled with those little shell animals above-mentioned.

Latham regards this bird as indigenous to Great Britain, for it is seen here all the year round. The nest is made of soft and downy materials, and it is suspended between three reeds, which the birds have the dexterity to draw together. It is rather of a close texture, and composed of the tops of dry grass, mixed with reeds and rushes, and interspersed with small leaves. The eggs are four or five, of a reddish white, spotted with brown. If we may judge of this bird when at liberty from its behaviour in captivity, we should say that its manners are more mild and social than those of the other titmice. The male and female show much attachment for each other, and will even extend their cares to young canary-birds. It is even said, that when these birds repose, the male will cover his companion with his wings. Such a disposition, joined to some generic differences, seems to separate them from the other titmice. The latter, it is true, if we may judge from their frequent cries of appeal when dispersed, seem attached to the society of their fellows. They appear, however, to fear approaching each other too nearly; judging of the disposition of others by their own, they are cautious and distrustful—such, indeed, is the usual spirit of associations among the mischievous and wicked of all species.

The bearded titmice are found in Denmark and Sweden, but rarely. They are common in the neighbourhood of the Caspian Sea, and the Palus Mæotis, where they inhabit the reeds. But they are not found in the more elevated latitudes of Asia.

The *Penduline Titmouse*, sometimes called the *Remiz*, is found in Poland, Italy, Siberia, and the south of France. It frequents aquatic situations, suspends its nest to the extremity of a flexible branch, hanging above the water, attaching it with hemp, flax, or some material capable of sustaining it in the air. It gives it the form of a purse, of a sack, or bag-pipe. It makes an opening on the side, usually on that which is turned towards the water. This nest is composed of the down of willow-flowers, of poplar, &c. It is interlaced with small roots, and forms a close tissue almost as substantial as pasteboard. A bed of the same down, but finer, furnishes the inside. The eggs are four or five, snow-white, and about the size of those of the wren. The female has two broods in the year. It is pretended that this titmouse is so cunning, that it cannot be taken in any snare.

The Buntings *(Emberiza)*, as we have seen, are distinguished from other passerine birds, principally by their conical, short, and straight bill, and by the addition of a knob in the roof of the upper mandible, which is made use of by the bird as an anvil on which to break and comminute its food. This apparatus is sufficient to lead the observing naturalist *per saltem*, as it were, to the conclusion that this genus of birds must be granivorous. It is true, indeed, that very many birds are enabled to crack and open nuts and hard seeds, without the aid of that extra provision with which the buntings are furnished: and this is one of the countless instances which might be adduced to display the various means employed by Nature to attain one and the same end. How different, for instance, are the means by which the several classes of animals attain the common object of locomotion, and how various are the modifications of those means in the respective genera. The buntings, however, do not feed exclusively on vegetable matter; like most of their order, they subsist also partially on insects and worms.

The *Yellow Bunting* (*E. Citrinella*). This common species, in our own country, is known to every one under the name of the Yellow-hammer. The yellow on the crown of the head is sometimes replaced by olive-green: and this, as well as other occasional deviations from the ordinary gamboge yellow of this bird, would in all probability have induced the erroneous multiplication of species, had the yellow bunting and its incidents been less universally known.

This bird builds in a careless manner, on the ground, or towards the bottom of a small bush. The exterior of the nest consists of straw, moss, dried leaves, and stalks; and within is a little wool. Notwithstanding the carelessness of its nidification, however, few birds display stronger attachment to the young and to their eggs, than this; so much so, as to be not unfrequently taken by the hand, on the nest, rather than abandon its offspring in time to save itself. The eggs are in general about five in number, and are whitish, with red streaks.

This bird seems common to almost the whole of Europe, though Dr. Flemming excludes it from Orkney. In spring and summer, it is found in the hedges and borders of copses, rarely penetrating far in woods; and in autumn, they may be observed passing in small flocks toward the south.

The *Foolish Bunting* frequents the warmer situations of Europe, and lives solitarily in mountainous districts. It is said to have gained, deservedly, its epithet, from the ease with which it falls into every kind of snare.

The *Cirl Bunting* may be considered a British species, as it is not uncommon in company with the yellow bunting and the chaffinch on the southern coast of Devonshire. A straggler has, indeed, been killed in Scotland.

The *Reed Bunting (E. Schœniculus)* is about the size of the yellow bunting, and is common in this country. It constructs its nest in grass or furze, near the ground, and has been said to attach it to three or four reeds above the water, whence its name. The eggs are four or five in number, bluish white, spotted, and varied with brown. "I have now and then," says Dr. Latham, "seen this bird in the hedges, or the high roads; but the chief resort is near the water; and that it, among other things, feeds on the seeds of the reed, is clear, as I have found them in the stomach." Though not uncommon, they are not found in large flocks. Though this species is said to be the best songster of the genus, its musical pretensions seem by no means to be boasted of. It is perennial in this country, though said to migrate in other parts of Europe.

The *Common Bunting (E. Miliaria)*. This species is rather larger than the yellow bunting, and is much less common here. While in France, they are merely occasional residents, and arrive there in the spring, from the south, shortly after the swallows, and quit that country again in the beginning of autumn, they are found here during the whole year, and congregate in winter in large flocks, when they are frequently caught in numbers, and sold under the name of bunting larks, ebbs, or corn bunting. They nestle on or near the ground, have four dirty-white eggs, spotted and streaked with brown; and the young have a reddish tinge.

During incubation, the male is generally found perched on a branch not far distant from his mate, constantly uttering a tremulous kind of shriek, several times repeated, with short intervals. Their unavailing anxiety to protect their eggs and young, frequently leads to the spot where they are deposited, which the simple birds are so unwilling to forsake, and, in their anxiety, so easily betray.

The *Ortolan Bunting* (*E. Hortulana*) is never known to visit this country. This bird, whose flesh is very highly esteemed, and which is consequently much sought after, appears to be confined to the southern parts of Europe, where it is found at all seasons. All the individuals of the species are not, however, confined to one locality the whole year; for a few of them quit the south in the spring, and visit for a time the intermediate latitudes of Europe. Even these, however, do not breed in all the countries they visit, as their nests are said to be found only in Germany, and Lorraine and Burgundy, in France. It is commonly near the stem of the vines that they build their ill-constructed nest, in which the female deposits four or five eggs. In Lorraine, they are said to build in the corn fields.

When these birds first arrive in France, they are far from fat; but human ingenuity soon makes them fit for the table: they are fatted by inclosing a number of them in a dark chamber, in which is placed a lanthorn, surrounded plentifully with oats and millet. The darkness seems to have the effect of confining the whole attention of the birds to their favourite food, thus placed within view; and it is said they will thus die of suffocation from their own fat, if left entirely

COMMANDING BUNTING.

EMB. GUBERNATRIX.

London, Published by Whittaker & C° Ave Maria Lane, June 1829.

to themselves. Another mode is, by confining them in cages, which admit a little light only to the box containing the food. In this state, the ortolan bunting is said to be one of the most exquisite morsels known for the table.

Of the *Commanding Bunting (Emb. Gubernatrix)* we insert a figure (*male*) from the "Planches Coloriées" of M. Temminck. A description of it will be seen in the text. Its habits are unknown.

Among the buntings, distinguished by an elongated claw to the thumb (PLECTROPHANES, Meyer), we shall notice only the *snow bunting*, as it is found in the northern parts of Great Britain, and is called in Scotland the snow flake. These birds appear there in large flocks, at the commencement of frost, and are feared by many as the harbingers of hard weather; they are about the size of the chaffinch, black above, with a white rump, crown, and forehead. They nestle in holes in rocks, and produce five white eggs, with dusky spots.

They are found in all the northern latitudes, as high as navigators have penetrated; nor is it at all apparent by what means they find food in these inhospitable regions. The higher the degree of latitude in which they are found, the whiter, as it appears, becomes their plumage; this tendency, which we have had frequent occasion to notice, among the mammalia, as well as in the present class, has led to the conclusion that there are many varieties of this species. It breeds in Greenland, visits this country in harvest, and retires in spring. As the winter advances, it approaches the corn-yards, and feeds with the sparrows and finches. In Zetland it is called oat-fowl, from the preference which it gives to that kind of grain.

We now come to the extensive genus of the FRINGILLÆ, all of which agree in the following characters:—

The bill is less thick than the head, straight-edged, conical, and pointed. The upper mandible covers the edges of the lower, is straight, seldom ever inclined towards the end; the palate is hollow, and longitudinally striated; the nostrils are round, covered entirely or partially with feathers, very short, and directed forwards. The tongue is thick, rounded, compressed at the point, and bifid; the toes are four in number, three in front, one behind; the external ones are united at the base, the internal free. The four first remiges are nearly equal with each other, and are the longest of all.

An immense number of birds, as may be seen by our tabular view, are united together under this division, because they present a common analogy in the form of the bill. All these birds strip the grains or seeds on which they feed of their pericarpe, previous to swallowing them. They have a crop, in which the food is macerated before it passes into the gizzard, and all, with the exception of one, *the widow-bird (Emberiza longicauda)*, are monogamous; but their mode of life, instincts, and manners, not being the same in all, have given rise to a natural division into small families.

The species which live between the tropics and the neighbouring regions are sedentary, while among those of the temperate and frozen zones some abandon their native country on the approach of winter, to seek in more southern climates the food of which the inclement season has deprived them in their own. Some, at this period, only quit their mountains to descend into the plains, and remove for a greater or less distance from their summer domicile, according to the greater or less severity of the winter. Some of our finches move still farther south, to give place to others of their own species which come from the north, to pass with us the winter season. The linnets, goldfinches, and sparrows proper never quit us.

Although all these birds are granivorous, there are some among them which also eat insects; such are the sparrows, finches, &c.; but, generally speaking, they only use them to feed their young: and when the bill of the latter is grown sufficiently strong to bruise the grain, they no longer show a preference for the insectivorous regimen. Goldfinches, linnets, and canaries do not appear ever to touch insects; neither do the bullfinch and greenfinch; they feed their young with the tender seeds of anagallis, groundsil, and other precocious plants. Those which are purely granivorous always feed the young from their own crop, when the food is macerated; but the entomophagous kinds carry the insect in the bill, or at the entrance of the œsophagus, to their little ones.

The species belonging to temperate and polar regions have but one season of reproduction in the year; but those of the torrid zone have many; some nestle in bushes, others on trees, and many of the last give to their nests a very elegant form. The house and hedge sparrows construct their nests rudely, in the holes of walls and trees. The fringillæ have rarely but a single brood, having often two, and sometimes even three or four, according to the length of the fine season in the countries which they inhabit. All these birds, with the exception of house and hedge sparrows and some foreign species, have a song more or less agreeable, and, in certain of them, it is scarcely inferior to that of the nightingale itself; they all are easily accustomed to a state of captivity, and many of them constitute a most attractive feature in aviaries.

The WEAVERS (PLOCEUS), which, in the "*Règne Animal*" form the first subdivision of the fringillæ, are placed in a separate genus by M. Vieillot, and in the same family with the orioles. Their bill is robust, advancing on the forehead in the form of an acute angle; it is longi-conical, convex above, a little compressed at the sides, entire, nearly straight, sharp, and sometimes a little gibbous. The lower mandible has its edges bent inwards; the nostrils are oblong, and covered with a membrane; the tongue is cartilaginous, and fringed at the point; the wings have, in many species, a bastard quill, and the second and third remex are the longest of all.

Most of these birds are found in Africa or the East-Indies. Their generic name is derived from the wonderful art with which they construct their nests, interlacing them with blades of grass; in which talent they resemble the divisions of carirostral birds distinguished by the names of cassiques, troupials, and carouges, whose place they appear to hold in the old continent.

The *Toucnam-courvi*, or *Philippine-grosbeak*, is one of the most remarkable of these birds, for the art with which it constructs its nest. It suspends it to the extremity of the branches by its upper part, composes it of small fibres of leaves, interlaced one with another, and gives it the form of a sack, swelled and rounded in the middle, the aperture of which is placed on one side; to this aperture is fitted a long canal, composed of similar fibres, and turning towards the bottom of the nest, with an aperture underneath, so that the real entrance to the nest does not appear at all.

The *Baglafecht*, or *Abyssinian grosbeak*, has been represented as a variety of the last; but it more probably constitutes a distinct species. It is a native of Abyssinia, and gives to its nest a different form from that of the toucnam-courvi, and displays rather more industry in its precautions for sheltering its offspring from the humidity of the weather and the voracity of its enemies. It rolls its nest in a spiral form, not unlike the shell of the nautilus, suspends it to the extremity of some little branch, over a placid stream, and fixes the entrance in the lower part; but the aperture is always on the eastern side, in opposition to the direction of the rain.

The *Cap-more (Ploceus textor)*, is a native of Senegal, and also of the kingdoms of Congo and Cacoucongo. Like most of the birds inhabiting these burning regions of Africa, its livery varies according to the season. In spring, for instance, its head is covered with a sort of capouche, of a reddish brown, which, in the after season, changes to a yellow; its song is peculiar, and very lively. Some of these birds, when seen alive in France, evinced a disposition to nestle, though unexcited by the presence of the female; they constructed nests with blades of grass or reeds, interlaced in the wirage of their cage. It is probable, that with care, and the production of a proper degree of warmth, they might be made to multiply, even in this climate.

The *Nelicourvi (Ploceus Pensilis)* does not display less ingenuity and industry in the formation of its nest than the baglafecht, or Abyssinian grosbeak, and the toucnam-courvi; it composes it of straws and reeds, interlaced with much dexterity, and suspends it to a flexible branch, on the edge of a rivulet; it forms a pouch in the upper part, in which the eggs are deposited, and fits to this an elongated tunnel, leading downwards, at the end of which the entrance is situated. When the second brood takes place, the bird attaches a new nest to the former, and so on successively; five or six hundred of these nests may be sometimes seen, thus suspended to a single tree. The female lays only three eggs.

Ploceus Cristatus, the *Malimbic Tanager*, of Latham, inhabits Malimba, a country on the eastern coast of Africa, in the kingdom of Congo, where it was first observed by Perrein. These birds usually sojourn on trees bearing figs, which exactly resemble those of Europe, and place their nests on such of the branches as form a triangle. The nest is of a round form, with the aperture worked on the side; the exterior is composed of fine plants, arranged ingeniously, and the interior is furnished with cotton. The eggs are from three to five, of a greyish colour, and the male and female partake of the incubation. It is in the months of October and November that these birds are found in Malimba; they remain on the fig-trees abovementioned only while they are loaded with fruit; and when that disappears, the birds disappear also, and do not return till the following year.

The *Worabee*, or *Black-collared Finch*, also exhibits wonderful industry in the making of its nest, and great foresight to protect its young from rain, and from the attacks of little animals. The form of this nest, according to Buffon, is pyramidal. It is always suspended above the water, and at the extremity of a little branch. The aperture is on one of the faces of the pyramid, and usually turned towards the east. The cavity of this pyramid is separated into two parts by a partition, thus forming, as it were, two chambers. The first, in which is the entrance to the nest, is a sort of vestibule, where the bird first introduces itself. Then he climbs up the intermediate partition, and descends down to the bottom of the second chamber, where the eggs are deposited. This species is a native of Senegal and Abyssinia.

Of the *Textor Alecto*, described p. 133 of the present volume, the opposite is a figure from M. Temminck. It is a recently discovered species, of whose habits we are wholly ignorant.

The *Sociable Grosbeak* is left by M. Vieillot in the genus Coccothraustes, although he confesses that its proper place is here. These birds are found in India, and the interior of the Cape. They unite in numerous flocks, often to the number of eight hundred or a thousand; select a large mimosa, or an aloe-tree, to establish their habitation in, at which they work in concert. They construct it with reeds and other fibrous plants tissued together. This habitation is divided into compartments, or small cells, and has many issues irregularly placed, or rather it is composed of as many nests as there are couples, about two or three inches distant from each other. Each year the total mass augments with the number of new couples, until the tree can sustain no more. The plants which they employ are termed by the Cape Colonists, *Booshmanées-grass*. As the wings and feet of insects are found in these nests, it would appear that this bird uses food of this kind. These grosbeaks live in peace

THE WEAVER ALECTO.

TEXTOR ALECTO. Temminck.

London Published by G.B. Whittaker March. 1829.

and concord, without tumult or quarrel. But in the small parrots, which are also republicans, they find dangerous enemies, who sometimes chase them from their dwelling, and take possession of it for themselves. These collections of nests are very numerous, and are found in retired places, sheltered from high winds, and at the back of mountains.

The *Surinam Crow*, as it has been called, is placed by M. Vieillot in his genus *Quiscalus*, of the corvine family. When the male is in his perfect plumage, he presents to the eye all the colours of the prism in their utmost splendour. The richest and most brilliant reflections of blue, purple, violet, green, and gold, play upon a ground of velvet black. In the female and the young, the colours are, according to the usual law of nature in the feathered tribes, of a duller cast in general.

These birds in their mode of life have some analogy with our own rooks. Like these, they delight all the year round in the society of their fellows. They fix their nests on trees, particularly pines, near each other. Sometimes fifteen or twenty may be seen on the same tree. The exterior of the nest is composed of stalks and roots of a species of plant full of knots, cemented together with clay. The interior is composed of a very fine kind of rush, and horse-hair. The eggs are five or six, of a bluish olive, sown with broad spots and stripes, some black or sombre brown, others of a weaker tint.

These birds are seen at times in the interior of woods; but they usually remain on the borders, from which they spread into saline marshes, meadows, cultivated grounds, and rural habitations, to seek their food, which consists of worms, insects, berries, and grains. Being of a very social disposition, as has been already mentioned, they remain the whole year in flocks, which are sometimes so numerous as to obscure the entire atmosphere. They inhabit the New World from the Antilles to Hudson's Bay; but on the approach of frost they quit the northern climates. They are often seen, like our pies and crows, in the train of the plough, to collect the worms and larvæ turned up by the share.

They are heard to sing only in the spring season. Their song is sonorous, and not destitute of melody, though of a melancholy cast. Of all the migrating birds of North America, they are the last to quit the central parts of the United States. Their departure takes place in the month of November. It would seem as if they did not remove to any great distance, as they re-appear in the month of February. They then frequent the saline marshes, and feed on the grains of the *Zizania-aquatica*; and in the month of March they retire into coppices and orchards bordering on rural habitations. They are often seen feeding at this time in the front of barns, and even come to the doors of houses, to partake of the food distributed to the poultry.

When the Europeans first established themselves in North America, these birds, and some of the Xanthorni, committed such ravages in the corn-fields, that a price was set upon their heads. They were easily exterminated, for they are not distrustful; and the more numerous they are, the more easily do they suffer themselves to be approached. But, from their nearly total destruction, an evil resulted which was not foreseen. The corn and pastures were devoured by worms and insects; this forced the settlers to spare the birds, that they might remove the other scourge. The damage which they still do, being less apparent, as the country is more cultivated, and their flesh being hard and dry, they are seldom hunted at present, except for the purposes of amusement.

We now come to the SPARROWS, properly so called.

It cannot be deemed necessary for our purposes to write a description of a bird so common among us as the *House*

Sparrow—a bird which the citizen lodges within his walls, under the roofs of his houses, which he meets at every step in his daily walks, which abounds equally throughout the country, which partakes freely of the grain distributed to the domestic fowls, and, in fine, a bird which the agriculturist has marked as one of his most active and persevering enemies. At the same time it may be noticed, that naturalists are often too careless, and even utterly negligent, in describing those objects which are familiar to themselves, and to their countrymen, without remembering that the objects most common among ourselves, are frequently strangers to another climate. The natural historian should generalize his views in proportion as he hopes that his writings shall be extended, and render the latter suitable to every age and every country. But as our pretensions are of a far humbler kind, and as the text itself generally furnishes sufficient indications of external description for all useful ends, we shall hold ourselves dispensed from all such minute details—details that, after all, are rarely consulted, and still more rarely supportable.

The large head, thick and short bill, and animated eyes of the sparrow, give to its physiognomy a characteristic of something like impudence, if we may with propriety apply to any lower animal those moral peculiarities which are the result of spontaneous volition in free and responsible agents. From the shortness of the neck, the bulk of the body in proportion to its length, and which bulk is apparently increased by the moderate breadth of the tail, there is nothing of lightness or elegance in the form of this bird. Its motions, too, although precipitate, are quite destitute of grace. Its monotonous and incessantly-repeated cry, is fatiguing to the ear, and pursues us perpetually around our houses, and in our gardens.

This species would seem, in a very extraordinary manner, to have changed its character, and lost its originally wild nature. It has become almost domestic, and lives, as it were, entirely in the society of man. The sparrows are very troublesome lodgers, and importunate guests: impudent parasites, in fact, which will share with us, whether we will or no, our grains, our fruits, and our habitations. What could have been the situation of these birds before man had formed large societies, before he had cultivated the ground, so as to produce abundant harvests, may be a very curious, though very futile question. They have never been found in the truly wild state, abandoned to their own resources. None have ever been taken which did not exhibit a very decided mark of the domestication characteristic of the whole race. We may, however, suspect, with much probability, that this species was much less numerous, in the earlier ages of the world, than it is at present.

The habit of living in the midst of us, has brought the instinct of the sparrows to a very great degree of perfection; perhaps we should express ourselves better, were we to say, that to their natural instinct it has superadded all the intelligence that they are permitted to acquire, through the medium of observation, stimulated by their natural wants. They know, perfectly well, how to accommodate their manners and habits to situations, times, and circumstances. They can even, in some sort, vary their language; and as they are extremely loquacious, it is not at all difficult to distinguish in their cry, the expressions of appeal, of fear, of anger, of pleasure. In the midst of an association, which they have formed against the inclination of one of the parties, and that, too, the most powerful, for their own advantage, and to the detriment, or at least to the apparent detriment, of those with whom this compulsory union is established, the sparrows have contrived to preserve their independence. Bolder than the generality of birds, they evince no fear of man; they surround him in cities, in the country beset his path, and will scarcely turn out of it to let him pass, either on the high-roads, or still more especially in public walks, where they enjoy an almost undisturbed security. They are not constrained by his presence, nor distracted from their search for food, from the arrangement of their nests, from the cares which they bestow upon their young, from their combats, or their pleasures. They are in no wise subjected, and their familiarity partakes of the character of insolence.

They are not less numerous in towns than in the country. They lodge and nestle under the tiles of houses, and in the holes of walls. Though many of them may be seen together in one place, they form no society, among themselves, in summer. They are often alone, or in couples. They are a small community, always in motion, the individuals of which cross each other's path incessantly, engaged in satisfying their appetites, thinking only of themselves, and paying no attention to the common interest—too faithful an image of the inhabitants of those same cities, whom they have chosen as their hosts.

In the evening, in summer, they generally assemble on large trees, for the purpose of squalling in unison. When this concert is more noisy and prolonged than usual, it may generally be considered in the country as a sign of fine weather on the following day. In summer, also, the sparrows may be seen, in troops, on such hedges as border those tracts of cultivated land, where the harvest is ripening; but this is a mere accidental assemblage, formed by the desire of booty, and dissipated when there is no more to plunder. If a shot, or any other noise, should suddenly arouse this band of robbers, they do not fly to any great distance, but speedily return to the post, from which they carry on their depredations.

The same family, however, will remain assembled together for some time. The young follow their mother, and may be killed, one after another, with a cross-bow, providing that she be taken down first. The young will then not fly away, but crowd together the more of them that fall; but if the mother should be missed, she will be off, and take the young ones with her.

The flight of the sparrows is short and difficult. They cannot raise themselves to any height, and when they depart in flocks, it is always at once, quickly, and with considerable noise. They are not migrating birds. They do not change their district, and there they may be observed to pursue the maturity of the different kinds of grain on which they feed. They disdain to fix in barren countries, and flock to those where the harvest is most abundant. We may judge, with certainty, of the fecundity of a country, by where the sparrows are most numerous. They are sometimes found in the most secluded and solitary spots, where a farm, surrounded by cultivated fields, and provided with a poultry-yard, and a pigeon-house, offers them an easy and abundant subsistence.

The sparrows are of a robust constitution, and can support, with equal ease, the heat of the most burning climates, and the extreme cold of northern regions. They are found extended through Greece, and northern Africa; and, again, on the other hand, they are found as far north as Siberia. Though common enough through a considerable part of Africa, they have not been seen along the western coast of that continent. This cannot be attributed to the heat of the climate, since they can endure that of Egypt; but it is the difference of the alimentary plants in these regions, which causes this difference. Corn, and productions of a similar nature, are cultivated in Egypt, as well as in Syria and Barbary. They cease to be so about the environs of the White Cape. They are replaced by other nutritive plants, among the negroes who inhabit to the south of this promontory, and the seeds of such plants form no suitable aliment for the sparrows; so that it appears, if these birds do not frequent all corn countries, at least it is certain, they do not approach those where corn, of some kind or other, is not cultivated. A fact, comparatively recently made known, comes in to confirm these observations. In the voyage of Commodore Billings to the north of Asiatic Russia, and the Icy Sea, &c. we read that the banks of the Pellidoui, a river of Siberia, which falls into the Lena, are famous as well for the kind of animals found there, as because they are the last spot in that quarter which produces corn. The sparrows and pies proceed no farther north than this. It is even particularly specified in the work just mentioned, that, at the time it was written, they had only been seen there for five years, exactly the time for which grain had commenced to be cultivated on these banks.

The sparrows are notorious for the warmth of their temperament, and the males fight desperately for the females, during the season of reproduction. These combats are so violent, that some of the parties frequently drop down dead. They employ hay and feathers in the construction of the nest, contenting themselves with arranging these materials negligently under the tiles, in the holes and crevices of walls. But they form a regular tissue of them, when they nestle in large trees, such as nut-trees, poplars, &c. They there give to the nest a rounded form, and cover the upper part exactly, leaving but one aperture below this cap. Some take possession of the nests of swallows, &c. The eggs are five, six, and sometimes seven or eight in number, of a whitish ash-colour, with abundance of brown spots. The little ones are born without feathers or down, and are all over red. Wherever they establish themselves to multiply their species, they do not appear at all affected by the noise about them, and to which they are accustomed from infancy.

Birds which come of their accord, like the sparrows, to form a sort of society with man, are of course gifted with all the dispositions for a more intimate association. The sparrows are easily reared in cages; accustom themselves to captivity, without difficulty; are sufficiently docile to obey the voice, to receive their food from the hand which presents it, to suffer themselves to be taken up, touched, and caressed. But their being tamed thus, proceeds from their natural boldness, and from the facility with which they find the means, in captivity, of satisfying their voracity. They are but little susceptible of affection. They do not understand, like the canary, how to excite caresses, to return them with sensibility, to exhibit symptoms of joy on the approach of a being whom they love, or of grief at their absence. Without anything like sentiment in their love with each other, it cannot be supposed that they are capable of friendship.

Various accounts have been given relative to the duration of life in the sparrows. Some allow them only two years of existence; others say four, and some even eight years. These assertions appear to be none of them well founded. The sparrows live much longer than is generally imagined. M. Sonnini knew one of these birds to live four-and-twenty years in captivity, and even then it died accidentally. The warmth of their temperament may, in a state of liberty, contribute to abridge their existence; but we may presume, with every appearance of reason, that they pass the bounds usually allotted to them by authors.

The gluttony of the sparrows is equal to their petulance. They feed alike on the first fruits which ripen in orchards, on the grain sown in the fields, and on those which the farmer has gathered into his barn and storehouses. They also eat caterpillars, locusts, flies, and various other insects. But this taste, which is only secondary with the sparrows, renders them still more pernicious to the agriculturist, for they also devour bees. It appears, therefore, to be erroneous, as some writers on rural economy assert, that the number of insects destroyed by sparrows, compensates for the devastations they commit on grain and fruit.

These birds, harsh as it may seem to say so, do little but evil to man during their lives, and are of no use to him after their death. Their flesh is hard and bitter, and the medicinal properties which were anciently attributed to some of their parts, are merely imaginary.

Rougier de la Bergerie, a French writer on rural economy, has made an approximative calculation of what the sparrows cost, annually, to France. If their number be reduced merely to ten millions, a reduction much below the reality, it follows, that each of them eating a bushel of grain, weighing twenty pounds, ten millions of bushels will thus be withdrawn from the consumption and commerce of men; and, only reckoning the price of a bushel to be twenty sous, no less a sum than ten millions of francs per annum, will be withdrawn from agricultural produce. This calculation of an able agriculturist is confirmed by observation. The quantity of grain eaten by these birds, may be easily ascertained by those who bring them up in cages; and M. Sonnini, from whom we borrow these observations, says, that he found two-and-twenty grains of wheat in the stomach of a sparrow just killed.

It is proper, however, to hear the other side of the question, and see what can be said in favour of the sparrow. A countryman of our own, Mr. Bradley, in his General Treatise on Husbandry and Gardening, shows, that a pair of sparrows, during the time they have their young ones to feed, destroy, on an average, every week, about three thousand three hundred and sixty caterpillars. This calculation was founded on actual observation. He discovered that the two parents carried to the nest forty caterpillars in an hour. He supposed the sparrows to enter the nest only during twelve hours in the day, which would cause a daily consumption of four hundred and eighty caterpillars; and this average gives three thousand three hundred and sixty caterpillars extirpated, weekly, from a garden. But the utility of these birds is not limited to this circumstance alone; for they likewise feed their young ones with butterflies, and other winged insects, each of which, if not destroyed, would be the parent of hundreds of caterpillars.

On this calculation of Mr. Bradley, we shall only observe, that the premises on which it is grounded, appear, to say the least of them, suspicious. He observed the sparrows for one hour only: they must have consumed the entire of that hour in going from and to the nest, in bringing such a number of caterpillars. Now, it really appears too much to suppose, that the birds consumed twelve hours out of the twenty-four in such operations. Besides, the number of caterpillars, supposed to be daily eaten by the young, which, probably, were not more than three or four in number, shocks probability, even taking the assistance of the parents into consideration. The caterpillar in general is not so very small a morsel, in proportion to the size of the birds, and it would appear that such a quantity must overload their stomachs, and occasion death. In fact, all such calculations must of necessity, be liable to great error; yet we cannot help thinking, on the whole that that of the Frenchman has the least fallacy about it of the two.

The doctrine of compensation itself, is liable to much difficulty. Birds, by destroying insects, make up for the damage they themselves do to the fruits of the earth. Granted. But let us carry this a little farther. It is allowed on all hands, and, indeed, this defence of their destroyers presupposes it, that insects do, even in their diminished numbers, unqualified mischief. Where is the compensation on their part? They set limits, say some, to the exuberance of the vegetable world, which would otherwise be too great. But these are dreams, and very presumptuous ones, too; for they rest on the supposition that this earth, and all its productions, animal and vegetable, were created for man alone. In pretending to defend the Divinity, we exalt ourselves: things were not made solely with a view to his advantage, nor is the Great Governor of the Universe employed in adjusting a nice balance among them, on his account. Myriads of living beings have been produced, and each has a right to subsistence as well as man. Correcting the exuberance of the vegetable system, indeed! Idle theory. Relieving plants of their superfluous juices! Pray what is the particular use of lice and fleas? Are they intended to relieve the plethora of the animal system?

In a wild state, the note of the sparrow is only a chirp. This does not arise, however, from want of powers. The sparrow can be educated to imitate the song of the linnet and goldfinch, or both.

Mr. Smellie relates a pleasing anecdote of the affection of these birds towards their offspring.

"When I was a boy," (says this gentleman) "I carried off a nest of young sparrows, about a mile from my place of residence. After the nest was completely moved, and while I was marching home with them in triumph, I perceived, with some degree of astonishment, both the parents following me at some distance, and observing my motions in perfect silence. A thought then struck me, that they might follow me home, and feed the young according to their usual manner. When just entering the door, I held up the nest, and made the young ones utter the cry which is expressive of the desire of food. I immediately put the nest and the young in the corner of a wire cage, and placed it on the outside of a window. I chose a situation in a room, where I could perceive all that should happen, without being myself seen. The young animals soon cried for food. In a short time both parents, having their bills filled with small caterpillars, came to the cage, and after chatting a little, as we should do with a friend, through the lattice of a prison, gave a small worm to each. This parental intercourse continued regularly for some time, till the young ones were completely fledged, and had acquired a considerable degree of strength. I then took one of the strongest of them, and placed him on the outside of the cage, in order to observe the conduct of the parents, after one of their offspring was emancipated. In a few minutes, both parents arrived as usual, loaded with food. They no sooner perceived that one of their children had escaped from prison, than they fluttered about, and made a thousand noisy demonstrations of joy, both with their wings and their voices. These tumultuous expressions of unexpected happiness, at last gave place to a more calm and soothing conversation. By their voices, and their movements, it was evident that they earnestly entreated him to follow them and to fly from his present dangerous state. He seemed to be impatient to obey their mandates; but by his gestures, and the feeble sounds he uttered, he plainly expressed that he was afraid to try an exertion he had never before attempted. They, however, incessantly repeated their solicitations: by flying, alternately, from the cage to a neighbouring chimney-top, they endeavoured to show him how easily the journey was to be accomplished. He at last committed himself to the air, and alighted in safety. On his arrival, another scene of clamorous and active joy was exhibited. Next day I repeated the same experiment, by exposing another of the young ones on the top of the cage. I observed the same conduct with the remainder of the brood, which consisted of four. I need hardly add, that not one, either of the parents or children, ever again revisited the execrated cage."

An anecdote of this kind, well authenticated, is worth all the Greek and Latin of all the nomenclators, that ever barbarized language for the purpose of obscuring knowledge.

There are some accidental varieties of the sparrow. Such is the white sparrow, which has sometimes the plumage of a dirty white, sometimes as brilliant as snow, sometimes the head and neck of the same colour as the others, with the iris sometimes yellow and sometimes red. The young ones, which are white in infancy, become like the others, not unfrequently, on the first moulting. This colour is also acquired by age' and it is not uncommon to see some old sparrows, which are partially white. The black or blackish sparrow, the yellow sparrow, and the red, are also accidental varieties.

The *Tree-Finch (Fringilla Montana)* is remarkable for its perpetual movements when alighted on any place, for turning and moving the tail, raising and lowering it, fluttering, &c. continually. It is sometimes confounded with the common sparrow; but it is easy to distinguish it by its mode of life, size, and plumage. It seldom approaches our houses, but remains in the country, frequenting the sides of roads, and the banks of streams shaded with willows. It perches on trees and low shrubs. It is also found in woods, but more rarely.

This sparrow fixes its nest in the hollows of trees, and in the crevices of old walls, at no great distance from the ground. It constructs it of fine and dried plants, cattle hair, and feathers. Six eggs are the most that are laid, of a dirty white, spotted with brown. Noseman relates a singular fact, that out of these eggs there is always one smaller than the rest, and that the bird which proceeds from it is also much smaller than the rest of the brood. It is called in Holland the *little king*.

Though these birds have two or three broods in the year, they are less numerous than the sparrows. They assemble in flocks from the end of summer, remain together during the winter, and often unite in this season with flocks of finches, buntings, &c. They are less subtle than the sparrows, and give more easily into the snares which are laid for them; but they have less docility, and are never so familiar. The young ones taken from the nest may be reared with moistened bread; and when they can eat of themselves, they may be given the same seed as is given to the canaries and goldfinches. The tree-finch will live five or six years encaged. The song of the male is nothing to speak of; but the voice is naturally more agreeable than that of the sparrow. This bird is also less voracious, and does not do so much mischief to corn. It prefers berries, and wild grain, and also eats insects. This species is extended throughout all Europe, and is even found in Eastern Siberia.

On the rest of this division we find nothing respecting habits worthy of detaining the attention of the reader. We proceed to the FINCHES.

The *Chaffinch (Fr. Cœlebs)* is a species generally spread through Europe, from Sweden to the Gut of Gibraltar. It is even found on the coast of Africa. Some of these birds migrate in autumn, but this portion, it is said, is composed of the females alone, while the males remain during the winter in their native country. It is not impossible that there may be some mistake in this, and that the males and females may have been confounded together; for, from the moulting time to the month of February, and more particularly in autumn, the two sexes exhibit colours nearly similar. At all events, it is quite certain that many females remain as well as males, and united together, form, along with the tree-finches, green-finches, buntings, and other birds, those innumerable flocks which, during winter, are seen in our fields, and in the vine-grounds abroad; and which, when the earth is covered with snow, come before our barns, &c. to partake with the sparrows the aliment of our domestic fowls.

From the early days of spring, each couple isolate themselves. Some fix in our gardens and orchards, others retire into coppice-woods. They all animate the places which they inhabit by their gaiety, and by their song, which is by no means devoid of melody. Besides this song, which is tolerably varied, and consists of burdens more or less long, these birds have various well-known cries. That which the male and female use in the autumn, and during the bad season, is simple, and sharp. The male alone in the spring utters a plaintive accent, especially in the evening, and repeats it most frequently in rainy weather. This bird, taken from the nest, has the talent of learning foreign strains, and will imitate that of the canary, and in part that of the nightingale, &c., if kept near them. It can even be taught to articulate some words. It has also been remarked, that it sings better and longer if deprived of sight, and this observation has proved very fatal to these little prisoners, since they are blinded for the purpose of gratifying our taste for music. With the details of this cruel and disgraceful operation, we shall forbear to trouble our readers. Information of this kind is worse than useless. The poor victims of this cruelty having their attention no longer distracted by external objects, become indefatigable singers; but if the operation has not been very adroitly performed, they are subject to a constant rotation of the head, very disagreeable to behold. It is seldom performed, therefore, except with those chaffinches which are used as calls or decoys to attract the wild ones into the nets. But this means is not necessary to produce good appellants. It is sufficient to put them into the moulting, which is done thus, and the same method will answer for any other birds destined for a similar purpose. Towards the end of April, you take two or three of each species, and much more chaffinches than others, which are tamed by gradual diminution of daylight before they are plunged into utter darkness. They are finally enclosed in an obscure chamber, or in a chest. This preparation demands at least fifteen days. You begin at first by keeping the door and windows half closed, and continue by degrees to deprive them of light, until at last a complete obscurity reigns in the apartment. Every singing bird should be carefully removed from the neighbourhood; the birds should be cleaned every day, fresh food given to them, and the water changed, which should be kept in a larger vessel than usual. But this task should never be performed except in the evening by candle-light. If this be done in the chamber where they are kept, the cages are attached to the wall, one near the other, or suspended with rings to a perch which is placed across the middle of the chamber. If there be any among them that sing, their tails mnst be plucked out. They are kept thus until the month of August, at the least, at which period they are taken out of the obscure chamber. It is necessary to do this cautiously, and give them light by degrees, in the same proportions as it had been before withdrawn. But first, it is necessary to purge them, which should have been also done on their entrance into this sort of discipline. This purgation is accomplished by giving them for three or four days some beet-root, sugar well strained and clarified, with a little brown sugar in their water. Then they are left some days shut up in the light chamber, before they are exposed to the air. Some beet-leaves are given them to eat, and a piece of plaster is left in their cage. The birds destined to undergo this should be put in the cage in the month of October, to have time to separate the good singers from the bad. In fact, those which do not sing from this time unto the end of March, are unfit for the purpose. They must also be accustomed to eat grass, because without that they would languish in the moulting, during which the beet-root must also be administered three or four times. To habituate them to this diet, their usual food should be withdrawn for three or four hours every morning, and cabbage-leaves and lettuce substituted for it. After they are put into the air, they must not be exposed to the sun for twelve or fifteen days.

The chaffinch commences to sing at an early hour in the morning. It is generally heard from the beginning of the fine days in February, and continues to sing until towards the end of the summer solstice. It is a bird of a very lively character, always in motion; and from its gestures and song, the very emblem of gaiety. The male is very jealous; and having once chosen his companion, and occupied a certain district, he will suffer no others in his neighbourhood; and if any one intrude, a desperate battle takes place, and continues until the weaker either falls or retires. He never quits his female during the time she is hatching, remains during the night very close to the nest; and if he removes a little during the day, it is only for the purpose of providing food, which he shares with her on his return. The female alone works at the construction of the nest, to which she gives, with a solid texture, so elegant a form, that it is one of the most beautiful nests made by any of our European birds. She places it on trees and shrubs of the most tufted kind, and even on fruit-trees in our gardens and orchards. It has been remarked that in woods, this nest is situated very high: while in orchards, its elevation does not exceed the ordinary stature of man; but she conceals it so well, that it is frequently passed without being perceived. Various white and green mosses, and little roots, are covered on the exterior by a lichen, similar to the branches on which the nest is placed. The interior is furnished with wool, horse-hair, and feathers, connected together with spiders' webs. The female deposits here from four to six eggs, of a reddish grey, sown with blackish spots, more confluent at the thick end. The incubation, which the male does not partake, lasts for thirteen days, and the little ones are born covered with down. The father and mother feed them at first with insects and caterpillars, and then unite to this aliment the small seeds of plants; and when they can manage for themselves, they feed on rape-seed, millet, hempseed, corn, &c., the husks of which they can remove with much dexterity, so as to arrive at the farinaceous substance.

Those which are destined for the cage, should be taken from the nest; for when adult, they are with difficulty reconciled to captivity, refusing to eat at first, or perhaps at all, continually striking the bars of the cage with their bills, and frequently suffering themselves to perish. They may be brought up like the canary. As at this age there is no external difference between the sexes, the males may be known, after eating of their own accord for about fifteen days, by their commencing to warble. It is pretended that by giving them a little bread, cheese, or milk, they may be made good singers: but the cheese must not be salted. Others give them meal-worms, and even locusts or grasshoppers. They may be fed also with different seeds; but hemp-seed, though they are very fond of it, is pernicious to them, as well as to many other small granivorous birds. Therefore, they should have but little of it. As they are very fond of bathing, their water should be renewed daily, and given to them in abundance.

The *Mountain Finch* arrives here and on the continent generally in autumn, and passes the winter with us, departing again in spring. This species assembles in flocks, more or less numerous, and unites with the chaffinch and other small granivorous birds, retiring in the evening into the forests. They are easily distinguished from the others; for they fly, settle, and rise again in close compacted bodies, and utter a cry that has some resemblance to that of a cat. M. Lottinger, who has observed them much, in Lorraine, assures us that the females alone migrate, while the males remain in the Vosges. But this assertion cannot be generalized: for numerous flocks are to be seen elsewhere, composed both of males and females. In autumn it is certainly very difficult to distinguish one from the other, the plumage being so nearly similar, especially in the young ones of the existing year—but in the early days of winter, the characteristic colours of the male begin to appear. Besides the cry just mentioned, these birds have another, which is never heard but when they are settled on the ground. It somewhat resembles that of the stone-chat, but is not so loud and distinct. Their song is feeble and monotonous; it is a slight warbling, which cannot be heard at any distance.

The mountain-finch is more docile than the chaffinch, and more easily taken in snares, and sooner reconciled to captivity. It retires northward generally when the frosty season is over, though it sometimes will remain until the end of March: but it then becomes a hurtful animal, for, like the bulfinch, it destroys the young buds of fruit-trees, especially those of the plum. According to travellers, it appears that it nestles in the territories of the Luxembourg, and in the forests of Northland. Its nest is situated pretty high on those fir-trees which have most branches. It begins to work on this nest about the end of April, constructs it externally with the long moss of these trees, and furnishes it within with horse-hair, wool, and feathers. It lays four or five eggs. It is probable, from their great number, that these birds have several broods in the year.

We now pass on to the LINNETS and GOLD-FINCHES.

The GOLD-FINCH *(Fringilla Carduclis)* is one the best known, and most beautiful of our native birds. To a graceful, well-turned form, and a brilliant plumage, it unites dexterity, docility, and a pleasing song. Such is the harmonious mixture of its colours, that, though so common, it never ceases to delight the eye. Nothing, in fact, is wanting to this charming bird, but to be the native of a foreign country, to render it an object of the greatest research and admiration.

From the earliest days of spring, the enchanting voice of the male begins to be heard; but it is in the month of May that it puts forth its sweetest strains. Perched on a tree of moderate height, particularly on a fruit-tree, of which these birds are very fond, it makes the orchard echo with its song, from the point of day to the setting of the sun. It continues thus until the month of August, with the interruption, however, occasioned by the care of its young: for such is its attachment, that at this period all its moments are absorbed by its paternal duties. It feeds the young with tender seeds, such as those of groundsel, lettuce, and other plants. It is said, that it also gives them caterpillars, small scarabœi, and other insects; but it appears more probable that the gold-finches are simply granivorous, like the linnet, the canary, &c. It is on this account that they nestle later than the sparrows, the buntings, and the chaffinches, which rear their young on insects, and do not disgorge the food for them. The gold-finch, when its young are more advanced in age, gives to them grains more difficult of digestion, but never without softening them in its crop, and disgorging them like the canaries. It is so much attached to its progeny, that if shut up with them in a cage, it will continue to take care of them at the very epoch when liberty is so dear to other birds, that few of them survive its loss. But to manage this properly with the gold-finch, it must receive abundance of groundsel, &c., and particularly the seed of the thistle, which is its favourite food, and from which its French name (*Chardonneret*) is derived. It is also sometimes called, in our language, the *Thistle-finch*. The fowlers, accordingly, who lay various snares for these birds, make use of thistle-seed as their bait.

While the female is hatching, the male remains near her, singing on a neighbouring tree. He rarely moves, unless disturbed; then he flies away, but for a very short time, for this is only a feint on his part, to prevent the nest from being discovered; he is soon seen to return. The female shows a still greater attachment for her young; nothing can distract her from the business of incubation. Her constancy is admirable. For the preservation of her eggs, especially at the moment they are about to disclose the young, she braves every thing, and endures every thing. The impetuous wind, the continued rain, the hail-storm, can neither terrify nor tire her. The male never quits her; he accompanies her in all those courses which the exigency of food or the construction of the nest may require; but he does not partake either of the latter labour, or of that of incubation. He simply watches for her safety, while she is on the ground, either seeking food, or materials to build her nest, and is always perched on the nearest tree. This female gives to her nest a greater solidity, a better rounded, and even a more elegant form than the chaffinch. She places it usually on fruit-trees, and chooses the weakest branches; it is also found in coppices and thorny bushes. She employs, for the external part, little roots, fine moss, and the down of certain plants, which she covers again with lichens. The interior is composed of dried plants, of horse-hair, wool, and the most downy feathers; on this bed she deposits five or six white eggs, spotted with brown towards the thick end.

Though the goldfinches do not construct their nests until the middle of spring, they have yet three broods, the last of which takes place in August. The young cannot suffice for themselves for some time, even after quitting the nest; accordingly there is much patience requisite to rear them artificially. The best are said to be those which are born in thorny bushes, and belong to the last brood. They are, it is said, more gay, and sing better than the others. They must be taken in the nest, when all their feathers are disclosed, and fed with the following composition: some simnel, peeled almonds, melon-seeds, or nuts, or marchpane, should be pounded together. Of the paste resulting from this mixture, little balls should be made like small grains of vetches. They should be given one by one, with a skewer, about three or four in succession at one time, to each young bird, to which you afterwards present the other end of the skewer, with a little cotton on it, steeped in water. When they begin to eat alone, they may be nourished with hempseed, bruised with melon-seed; and when they are strong, they may receive the hempseed alone. This complicated paste which we have mentioned, and which is not easy to make, might be replaced by another, less difficult of composition. This is composed of hempseed and rapeseed, well bruised, crumb of bread, and yolk of egg. The whole is steeped in a little water, and given by billsfull. When they can eat alone, the hempseed may be removed from this composition, and replaced by millet, especially if the goldfinches are intended to be coupled with canaries. With this diet these birds enjoy better health, and live longer.

Olina tells us, that the young goldfinches which are placed near linnets and canaries, will appropriate their notes. Others say, that they show a greater disposition to copy those of the wren.

Some fowlers pretend, that among the goldfinches taken in nets, the best are those which have the six intermediate quills of the tail terminated with white. Some of them have eight quills, so terminated, and some four. These last, say the fowlers, sing badly; but all these distinctions are without foundation, and only turn to the account of the dealers in birds, because they sell the first at double the price of the others. But they also take care to mention that these spots vary in the same individual during the summer, so that one of one kind may be easily transformed into another. On the wild birds all these spots disappear, in a great measure, from the month of June to September; then all the quills, with the exception of the lateral ones, are black. It is the same with the spots which are on the quills of the wings: often in September, no further trace of them appears; but they come again with the new feathers. This progressive change does not take place so completely with birds educated in the cage; there always remain some white spots on the quills of the wings and tail.

The goldfinch is very easily reconciled to captivity, and even becomes quite familiar. From its activity and docility it may be taught a wonderful degree of precision in its movements; it will counterfeit death, and perform a great variety of other movements with the greatest dexterity; it can be taught to fire a cracker, and draw up small cups, containing its food and drink.

The mode in which it is made to perform the last operation

is curious. The bird must be equipped for the purpose; this equipment consists of a little band of soft leather, about two lines in breadth, pierced with four holes, through which the wings and feet are passed, and the two ends of which joining under the belly, are maintained by a ring to which a chain is attached; at the other end of this chain is a ring, which passes into the semicircle of wood which serves the bird for a perch, and the two ends of which are placed in the bottom board; on this board is a small looking-glass, placed in face of the circle, and below this circle is another of greater diameter, that the bird may mount and descend at pleasure. The two cups are suspended by a little chain to the upper circle; in one is the food and the other the drink, and they are so arranged that one cannot be lowered without raising the other; thus, a degree of industry is necessary for the bird to draw to himself the vessel he wants. The want of society appears to be of the first necessity to the goldfinch, which delights in that of its own species. It is on this account that it is fond of looking at itself in the glass, and is often seen to take its hempseed, grain by grain, and go and eat it before the mirror, thinking, doubtless, that it is eating in company.

Sometimes the mirror is omitted, and its place supplied by a small measure, closed on all sides, with the exception of a small aperture in the front, which is shut with a lid, so arranged, that it will obey the slightest touch, and shut of itself. To make the bird know where its aliment is, the lid is first held half open, then shut about three-fourths, it is then completely closed; and the bird, knowing where its food is, uses all its address to open, and holds it open with his feet. The water is placed in a little cup, attached by a chain to one of the circles; the bird draws it to him, seizing the chain with his bill, and holding it under his feet until he has quenched his thirst.

"Some years ago, the Sieur Roman exhibited in this

country the wonderful performances of his birds. These were goldfinches, linnets, and canary birds. One appeared dead, and was held up by the tail, or claw, without exhibiting any signs of life. A second stood on its head, with its claws in the air. A third imitated a Dutch milkmaid going to market, with pails on its shoulders. A fourth mimicked a Venetian girl, looking out at a window. A fifth appeared as a soldier, and mounted guard as a sentinel. The sixth was a cannoneer, with a cap on its head, a firelock on its shoulder, and a firelock in its claw, and discharged a small cannon. The same bird also acted as if it had been wounded; it was wheeled in a little barrow, to convey it (as it were) to the hospital, after which it flew away, before the company. The seventh turned a kind of windmill; and the last bird stood in the midst of some fire-works, which were discharged all round it, and this without exhibiting the least sign of fear."*

The goldfinch, naturally active and laborious, is fond of occupation in its prison, and if it has not some poppy-heads, hemp-stalks, and those of lettuce, to pick, for the purpose of keeping it in action, it will remove every thing that it finds. A single goldfinch, in an aviary where canaries are hatching, if he be without a female, is sufficient to make all the broods fail; he will fight with the males, disturb the females, destroy the nests, and break the eggs. These birds, however, though so lively and petulant, live in peace with each other, excepting a few quarrels about the perch and their food; all of them try to get possession of the highest perch in the aviary, for the purpose of sleeping, and the first who obtains it will not suffer the others to approach. It is necessary to place all the perches at a similar height, to isolate each from the other, and make every one only of length sufficient for a single bird.

* Bigland's Animal Biography.

Though the goldfinches will couple in aviaries, the union is rare, and the result seldom fruitful. It is true, that little attention is given to this point, in consequence of the facility of procuring these birds at every season, in any number desired.

A single female is said to be sufficient for one male, and both should have a large cage to themselvess. In captivity the male goldfinch will pair more readily with a female of another species than his own; for example, with the female canary; but the female goldfinch will seldom couple with the male canary. This union is not the result of any conformity of song, and still less of plumage; it proceeds from the goldfinch disgorging the food found in its crop in the same way as the canaries do; he pleases the female canary by this, and feeds her while she is hatching. This is not to be expected of the bunting or chaffinch, or any other bird that brings its bill full of food to the female and the young. This supplies a good rule for those who desire to make birds of different races couple together. Though the broods sometimes succeed from a female canary and a wild goldfinch, *i. e.* one caught in a net, yet it is better to educate them together, to accustom the goldfinch to the same food as the canary, and not to pair them until the end of two years. It will also be much better if the female canary has never paired with one of her own species, and be kept in the spring from seeing or hearing any of them, that she may totally forget them. Her first eggs will often prove fruitless, especially in the earliest days of spring, which is not the period of reproduction with the goldfinches; but the second brood will succeed, and the male goldfinch then becomes more assiduous and attentive than even the canary itself. He shares all the cares of the household, remains almost continually on the edge of the nest, and frequently feeds the female while she is hatching; he also assists in the rearing of the young.

The bill of the goldfinch is liable, especially in captivity, to elongate, sometimes to such a degree that one mandible so far passes the other, that the bird cannot take up its aliment· If the mandibles elongate equally, another inconvenience results; in feeding the young or the female, the male is apt to wound them grievously. To prevent this, the mandibles should be equalized and blunted with scissors.

The mules from the goldfinch and canary are more robust than the latter, and live longer. Their song is also more brilliant; but Buffon says, that they imitate airs with difficulty. Others, on the contrary, pretend that they can very easily be taught by the bird-organ and flageolet. These mules resemble the male in the form of the bill, and the colours of the head and wings, and the female in the rest of the body. Some beautiful varieties result from this alliance. M. Vieillot once caught a mule, which he conceives was the produce of a male greenfinch and female goldfinch, judging from its size, colours, and song. This bird did not appear to be the result of any forced union; it always remained extremely wild, and by no means familiarized with the cage—a seeming confirmation of the last remark. It was brought, notwithstanding, to couple with a female canary; but nothing resulted from the union. Some, however, say that these females are not unproductive, and that the second generation insensibly approaches the characters of the male; but this second generation must be marvellously rare, for no authentic proof appears of its having ever been witnessed. These mules, however, pair very readily with each other or with canaries; but the eggs produced are not fecundated. The female mules construct their nests much better than the canaries; and are such excellent nurses that they may be frequently substituted for the others, when the latter are sick, or are bad mothers.

In autumn the goldfinches assemble together, live, during

winter, in numerous flocks, and frequent those places where thistles and wild endive grow. During the severe cold, they shelter themselves in thick bushes; but they seldom recede far from the places where their food is found. Sometimes they mingle with other granivorous birds. Hempseed is the grain given to familiarize them with the cage; but it would be better to mingle millet and rape-seed with it, and to vary their aliment; thus the maladies might be avoided which attack them in captivity. This is a point not always properly attended to, for cage birds of all descriptions. Variation of food preserves them in good health, lengthens their days, and approximates them more to their natural state.

The maladies to which this bird is subject are the epilepsy, in which it frequently falls when apparently most vigorous, and singing with its full force; the *molten-grease*, or inflammation of the abdomen, and the moulting, which frequently proves mortal. The epilepsy is said by some writers to proceed from a very small worm in the thigh, lodged between the skin and flesh; sometimes this worm goes away of itself, and sometimes the bird draws it out with its bill, when it can lay hold of it. It seems more probable that this disease is attributable to the exclusive use of hempseed, for it also attacks canaries and bulfinches, when they are limited to this aliment. On the other hand, the goldfinch, when totally deprived of this kind of grain, is seldom visited by this distressing malady. When the bird is attacked by it, it is so violent and dangerous, as generally to prove mortal in less than a quarter of an hour. After some very precipitate movements, the bird falls, and lies extended in its cage with its feet in the air. Without the promptest assistance it soon dies. The most certain remedy is instantly to take the bird, and cut its nails with scissors, particularly the hinder one; some drops of blood will follow; and then dip the feet several times in white wine made lukewarm. If it is winter time, the bird should also be made to swallow a few drops of it, to which a little melted sugar is added. This remedy is very efficacious; the bird resumes new strength, and in a few hours enjoys as good health as he had before. It is recommended never to leave the goldfinches without a piece of plaster in their cage, suspended so that they may pick at it with facility. When these birds are properly taken care of, and attentively kept clean, they are not very subject to disease, and will live sixteen or eighteen years, and sometimes longer.

The species of the goldfinch is extended throughout the whole of Europe nearly, and through some parts of Asia and Africa. It is found in Greece, where it bears the name of *karedreno*; though no migrating bird, properly speaking, it does not remain all the year round on all the islands of the Archipelago. It prefers the largest, and also the lands of the neighbouring continent, because it doubtless finds there more safe and agreeable retreats.

Few species present more varieties than this; besides those which proceed from forced alliances, there are others attributable to aliment, to age, and to domestication.

There is one which is white where the others are red, namely, on the forehead and eyebrows, which colour also prevails on the top of the head, instead of black. On some the red is shaded with yellow, and the black appears through these colours. A goldfinch, with the head striped with red and yellow, has been found in America. One with the cap altogether black has but a few red spots on the forehead; the back and chest are of a yellowish brown; the iris yellowish, and the bill and feet flesh-colour. The whitish goldfinch has the tail and wings of an ashen-brown, the upper and under parts of the body whitish, and the yellow of the wings pale. Some varieties are totally white, and others, among which are the handsomest races, have the head red and the wings bordered with yellow. On the bodies of many the tints are more or less mingled with white. Among the black goldfinches some are entirely black; others more or less varied with this colour. These last varieties are chiefly attributable to food, especially to the exclusive use of hempseed. Still the colours are not fixed, for goldfinches have been known to resume their primitive tints after the moulting; and some which were even totally black, to retain very fine feathers of that hue. These changes from one moulting to another become still more palpable when millet or other grain is substituted for hempseed.

Of the other species of the goldfinch little can be said, except in the way of description; we shall proceed, therefore, at once to

The *Linnet.* To persons who have seen the numerous treatises extant on European birds, it must be a matter of surprise that any necessity should still exist of further investigation on this subject; nevertheless, it is most certain that many of the birds which are actually subjected to our daily view, do require further research.

Brisson, Mauduyt, Sonnini, and Frisch, have made two species of the linnet, properly so called, under the denominations of *grey* and *red;* Latham and Gmelin under those of *linota* and *cannabina.* Belon, Linnæus, Olina, Gesner, Montbeillard, Meyer, and Latham, in the second supplement to his synopsis, after the remarks of Bays and Montague, make but one. The Doctor, also, though he formally separates them in his *General History,* expresses there his sentiments to the same effect. As for M. Vieillot, he has no doubt of the identity of the red and grey linnet; and his opinion is confirmed by repeatedly multiplied and indefatigable observations. Both kinds, young and old, male and female, are grey in the back season, and resemble each other so much, that the sexes cannot be distinguished, except by the white border on the primary alar quills, which is more broad and brilliant in the male than in the female. The red colour, which characterizes the male during summer, commences to appear towards the end of autumn; but at this time it is tarnished, and occupies only the middle portion of the feathers, the extremity of which is of a reddish grey, so that it can only be perceived by raising them up. In proportion as the spring approaches, this colour extends and grows brighter, and towards the month of May becomes very brilliant in the male of two years old; less pure and less extended in the bird of the first year; and among the old ones it sometimes assumes an orange shade. Of course, the linnets which remain grey must be only females; and it does not appear that any well-authenticated instance of a male of this hue at such periods has been found.

It is only on the head and chest that their plumage suffers the variations we have alluded to. The occiput and nape become of a clear ashen, from the grey and reddish, which they were immediately after the moulting. The morone-brown of the feathers of the back assumes a finer and more decided colour. The croup passes from grey and reddish white, to blackish, and pure white. Such is the case with the male linnets in a state of liberty; but it is quite different with those confined in a cage, or even in an aviary, exposed always to the air. The red disappears; the brown morone remains dull; the grey of the occiput and nape preserve the reddish tint. The principal attributes, which in this state distinguish the male from the female, are the colours of the summit of the head, which is of a tarnished red towards the middle of the feather, and in the white of the alar quills, which is more extended in the male than in the female. No arguments, derived from size and proportions, appear to be more convincing in favour of the separation of species, than those drawn from colours.

There are some accidental varieties of the linnet, as for instance, some entirely albino individuals; others, which have only the head, wings, and tail, white. On others, white is the predominant colour, but the alar and caudal quills are black, edged with white, and some vestiges of grey on the wing-coverts.

The male neither partakes of the labour of constructing the nest, nor of that of incubation, but employs himself in several little attentions to the female, brings her food, which he disgorges, like the canary, cheers the monotony of her situation, by an agreeable song, incessantly repeated during the time of hatching, and watches for her safety. When any thing offensive appears, he puts forth a plaintive cry, flutters from bush to bush, removes for a moment, but only to re-appear directly. The more his companion is approached, the more his cries redouble. Then the female, warned by his complaints, and pressed by danger, quits the nest. Both she and the male fly away, and do not return for an hour; but when the eggs are ready to open, they return sooner. Both parents exhibit much affection for their young. They nourish them with tender seeds, prepared in their crop, and disgorged by the bill. These linnets have usually two or three broods, and sometimes four, if the first be disturbed. After hatching time, they assemble in numerous flocks, quit the high countries, and descend into the plains. It is at this time that snares are laid for them, and they are caught in great numbers. As these birds soon grow fat, when they have abundant aliment, their flesh acquires a savour which makes it in request in some countries, especially in the southern provinces of France. From this circumstance they have received, in Provence, the name of *bec-figue d'hiver*.

Montbeillard observes, that there is a great analogy between the linnet and the canary, and he is right. Their habits and nature are extremely similar, and of all birds the linnet is that which most readily couples with the canary; but it is difficult to believe that the produce of this alliance is more fruitful than that of the canary with the goldfinch or the siskin. Though repeated trials have been made, it does not appear that fruitful eggs have been obtained, either from a female canary, paired with a linnet-mule, or a female mule with a male canary, or from the mules paired together.

Although the linnet is one of the commonest of our small granivorous birds, and though it preserves no brilliant colours in captivity to render its possession desirable, it is not less in request than the brilliant goldfinch and charming bulfinch. Though sombre tints replace the fine red colour of its head and chest, and though its entire plumage changes to a tarnished brown, or dirty white, the linnet does not the less merit our attention, nor the less contribute to our amusement. Its qualities are truly interesting. Its natural disposition is docile, and susceptible of attachment; its song is agreeable, and the flexibility of its throat enables it to imitate with facility, the different airs which it is attempted to be taught. It can even be taught to repeat many words, distinctly, in different languages, and it pronounces them with an accent that would actually lead one to suppose that it understood their meaning. The tender attachment of which these birds are susceptible is astonishing; so much so, that they often become troublesome in their caresses. They can perfectly well distinguish the persons who take care of them. They will come and perch upon them, overwhelm them with caresses, and even seem to express their affection by their looks. They can also imitate and unite to the varied modulations of their own voice, the strains of other birds, which they are in the habit of hearing. If a very young linnet be brought up with a chaffinch, a lark, or a nightingale, it will learn to sing like them. But it will in most cases totally lose its native song, and preserve nothing but its little cry of appeal.

The linnets intended to be intructed in foreign strains, should be taken from the nest when the feathers begin to shoot. If taken adult, they will seldom profit by their lessons, though they will become both familiar and caressing. Different modes of instruction have been pointed out for them—such as whistling to them in the evening by candle-light, taking care to articulate the notes distinctly. Sometimes, to put them in train, they are taken on the finger, a mirror is presented to them, in which they think that they see another bird of their own species, which illusion is said to produce a sort of emulation, making them sing with more animation, and expediting their progress; but these precautions are not absolutely necessary, for the best instructed linnets are often brought up by cobblers, who whistle to them without interrupting their work. It has been remarked of the linnets, and it is true of many other singing birds, that they sing more in a small cage than a large one.

This bird lives a long time in captivity, if well taken care of. Sonnini quotes an instance of one that lived forty years, and might have lived longer had it not perished by accident. This was a bird of the most extraordinary amiableness and docility. It was in the habit of calling many persons of the house by their name, and very distinctly. It whistled five airs, perfectly, from the bird-organ.

The linnets have the advantage of singing all the year round, and they may be taught a variety of tricks, like the siskin, and the goldfinch.

When it is intended to educate young linnets, the males must be chosen; for the females neither sing, nor can be taught to sing. The males may be known by the white colour on the wings, which is more pure and extended in them than in the females. They are fed at first with oatmeal, and rapeseed bruised in milk or water; some use crumb of bread instead, and the yolk of a hard egg. They are given their bills full, like the canaries, and must be kept clean and warm. To render them more familiar, they are fed with the hand, and chirped to. When they can feed alone, the rape-seed is given to them entire, but softened in water, so that they can break it more easily. Their aliment must also be varied with millet, radish, cabbage, lettuce seeds, and plantain, and sometimes with bruised melon seeds, and from time to time with barberries, marchpane, and anagallis. Hemp-seed should be given to them very sparingly, because it fattens them too much, and they either die or cease to sing. Many persons give them nothing but rapeseed, but the same inconvenience results from this. The more their food is varied, the fewer maladies they will have. A small piece of plaster or chalk should also be put into their cage to prevent constipation, to which they are subject. They are also subject to a curious disease, which the French call *subtile*, and for which the chalk or plaster is a remedy. This disease is indicated by their melancholy, their silence, and their stiff and bristling feathers. When it has made some progress, their bills become hard, the veins are thick and red, the breast is swelled, the feet are also swelled and callous, and they can scarcely sustain themselves upright. Linnets are subject to epilepsy, for which chalk is also prescribed, and a disease called *the buttom*, which is a small tumour, and generally considered incurable. It is recommended, however, to pierce it quickly, and staunch the wound with wine. Besides all these complaints, which are for the most part the result of captivity, they also suffer from the asthma, the symptom of which is, striking angrily with their bills. A little oxymel should then be put in their water, and their food changed for some days, giving them some tender wild endive, pounded with barberries, or cabbage, if this malady attack them during winter. Nothing is better to keep them cheerful and in good health, than to give them occasionally red currants. As nothing should be neglected to preserve a bird, to rear which so much pains are taken, it should be kept as nearly as possible in its natural state. These birds are pulverators, and a bed of fine sand, renewed from time to time, should be kept at the bottom of their cage, and as they are fond of bathing, they should also have a small bath, into which fresh water is put every day.

The linnets unite in flocks towards the end of September, and remain so during the winter. They fly in close bodies, rise and fall together, and fix on the same trees. Their flight is continuous, and not performed by repeated springs, like that of the sparrows. On the ground they proceed by hopping. They pass the night in the woods, and choose for an asylum the trees, whose leaves, though dry, are not yet fallen, such as oaks, &c. They frequent fallow, and cultivated lands, living on various little grains; they also pick the buds of poplars, of linden trees, and birch, as do the bulfinches, &c. Another of their aliments is indicated by their name—this is linseed. In short, all kinds of grain suits them; but it does not appear that they ever touch insects. It is certain that they do not take them to their young, as the birds which use both grain and insects do. Towards the commencement of spring, they are heard to sing altogether at a time, and their concert is always preceded by a sort of prelude. At this time they pair. Once having made their choice, each pair isolates, and chooses a particular district, from which it never departs during the summer.

Linnets are common in France, in Italy, in the Levant, in Germany, and in the southern parts of Russia.

The *Siskins* (Fr. *Spinus)* are birds of passage, and fly so high that they may be heard before they are seen. They are very numerous in the southern provinces of Russia, and common enough in this country during the winter; they are fond of places where the alder-tree abounds. They arrive in

France about the time of the vintage, then proceed farther south, and reappear when the trees are in flower; but in summer they are not seen. In all probability they then voyage northwards, or return into thick forests on the lofty mountains. This is affirmed by Sonnini in his edition of *Buffon's Natural History*. "I know," says he, "to a certainty, that the siskins nestle on the highest mountains of the Vosges, in Lorraine, and particularly on that one which is called the Donon; they pass into the plain in spring, and afterwards return to this chain of mountains in like manner as they do in Switzerland and Franche-Comté. They descend after the hatching in September and October." The individuals which nestle in the north do so on the tops of pines and fir-trees; the brood is four or five eggs, of a greyish white, spotted with red.

The siskins, in their habits, have very considerable relations with the linnet: they give a preference to the seeds of the alder-tree; they often dispute with the goldfinches for the seed of the thistle. Hempseed is for them an aliment of choice; but they appear, especially in captivity, to be greater consumers of it than they really are, from a habit which they have got of breaking more grains than they eat. In their passage in Germany, in October, they considerably damage the hop-grounds, by eating the seeds. In France, also, they do considerable prejudice to the apple-trees, by picking at the flowers.

The song of the siskin is by no means disagreeable, but very inferior to that of the goldfinch; it is said to possess the faculty of imitating the song of the canary, linnet, &c. if taken very young, and placed within hearing of these birds. It has, moreover, a note of appeal peculiar to itself. Even when taken adult, it is easily tamed, and becomes almost as mild as a canary. It is very docile, and can be taught to hoist up its food and water in the mode we have already described for the goldfinch, and can be even accustomed to come and perch upon the hand, at the sound of a little bell. The mode of teaching it is, by ringing the bell every time that you are about to feed the bird. Lively and gay, it is always the first to awake in the aviary, and the first to warble and set the others agoing. The manners of this little captive are so mild, that it never seeks to quarrel with any of its companions, and it yields promptly when it is threatened. Placed in an aviary where there are several birds of different species, it will conceive an affection for some particular one, and often feed it; but it gives the preference to those of its own race, whether males or females.

It has been remarked, that there is a great sympathy between the siskin and the canary. It is so great, that if a siskin be let loose in a place where there are canaries with other birds, it will fly immediately to the former, and come as nearly to them as possible, and they will also seek its society with eagerness. The male or female siskin will easily pair with the canary, the female, as is asserted, particularly; but the male siskin, when he has once coupled with the female canary, will assist her in all her labours, with much zeal, aiding in the construction of the nest, carrying the materials, and even making use of them. He continually feeds her while she is hatching, disgorging the food from his crop; in spite of all this, however, the eggs are frequently fruitless. The mules which do sometimes result from the union partake, in their appearance, of the characters of both parents.

In captivity these birds will live for ten years, and are very little subject to diseases, excepting the *molten-grease*, which is brought on by feeding them exclusively on hempseed; they should, therefore, be accustomed, in preference, to rapeseed and millet.

A variety of the siskin has been observed with the head yellowish and the rest of the plumage black. To make these birds entirely black, it will be sufficient to feed them constantly on hempseed. This variety is said to be found in Silesia.

The *Citril Finch* is found in all Italy, Greece, Turkey, Austria, Provence, Languedoc, Spain, Portugal, and sometimes in Lorraine. The male has an agreeable and varied song, but not so fine and clear as that of the canary. In Italy this species makes its nest not only in the country, but oftentimes in gardens on tufted trees, particularly on the cypress, and constructs it of wool, horse hair, and feathers. The eggs are four or five: the male easily pairs with the female canary, and the mules have been found productive. The Count de Riocourt had for many years several of these mules, which coupled with female canaries, and the young produced new generations. The same fact occurred with Mr. Vieillot. Some resembled the father and some the mother, without reference to the distinction of sex in themselves; while, on the contrary, the alliance with the female canary, goldfinch, bullfinch, greenfinch, &c. gave rise to a mixture of colours in the offspring, none of which perfectly resembled either father or mother. Another remark, very essential, is, that if the alliance of the male citril with the female canary gives rise to new generations, by the reproduction of the young, the female citril will reject, in captivity, the advances of the male canary, and even those of the male of her own species, and that the same is the case with the female mules of the first alliance. They refuse all pairing either with the male mules, the citril, or the canary, which proves that the original type is more firm with them than with the males. Notwithstanding this, M. Vieillot is inclined to consider the citril and the canary not as two distinct species, but as two races springing from the same stock, one of which has fixed in Europe and the other in the Canaries, and whose differences are attributable to localities. The fecundity of mules, from the alliance of canaries with goldfinches, linnets, &c. does not appear to be proved by a single fact, which is natural enough, as the species are so evidently distinct, and proceeding from different stocks; and the same observation may be extended to all other birds whose union produces unfruitful mules.

The Canary.—There is a considerable degree of interest and charm in every thing relating to this bird of the Hesperides; an elegant form, beautiful plumage, melodious voice, and a disposition amiable, docile, and familiar; it unites all the pleasing qualities and all the little talents which are isolated in other birds. This delightful bird particularly constitutes the amusement of the young female, who is well employed in the developement of its mild and social habits. Care, attention, and caresses are lavished on it, and nothing, in short, is spared in bringing it up. Its infancy and education give rise to some little trouble and embarrassment; but it does not prove ungrateful. It gives daily proofs of its susceptibility of gratitude, and of attachment. In the evening its adieus are caresses, and it scarcely wakes in the morning before its benefactress becomes the object of its first regard; it claps its little wings at sight of her, flies to her, and expresses, by a number of gentle gestures and semitones, the sentiments of affection with which it is inspired.

The docility of the canary is such, that it has been known, at the command of its instructress, to fly to the head of a cat, sing there with full throat, and receive caresses from its natural enemy. The liberty granted to other birds during the period of their amours, generally puts an end to their attachment to us; not so with the canary. Two of these familiar birds, male and female, having escaped from their aviary, fixed themselves in a thicket at some distance, and nestled there. The male was in the habit of coming regularly twice a-day to sing near his former dwelling, and to gorge himself with food, for the purpose of partaking it with his companion; but he did not suffer himself to be taken, although he approached very near his mistress, and appeared to take pleasure in fluttering over her, and repeating the lesson which she had taught him. At last he ceased, all of a sudden, to make these little journeys. The lady, uneasy about the birds, having sought them everywhere, without success, believed that they had fallen victims to some bird of prey; but ten days after these fruitless researches, the couple reappeared, accompanied by their young family, and established themselves in their ancient domicile, where they resumed all their former familiarity; "as if," says the writer from whom these observations are taken, "they were endeavouring to make atonement for the pain which their temporary absence had occasioned."

The canary is as docile as familiar. One of these birds has been seen, at a signal, to seize a match in its claw, light it, set fire to a little cannon, fall, as if dead, at the explosion, rise suddenly again, and place itself as if on duty.

Bingley in his "Animal Biography," says that "In the month of May 1820, a Frenchman, whose name was Dujon, exhibited in London twenty-four canary-birds, many of which, he said, were from eighteen to twenty-five years of age. Some of these balanced themselves, head downwards, on their shoulders, having their legs and tail in the air. One of them, taking a slender stick in its claws, passed its head between its legs, and suffered itself to be turned round, as if in the act of being roasted. Another balanced itself, and was swung backward and forward on a kind of slack-rope. A third was dressed in military uniform, having a cap on its head, wearing a sword and pouch, and carrying a firelock in one claw; after some time sitting upright, this bird, at the word of command, freed itself from its dress, and flew away to the cage. A fourth suffered itself to be shot at, and falling down, as if dead, was put into a little wheelbarrow, and wheeled away by one of its comrades, and several of the birds were at the same time placed on a little firework, and continued there quietly, and without alarm, until it was discharged." Such facts prove the wonderful intelligence of the birds, and the exceeding patience of those who instruct them.

If the young beauty derives amusement from this charming bird, and observes in its little household management the example of those delicate attentions which a growing family requires; if it charms the ennui of the cloister, and cheers the labours of the sedentary artist, it is not less agreeable to the aged, who find in its society a consolation in their sufferings; while its amiable liveliness and docility awaken in their hearts that gaiety which had been banished thence by the weight of years.

This little musician has at times its fits of sullenness and anger; but these are neither hurtful nor offensive. Still it ought to be managed with caution; for caresses, too much repeated, will on some occasions raise its anger to that pitch that it sometimes becomes the victim of it. Its vocal powers are so flexible, that it will learn to speak, and to whistle the most melodious airs. Words and phrases of the most tender kind are those which it learns to repeat and pronounce with the greatest facility. Of all birds it contributes most to the pleasures of society, and takes the greatest share in them. The nightingale astonishes us by the resources of its incomparable organ, interests by the variety of its tones, and even ravishes the ear by its brilliant and rapid execution; but, proud of its talent, it seems to disdain everything foreign, or, at all events, it is with the utmost difficulty that it is brought to learn any lessons. Moreover, the charm of its voice continues only during a few months; and to enjoy it in all its brilliancy and perfection, it must be heard in the woods, in the silence of night. Once become our prisoner, and shut up in our apartments, its song loses its melody, because it is too powerful for the sphere which it fills, and its accents acquire a harshness which is fatiguing. The linnet, the goldfinch, the bullfinch, willingly receive instruction; but the canary has a better ear, more facility of imitation, and a better memory. Its natural disposition is also more docile and amiable. Its song, which is a model of grace, is heard at all times, and charms us when all else is silent in nature. It is of all birds that which is educated with most pleasure, for its education is the most easy and the most successful. The canary acquires, by domestication, colours more pure and brilliant than it possesses in a state of nature, and its song is also infinitely superior; it has been embellished and made more perfect by borrowing foreign strains. Some have acquired the notes of the titlark, and others those of the nightingale, and all possess the pure, soft, melodious *timbre*, which the wild canary never possesses. Mr. Barrington saw two of these birds which came from the Canary Islands, neither of which had any song, and the same was the case with a great number of them brought over afterwards. Most of the birds imported from the Tyrol have been educated under parents whose progenitors were taught by the nightingale. Our English ones, however, partake more of the notes of the titlark. M. Vieillot had several of the native canaries alive, and he assures us of the inferiority of their song. Another fact, equally singular, is, that he could never bring them to couple, either together or with the domestic canaries. A description of these native canaries may not be improper here, as there is nothing of the kind in the text; and it will be useful in pointing out the differences occasioned by captivity and domestication.

Their size is the same as that of the domestic bird, but their form is thicker and more compact. The head is thicker; the feathers which cover it, as well as those of the top of the neck and back, are grey on the edges and brown in the middle: the crupper, sides of the head, forehead, throat, front of the neck, and chest, are of a yellow green, varied on the flanks with brown marks; a whitish tint prevails on the belly in its lower part, as well as on the smaller wing-coverts, and on the under tail-coverts; but the upper resemble the crupper. An embrowned shade prevails on the greater wing-coverts and quills, and on those of the tail. Their external edge is yellowish green. The bill is horn-coloured, and terminated with blackish; the feet are brown. The tints of the female are less lively.

Such is the canary-bird in its natural state, the type of those numerous varieties attributable to the varied influences of domestication.

It would be endless to enumerate all the changes which such influences produce. No less than twenty-eight perfect varieties are enumerated by connoisseurs, independently of the wild canary just described, and of that which is fully and entirely of a jonquil yellow, which is generally the most esteemed. To enter into descriptions of all those, would neither suit our limits nor our plan. But as the canary is a bird in such general estimation, we shall hazard a few more observations concerning it.

Great variations of disposition and temperament are observable among canary-birds. Some males are of a melancholic and sombre temperament, apparently bloated, singing rarely, and when they do sing, in a lugubrious tone. They take an infinite time in learning whatever is attempted to be taught them, never know it but imperfectly, and generally forget it after the first moulting, or the first malady with which they are seized. They are so much chagrined at being covered up from the light, during the period of instruction, that they often perish. To draw them out of this apathy, old canaries are brought as their instructors; and these being ardent and full of vivacity, rouse them up a little, and they begin to sing and show some symptoms of animation. These same individuals are naturally uncleanly. Their feet and tail are almost always dirty, their plumage neglected, and never smooth. Such males cannot possibly be agreeable to the females. Of a melancholic character, they cannot enliven them with their song, and seldom attempt it, even when the young ones are about to come forth. These last are seldom better than their sires. Besides, the least accident which happens in their little household, makes them taciturn, and saddens them to such a degree, that they are ready to die. Such birds should be rejected by those who are desirous of raising and educating a brood of canaries.

Others are of so mischievous a disposition that they will kill the female which is presented to them. But these *mauvais sujets* sometimes possess qualities which atone a little for their wickedness, such as a melodious song, fine plumage, and great familiarity. It may be remarked, that the more mild, free, and playful the canaries, either male or female, are with their master, the worse they discharge their domestic duties; something like the *bon-vivants* and *bel-esprits* of either sex in our own species. Such birds as these should be preserved, but not paired. There is, however, a method of taming the perverse character of a male of this kind. For this purpose, two females are selected, a year older than himself. These two females should be encaged together for some months, so as to become very familiar, and not to fight when they find themselves placed with a single male. About a month before the hatching time, they are let loose together in the same room; and when the coupling time arrives, the male is put in along with them. He will not fail to attack and begin to fight with them. But they will unite for their mutual defence, and end by subduing him, and compelling him to love. These forced alliances are sometimes known to succeed better than the others.

Other males again, are of so barbarous a disposition, that they will destroy the eggs, and often eat them in proportion as the female lays them; or, if these unnatural fathers do suffer the females to sit, scarcely do the little ones appear before they seize them with their bill, and drag them about the aviary until they are all dead. To remedy the former evil, the first egg which the female lays must be taken away, and one of ivory substituted in its place, and the same done for the second the following day, the moment it is laid, so that the male cannot break it; and this plan must be persevered in to the last. Then the female, having no further need of the male, he may be shut up in a separate cage, and placed in or near the aviary during the whole time that she is hatching. The eggs, according as they are taken away, should be placed in a little deal box, filled with vitreous sand, so as to preserve them freshly, and prevent them from being broken. As for the male, who does not touch the eggs, but kills the young, he also should be placed in a separate cage, and in a similar position, the evening before the young are about to be disclosed. There is no fear that the privation of his female will cause him any *ennui* or disgust, or that she will abandon her brood. If she be of a good race, she will educate them very well without his assistance. But as soon as the little ones are taken away for the purpose of feeding them by hand, the prisoner may be let loose, and restored to his female. The same process should be adopted at every hatching. One would think that canaries of such a disposition ought to be rejected altogether; but such means will succeed when persons are absolutely desirous to have a brood from them.

We may further remark, that among the canaries, some

individuals always remain wild, of a rude and fierce nature, and a tameless, independent character. They will not suffer themselves to be touched, nor caressed, nor governed, nor treated like the others. Such canaries would certainly thrive if they were at full liberty. A narrow prison, such as a cage, will not suit them. They should be kept in a large cabinet or an aviary, in the open air. However, if it be absolutely necessary to confine them in close quarters, they should not be meddled with, after they are fixed in one place, but simply furnished with the necessary food, and allowed to live according to their fancy.

Some males are of a weak habit, indifferent to the females, and always sick after the nestling. They should never be paired; for it is to be observed, that the little ones always resemble them. There are others which beat the female to make her come out of the nest, and hinder her from hatching. Those are the most robust, the best for song, and often the finest in their plumage, and the most docile and familiar. They should either have two females, or be treated like those which break the eggs or kill the young.

In fine, there are some canaries, which are always gay, always singing, of a mild character, and happy disposition, so familiar that they will take from the hand, and even from the mouth, all that is presented to them; good husbands, good fathers, susceptible, in short, of every good impression, and endowed with the best inclinations, they amuse the female incessantly with their song, and take so much care of her, that they feed her every moment, soothe her in the painful assiduity of hatching, seem to invite her to change her situation, often hatch themselves for some hours in the day, and feed the little ones the moment they are disclosed. Besides these excellent domestic qualities, they are susceptible of a more perfect education; they easily learn airs from the bird-organ or flageolet, and execute them in a more elevated key than the others. It is from such that the species must be judged, for they are by far the most common; and even the bad disposition of those which break the eggs or kill the young, is often only apparent: it proceeds from their amorous temperament—from the desire of enjoying the company of the female more fully and exclusively. The best mode of managing these, is to place them in an apartment exposed to the sun, and facing the east, in winter, and to have more females than males. Thus, while one hatches, they seek another. When the females are less numerous, the males will fight desperately; and it is said that when they see one male tormenting a female, the others will beat him for the purpose of taming down his ardency.

The same differences of character and temperament may be remarked among the females. The females of a particular race, called *agates*, from their colour, are the most feeble (as are also the males), and often die upon their eggs. They are full of whims, and often quit the little ones to give themselves to the male. Some females, such as the grey, are so idle, that their nests must be made for them; but they are generally good nurses.

These birds evince very strong natural sympathies, and sometimes antipathies, which nothing can subdue. The sympathy of a male may be known by putting him alone in an aviary, where there are many females, even of a colour dissimilar to his own. In a few hours he will make choice of one, and will not cease, for an instant, to shew his attachment by feeding her, while for the others he evinces the greatest possible indifference. He will even choose a female without seeing her. It is sufficient that he should hear her cry, and he will not cease to call her, although others may be with him in the same cage. This penchant sometimes becomes fatal to the male, who has been known to die of grief, if the female belonged to another master than his, and could not be procured for him. The same observation is also applicable to the females.

The males, however, exhibit more marks of natural antipathy than their companions, and cannot couple, indifferently, with all sorts of females. If the female which is presented to a male, does not suit his taste, all possible care and attention will be of no avail to reconcile him to her. They will quarrel every instant, and fight incessantly. Their antipathy continually grows stronger and stronger; so that this ill matched pair, if left together, will soon cease to eat, grow emaciated, and die. To assure one's self of this natural aversion, it will be only sufficient to separate them, to allow them to repose a few days, and then to let them both loose in a large aviary, where there are several males and females. They will soon be seen in a few days, each attaching itself to another, and pairing with as much promptitude with it, as if they had been always together. The mutual antipathy between the two former is still observed to exist, for if any dispute arises in the aviary, this couple is sure to be in the midst of it, fighting on opposite sides. This antipathy is very remarkable between canaries of different colours. If one of a brilliant colour has lost his female, he will display a most invincible aversion for a female of another and more sombre hue—such as the grey.

There are some canaries, but few in number, which do not sympathize with birds of their own species: their antipathy is so great, that they cannot be brought to pair with any, and will die sooner than couple. These individuals always remain sterile and inactive. There are more males than females thus constituted. They are usually, too, the best singers, and the most long-lived birds. Forced alliances, especially in such cases, ought always to be avoided; the only result will be fruitless eggs, and, not unfrequently, the destruction of the individuals so paired.

There are some, in fine, especially males, which have so great an aversion for their own species, that they cannot endure them even in the neighbourhood. If they but hear each other sing, they instantly give symptoms of the most violent rage, and endeavour, by all means, to escape from their respective cages, to tear one another to pieces. If they are not put at a sufficient distance to prevent hearing, they will fall sick, and irretrievably perish. This malady is the more difficult to cure, as its cause is frequently unknown. It is manifested by one canary's replying to another in the neighbourhood by clapping his wings, and shewing every symptom of fury.

The ardour of the male is shewn, as in all other birds, by the extension of the voice. In the female this is not the case. But love appears to be of a more permanent necessity with her, for she will fall sick and die, if deprived of her companion.

Birds, which occasionally evince such a decided antipathy among themselves, could hardly have been expected to sympathize with species totally different, such as the goldfinch, linnet, or siskin—still less with the bunting and the chaffinch, which feed in a different manner, and can neither administer food to the female canary, nor assist her in the education of the young. Yet so it is, that all these birds, however different, and apparently remote from the canaries, will produce with them, if proper pains be taken. It is always in the males that the most marked antipathy is observable; accordingly, the experiment is in general most likely to be successful with a male of a strange species and a female canary. Notwithstanding this, if it be possible to couple the male canary with a female goldfinch, linnet, or any other finch, the males will be handsomer, and will sing better, because the male is more predominant in the race than the female. When such alliances are projected, the canaries should be separated altogether from any of their own species. If a male be tried, one of two years old should be taken, which has never coupled with females of its own race. The same should be done with the female, according to some writers, but others say it is not absolutely necessary. It is probable that they will not couple for the first year, but that is no reason for despair. Difference of plumage, of cries, of song, certain disparities in habits and manners, are obstacles which nothing but the greatest ardour can overcome. They are also most difficult to be vanquished on the part of the male canary, on which account it is better to make use of females in such attempts; they will also, assuredly, produce with the males of all the birds above named; but the male canary is not equally certain with the females of the aforesaid species. The female canaries seldom produce with strange species, but from the age of one up to four years; but with their own males, they produce up to eight or nine years old—the streaked race excepted.

The siskin, the goldfinch, and the linnet, are those respecting which the production of the female with the male canary is best authenticated. If mules are desired from these birds, they must be taken on the nest, brought up by hand with the canaries, fed on the same aliment, and kept in the same aviary. The goldfinch, for example, which is generally chosen in preference, should be kept from hempseed, and accustomed, as soon as he is able to eat alone, to millet and rapeseed, the ordinary food of the canaries. Without this, a risk is run of losing one or the other, in changing their diet. If hempseed be suddenly taken from a goldfinch accustomed to it, to give him the ordinary food of canaries, the change will make him ill, and may cause his death. If, on the contrary, you leave him the hempseed, the female canary will eat so much of it, that she will get a fever, and probably die. What is said of the goldfinch is applicable to all other birds destined for the same purpose. It is also recommended, in the case of the goldfinch, to cut the extremity of his bill dexterously, for about the thickness of a halfpenny, or not quite so much. If some drops of blood should follow, there is no occasion for apprehension. It may be staunched with a little saliva, mixed with pulverized sugar. This operation, however, should only be performed on those goldfinches whose bill is very pointed, which often happens in captivity. This is absolutely necessary, because this bird, pursuiug the female, may wound her with his sharp bill, and prick the little ones in disgorging to them their food, which will destroy them. This inconvenience never takes place with goldfinches at liberty, for their bills are never so pointed, as the bills of the caged birds. If a female goldfinch is paired with a male canary, she should be two years old, for it is seldom that she lays in the first year. These birds, naturally wild, should be rendered as tame and familiar as the canaries, which may be accomplished by putting them in a low place, where there is plenty of company. It must not be imagined that all the mules which result from this alliance will be handsome. Of some, the plumage is of a very common kind, and the song very inferior. It would be useless to give any description of them, for they vary *ad infinitum*, and no description would suit any but the individual described. It is sufficient to say, that it is constantly observed that the mules resulting from these mixtures resemble the father in the head, tail, and limbs, and the mother in the rest of the body; and that the mules which come from the male linnet and female canary, have neither the white colour of the mother, nor the red of the father, as some have pretended.

The union of canaries with siskins, whether males or females, requires less attention. It is enough to let loose one or many of these birds, but always of the same sex, in a chamber, or large aviary, with canaries, and they will soon be seen to couple. We have said, of the same sex, because when the sexes are different the birds will naturally prefer their own species. The goldfinch, on the contrary, will only pair with the canary in a cage; to the linnet, greenfinch, and bullfinch, the cage and the aviary are indifferent. The commonest mules are produced from the linnet, the greenfinch, and the siskin, and the most esteemed of these, for song and beauty, are those from the male canary and a strange female. The mules from the greenfinch are in general of a bluish colour, and the males sing very badly, especially if the father be a greenfinch. The male mules from a linnet sing much better, but their plumage is very ordinary. Those of the siskin are small, and sing badly. Those from the bullfinch are susceptible of a perfect education, and their plumage is singular; but this alliance rarely thrives. The male feeds, it is true, like the canary, and pays much attention to the female. But she dislikes and flies from him. His cry, and the opening of his wide bill, frightens her. It is necessary to choose a vigorous female or male, which has been brought up with bullfinches, and has never coupled with a bird of its own species.

To have fine mules and good singers, they should be of the race of the goldfinch. This bird should be chosen robust, gay, ardent in singing, and of a fine plumage. A goldfinch even taken in the net will couple, but he must at least have passed a month with the canaries, and be accustomed to their food from the moment he is taken; for had he been previously fed on hempseed, and suddenly deprived of it, he would assuredly perish. After coupling, and when the young are produced, the goldfinch, whether cock or hen, should receive thistle-seed from time to time, for these birds are extremely fond of this seed, and it may be considered as their primitive and essential aliment in a wild state. Groundsel is also suitable to them, and may be substituted for the thistle-seed when the latter is not mature. If a linnet be chosen, it should be a male, for the experiment with a female is rarely successful.

Chaffinches and buntings are extremely difficult to make unite with canaries. M. Vieillot says, that he knew but of one example of a female of these species having produced fruitful eggs with a male canary. From these facts it would appear that the siskin, male or female, will produce equally with the cock or hen canary; that the hen canary produces easily with the goldfinch, less easily with the linnet; that it can produce, but not easily, with the male chaffinch, bunting, greenfinch, and sparrow, and, very rarely, with the male bullfinch; but the male canary will not produce easily, except with the female siskin, hardly with the goldfinch, and not at all with the others. It appears also, from observation, that of all birds coupled with the canary, the serinfinch, or green canary, as it is sometimes called, has the strongest voice, and is most vigorous and ardent for propagation. It would also appear that it is the only one whose mules are fertile, which argues a close affinity, if not identity, of species. The siskin and the goldfinch are neither so vigorous nor so vigilant.

The mules sing longer than the canaries, are of a more robust temperament, and their voice is stronger and more sonorous. But they learn foreign strains with greater difficulty, and always whistle them imperfectly. All the young mules should be placed under old canaries, of a fine voice, and fond of singing, to instruct them, and serve as music masters. The same thing should be done with the young canaries. There must be always, either in the aviary, or near it, three or four old canaries that are good singers. According to the observation of Père Bougot, there are more males, among the mules than females, for in eighteen young mules from a female canary and a goldfinch, there were sixteen males; but this fact requires further authentication, from repeated observations, before we can venture to generalize it. These mules, as we have said, are stronger, and have a more piercing voice and greater breadth than the pure species. They also live longer, if not employed in attempts to propagate.

It is pretended, that those bastard birds which come from the mixture of canaries with siskins, goldfinches, &c. are not sterile mules, but fertile mongrels, which can unite and produce not only with their paternal and maternal races, but also with each other, and give birth to fruitful individuals, the varieties of which may also mingle and be perpetuated. Sprenger assures us, after many observations, of the truth of this assertion. It is also the opinion of Hervieux, who has seen the father, the mother, and the young of this second race; and he tells us that "nature never produced any thing so fine in this kind." It would appear that this production, if real, depends on many circumstances, which it is impossible to ascertain, and still more so to point out precisely. M. Vieillot tried experiments in this way, and used every possible means, for more than twenty years, without success. He also consulted in Paris a great number of amateurs, and of bird-dealers, who might be relied on, who sell every year a great number of mules from the goldfinch and hen canary, either born in Paris, or brought from Amiens, where the handsomest are bred; and all certified that these mules were unfruitful, and that they never knew an example of the contrary, in spite of the reiterated attempts which they had made every year, but to no purpose, to produce one. The male mule will, it is true, couple with the hen canary, and *vice versa*, and also bestow all the necessary attentions; but nothing but barren eggs is the consequence. The result is similar from the junction of the mules themselves, and it is the same with those which proceed from the linnet, the siskin, the greenfinch, and the bullfinch, and the same remark may be applied to birds of every other order, genus, and species. It is the same with the mules of the white or collared turtle dove and the common species, with those of the cock pheasant and the common hen, the duck of India and our domestic breed.

For the breeding cage given to the canary, the most convenient is that which is long, wide in proportion, and of a good elevation; so that the bird may have room to fly and move about, by which means it becomes stronger and more robust. It should not have drawers at the two sides, like other bird-cages, for it is necessary that the little prisoner should always be plainly seen. The drawers should be placed below on the ground of the cage, and made to draw out together. They should, moreover, be grilled over inside the cage at certain intervals, so that the bird, being only able to pass its head, cannot overturn its food and drink. A cage thus constructed presents many advantages:—1. The bird cannot conceal itself from view by any movement. 2. It has not its food so constantly under view when it is perched upon the sticks: it eats less frequently, and grows less fat, which prevents diseases that result from over eating. 3. It is a great advantage to the birds, when indisposed, or any thing is the matter with their feet, to find their food without being obliged to mount on the perches, where frequently they cannot sustain themselves.

The best breeding cages are those which are constructed of oak or hazel wood, and the bottoms and drawers of which are all of a piece. Those of deal are cheaper, but have one great inconvenience: after a year's wear they are apt to warp, and afford a retreat for mites and bugs. The four faces should be of iron wire, with two doors at the two sides, as large as the middle one. Such a cage is preferable, because the birds may always be seen in it, in whatever position they may be. The two doors serve to facilitate the passage of the canaries from one cage to another without touching or frightening them, if the change be necessary for any purpose. With a number of cages of this construction, a large aviary may be formed by joining them together, and opening all the doors of communication. The birds also, by being thus discovered, become more familiar, and are secured from many little accidents which might occur to them in more obscure places.

The time for the first nestling should never be hurried. It is usual to allow these birds to couple about the 20th or 25th of March, and even sooner, but it is better to wait until the middle of April; for when they are put together in a cold place, they often grow disgusted with one another; and if by chance the females have eggs they abandon them, unless the season grows warmer: then an entire breed is lost by hurry and impatience.

In pairing them, you put at first a male and female in a small cage, which suits them better than a large one, as they will sooner form in it an intimate acquaintance. They should be left there eight or ten days, and then you will know whether they agree together. After fighting, which usually takes place between them during the first day or so of their domiciliation, they will be seen to form a friendship, by picking gently with their bills. You then transfer them to the breeding cage which is destined for them, and which is provided with every thing needful for their little household. Although these birds will hatch in any situation in which their domicile is placed, its best position is facing the east. The father and mother are more lively and in better health; and the little ones profit more in one day in this aspect, than in two in any other. A southern or western aspect heats their head, engenders a quantity of mites, and makes the females sweat, and suffocate their progeny. That of the north is prejudicial, for even in summer the wind which blows from this point causes death to the new-born young, and even to the old. An obscure place renders them melancholy, and gives rise to abscesses which destroy them.

In fine, they must be approximated as much as possible to a state of nature. In their native country the canaries remain on the borders of little streams, or humid ravines. They must not therefore be allowed to want water, both for drinking and bathing. As they are aborigines of a very mild climate, they should be sheltered from the rigour of winter: still, having been long naturalized here and on the continent, they can bear cold, and it will be sufficient to lodge them in a chamber without a fire; and it is not absolutely necessary that the windows should be glazed. They will thrive better so, than in a chamber with a fire.

The little ones produced from canaries of a uniform colour, resemble the parents. We must not expect from a male and female of a grey colour, young ones of any other. It is the same with all the other varieties, with the *is-abella*, the *white*, the *yellow*, the *agate*, &c., all produce their similars in colour. But when these different races are mixed, some very beautiful and rare birds are the result. The most complicated and elegant individual varieties may be produced *ad infinitum*.

The male is sometimes paired with two females; for this he must be strong, vigorous, and very lively. These qualities may be recognised by his incessant movements in his cage, never resting for a moment, and singing in a very elevated key for a long time together, and frequently. The choice being made, there are two small cages, in each of which is one female: they are placed so as to communicate together by one door, and the male is let in. A single cage may also be employed, but it should be large, and have a separation in the middle, that the two females may not see each other when hatching. Such unions take place naturally in a large aviary or cabinet. Four vigorous males will suffice for a dozen females.

The canaries usually receive, to make their nests, the hair

of deer, which has not been employed for any other purpose, moss, divided cotton, flax, dog's-grass, and a little dry and fine hay. Of these materials, there are scarcely more than two of which they can make use with advantage—the fine hay, to make the body of the nest, and a little moss, dried in the sun. When the nest is made, you may give them a pinch of deer's hair, but only during the first brood, because the heat is then not considerable; but this material must be avoided afterwards. This hair is too heating, and will make the female sweat, and suffocate the young, when they are just born. If the cotton and flax be not cut very finely, they are apt to embarrass the feet of the hatching bird; and if she rises with vivacity, she may carry off the eggs and nest along with her. Some use a sort of dog's-grass, which is a very proper material for the construction of the nest. The finest must be chosen, and well cleared of dust; it is best to wash it, and dry it in the sun, then cut and scatter it in the cage. The dog's-grass alone will be sufficient, and will give the nest a form and solidity not to be expected from other materials. Besides, the same will serve for several times: it only wants to be washed in boiling water, every time it is to be used.

Three kinds of little balls or baskets are given to the canaries as receptacles for their nests—of osier, of wood, and of clay. The first should be preferred, but it should not be too large; the two others heat the female too much. The nest, too, made in the wood, holds badly, and the male or female often drag it away with their claws, break the eggs, or overturn the young. But one pannier at a time is given to them, for if they receive two, they sometimes carry the materials into one, sometimes into another, and amuse themselves for some time before they seriously begin to make the nest. About ten days after the birth of the young, they may receive the second, which should be placed on the opposite side, because the females then begin to occupy themselves about a new brood, though they are still feeding the young of the former. Some varieties of the canary are so idle, that it is better to make the nest for them; if they do not find it to their taste, they have only the trouble of setting it to rights.

Too much precaution cannot be used in the choice of food for these little birds; to give them too much or too little, is equally injurious. One should also know the proper food to give them, and the proper time for feeding them, with a particular food; what is proper in one season, is often poison in another. When they can eat alone, their usual food is rapeseed, millet, grass-seed, and hempseed; they are mixed in the following proportions: half a pint of hempseed, as much of grass-seed, a pint of millet, and six pints of rapeseed, well fanned; the whole mixed together. This mixture should be kept in an oak-box, well closed, to prevent any dirt from falling into it; a sufficient quantity to last two days is put into their drawer. Some persons give them nothing but rapeseed, but this is not sufficiently nourishing; it makes them melancholy and thin, especially the young of the last brood, which are not so robust as the others. Some give them hempseed in abundance, particularly when they couple with goldfinches, siskins, &c.: but this aliment is actual poison for them when it is not mixed with others, and that in small quantity. Moreover, all these seeds should not be too fresh, otherwise they may produce maladies.

When the canaries are coupled, they give them, beside these grains, a little simnel, or some hard biscuit, particularly when it is observed that the female is ready to lay. They give them besides, for the first eight days, a considerable quantity of lettuce-seed; this purges them.

The most difficult time to manage the canaries is when they have young ones. The evening before they should appear, which will be on the thirteenth day of the female's hatching, the fine and sifted sand, which ought to have been put at the bottom of their cage on their first entrance, must be changed. This sand is useful, because if the female lays, as she will do sometimes, at the bottom of the cage, the egg is not injured; and also, as she often happens, in rising too quickly from the nest, to carry off the new-born young with her, they, falling on the fine sand, are not wounded. Having changed the sand, the perches must be cleaned, the drawer filled with fresh grain, the old being removed, and fresh water put into their trough, which must be well cleaned first. All this should be done at this time, that the birds may not be tormented during the first days after the birth of the young. They should also have a little biscuit, or something of the sort, pretty hard, to prevent them from eating too much. While this aliment lasts, nothing else is given to them. It may be succeeded by a kind of paste, composed of hard egg, white and yolk, hashed very small, with a morsel of simnel; the whole pressed with the hand, and put into one small saucer, and in another some rapeseed, steeped in water, or rather on which boiling water has been poured, to remove its roughness. Sugared biscuit, say the connoisseurs, ought to be rejected, as too heating, and either making the eggs unfruitful or the young feeble and delicate.

Many other precepts regarding food and treatment, are laid down by authors, but we omit their notice, in the fear of becoming tedious. It may be observed, that many of them should not be literally followed, being more prejudicial than useful to the health of our little prisoners. Too much care and attention are just as bad as negligence. A diet, properly regulated, of rapeseed and millet; water, once or twice a-day, in summer, and from one day to another, in winter; some green plant, from time to time; pounded oats: and, above all, careful cleanliness, will be found to suit them best.

There are some females which never lay ; others which have but one brood, and which having laid their first egg, rest, and do not lay the second for two or three days after. There are others which have but three, of three eggs each, laid in uninterrupted succession, i. e. without the interval of a day. A fourth kind, which is the most numerous, has four broods, of from four to five eggs each, but not invariably so ; others, the most fruitful of all, have five and even more, if allowed ; and not unfrequently of six or seven eggs each. When this last kind of canary hatches well, it is a perfect race.

It is proper to separate the bad eggs from the good ones, but it is impossible to judge with certainty on this point, until the female has sat for eight or nine days. Then each egg must be taken gently by the two ends and examined by day or candle light. If they appear muddy and heavy it is a sign they are good, and that the little ones are forming ; if on the other hand, they are as clear as on the day when the female began to hatch, it is an indication that they are bad. They should then be thrown away, as they can do nothing but uselessly fatigue the bird. By thus rejecting the bad eggs, two broods are easily formed out of three, when you have many canaries hatching at the same time. The female which shall find herself free, will set to work at a second nestling. In distributing these eggs from one female to another, care must be taken to observe that they are all good, for some females if they receive clear eggs will not fail to fling them out of the nest themselves, instead of sitting upon them. A greater inconvenience results, if the nest be too deep to allow her to do this. She will not cease to peck at them until they are broken, which hurts the other eggs, infects the nest, and destroys the entire brood. This is the case with the females with that arrangement of feathers which the French call *panachés*. As for the others, they will sit on clear as well as on full eggs.

The most robust female should always be preferred to cover the extraneous eggs. Some can cover five or six. Breeders recommend the substitution of ivory eggs, in the manner we have already described, in all cases, for by so doing, the young will all be disclosed together. The brood generally comes forth about the same hour, between six and seven in the morning, if the female enjoys uninterrupted health. If it be retarded but a single hour, it is a proof that she is not well, except in the case of the last egg, which is generally delayed for some hours, and sometimes for an entire day. This egg is always smaller than the rest, and we are assured that the young one from it always proves to be a male. The defenders of the above-mentioned practice, say, that if the eggs are left to the females, they will be hatched at different times, the first born will be stronger than those which come forth after, will consume all the aliment, and often crush or suffocate the last. Others find this practice to be contrary to the process of nature, say that it subtracts from the female a considerable portion of heat, and encumbers her at once with five or six young ones, which, coming all together, disturb her. They also assert, and, after all, this is a matter of experience, and not of theory, that letting the female alone is the best way to have the brood thrive. It is certain, that too much nicety and too much care are often found more prejudicial than useful. Nature should as much as possible be always followed.

The incubation lasts thirteen days. It may be retarded or advanced one day, from some casual circumstance. Heat accelerates the exclusion of the young. Cold retards it. Therefore in the month of April, the incubation lasts thirteen days and a half, or fourteen days, according to the temperature of the air. On the contrary, in the month of July and August, it sometimes happens that the young are excluded at the end of twelve days. It is said that thunder will turn the eggs, and often kill the little ones, in the seventh or eighth day of incubation; a bit of iron put into the nest is said to prevent this effect. The eggs ought never to be touched without the most urgent necessity, and young persons are too fond of doing this. It chills the eggs, and retards the birth of the young; frequently repeated, it will prevent it altogether.

It is seldom that canaries, brought up in an apartment, are sick previous to the formation of the brood. The males will be so sometimes, and that when their attention is most required. In this case, they should be taken out of the cage or cabinet, and placed apart in a small cage. As soon as the malady is discovered, the proper remedy should be applied. The sick bird should be put in the sun, and a little white wine blown over his body. This is good in all cases. If the malady continue, and the female evince chagrin for the loss of her partner, another should be substituted. Some females do not require this, and will nurse the young without the assistance of the male; but few will support the absence of the male for eight or ten days. The female, at all events, should be allowed to see him from time to time, by putting his little cage into the brood cage. The male should be suffered to repose for eight or ten days, and be fed with nothing but rapeseed, for his malady often proceeds from having eaten too much of the succulent food given abundantly to the birds at this particular time. After this, he may again be admitted to the female, and his behaviour will show whether he is cured or not; but if he be attacked again, he must be withdrawn altogether, for it is a proof that his temperament is too delicate. The female should then receive another male resembling the one she has lost; or, in default of that, one of her own race, for there is always more sympathy between those which resemble each other, than between others. To this however, the *is-abella* breed form an exception, for the males of that prefer females of another colour. The bird substituted, however, should not be one that is new to the cares of a family. If the female fall ill, the same course of treatment adopted with the male should be pursued. But if she be hatching, her eggs should be withdrawn, and given to other females, which are hatching at the same time, or nearly so, and also the little ones, if they are too young to be brought up by hand, even though the male might feed, because they are liable to perish from cold, for the want of a mother.

Accidents, such as breaking the eggs, will occur at times for want of caution. The female, instead of laying in the panier, will sometimes lay an egg in a corner of the cage; great care must be observed in transferring it to the nest.

The females are sometimes seized with an obstruction or difficulty in laying, which is attended with very grievous symptoms. They refuse to eat, and are sometimes so ill that they cannot stand on their feet. In this case a few drops of oil of sweet almonds should be introduced with the head of a large pin into the oviducts, which will facilitate the passage of the egg. If this do not succeed, the female should be made to swallow a few drops of the oil, which will have the effect of allaying her intense pain. She should then be put into a small warm cage, placed in the sun or near the fire until she has laid and resumed her former vigour. Her food then should be boiled grain, biscuit, or pinkseeds. If she still continue ill, the white wine should be used, as we have mentioned with the male, and a little of it, lukewarm and with sugar candy, should be given internally. Even when she is recovered, her eggs should not be left with her if she has laid any, for she will not return to the nest; they should be given to other females.

The female has sometimes a strange habit of plucking the feathers out of the young ones as they shoot out, about eight or ten days after their birth. In that case, the young should be taken away, if they can be fed by hand; if not, they should be placed along with the nest in a little cage in the centre of the other, between the bars of which sufficient space is left to allow the parents to feed them with facility.

The female will sometimes perspire over the young when they are but a few days old, or just born, which is very pernicious. This is observed by the feathers under her stomach and belly being wet. When this occurs, they should be given to another female. If, however, they are six days old, there is no danger.

Some females will lay three or four eggs, and then abandon them. This is ascertained by leaving the eggs two or three days in the cradle; and if the female does not return, they must be given to another. But on such occasions, as Hervieux remarks, the eggs are usually clear, which the female perceives, and therefore refuses to hatch. There are, however, some females—but this is very rare—that will never hatch in any case; such females should be allowed to lay, and their eggs transferred to others; still, they should be left for a day or two in the nest, to try them.

The claws of these birds should be carefully cut from time to time, otherwise they may chance to be broken. They should not, however, be cut more than half the length, or the bird cannot support itself on the perch.

There are some females which hatch very well, but will not feed the young. These should be immediately taken away and given to another female, whose young are of the same degree of strength. The same thing should be done when some in a brood are more advanced in age than others; for the stronger will often smother the weaker, and eat up all the food. Some varieties are more careless of the family than others, and from such the eggs should be withdrawn before the little ones are excluded, and given to more careful nurses. The grey canaries are the best for this, and their own eggs may be thrown, for the loss is not great, as they never produce but very common colours. The female mules are still better, as they are equally good nurses, and no risk is incurred in consequence of their own eggs being so very rarely productive. In the country you may even put the eggs of the canary into the nest of a wild goldfinch, removing its eggs, having first ascertained that they are in the same stage of incubation, which is easily done by breaking one of them. The goldfinch will hatch the others; and when the young are fit to be fed by hand, you may take them away. The adopted father and mother will even continue to feed them if you place the nest in a low cage in the neighbourhood; for this purpose the goldfinch is the best bird, for linnets and chaffinches will abandon their nest if the eggs be touched. The chaffinch indeed can distinguish strange eggs, and will fling them out of the nest. The greenfinch might certainly feed the young canaries, as it disgorges in the same manner as their own species; but the nature of its food is destructive to the canaries.

If a female fall sick or abandon the young a few days after they are excluded, and you have another female to give them to, you may take a nest of very young sparrows, and put a few of them in with the little canaries, to keep up their natural heat. They should be covered with a little soft lamb-skin, if the weather be cold, and fed carefully. As for the sparrows, their food should be of an ordinary kind, to prevent their growing too big.

Such are the most usual accidents with canaries in a brood-cage; but they are very rare if the birds are kept in a cabinet or large aviary.

Although moulting is one of the most dangerous maladies for the canaries, yet some males support it very well, and even sing a little every day; but the majority lose their voice, and some perish. The moulting proves mortal to most females that have attained the age of six or seven years. Males of the same age resist it better, and live three or four years longer. It is less dangerous to those birds that are kept in large aviaries with plenty of verdure, and thus approximate to a state of nature. But to those cooped up in a narrow prison, and fed with no variety of aliments, and rendered delicate by too much care, it becomes a most serious and often fatal disorder. Such are the consequences of all aberrations from nature. In the wild bird, the moulting is only a less perfect state of health; in the tame, a grievous malady, which can only be treated by palliatives, for no remedy has been found completely to remove it. It is, however, less to be feared in warm weather. The young moult six weeks after birth, generally speaking, but the weakest undergo this change first; the strongest sometimes do not moult for a month after them. These birds then grow heavy and melancholy, sleep during the day, often put their head under their wings, lose their down, but do not cast the quills of the wings and tail until the following year. They appear quite disgusted, eat but little, and do not even touch what they like best when in good health. The young of the latest brood suffer most, because they moult when the weather commences to be cold in September and October. They should be kept very warm, for a breath of air is sufficient to destroy those delicate birds which are born in our apartments. It is otherwise with those brought up in aviaries. They are more accustomed to changes of temperature, and fewer of them perish. Their temperament indeed is so robust, that they are scarcely sensible to cold, and may be seen in the depth of winter wallowing and bathing in the snow. The bath, by the way, is necessary for all the canaries, and in every season. They should always, therefore, be provided with plenty of water, which should be changed at least once a day.

As for the pastes, &c. used for canaries, we must refer the curious to such works as are expressly written on the subject. Properly speaking, these things form no part of natural history; but a few general observations respecting the diet of these birds may not be out of place.

Great judgment is required both in feeding the young birds, and in refusing them food. The least excess will destroy them, and the want of regularity render them meagre and emaciated, and unable to resist the effects of moulting. Even if they escape with life from this change, their constitutions are seriously injured, and both males and females become weak, languishing, dull, and unfruitful. Under a well regulated regimen, on the contrary, they will grow up as strong and robust when fed by hand, as under the care of their parents. To this care, however, they ought always to be left, except when destined to an artificial education. They should be fed almost every hour and a half, from half-past six in the morning till eight at night, receiving three or four bill-fulls each time, with a small and very smooth wooden skewer.

This feeding by hand is left off in about twenty-four or twenty-five days, as soon as the young birds begin to peck of themselves. They should be then kept in a cage without perches, furnished with a little hay, or dry moss, and their food, for the first month, should be composed of bruised hemp-seed, yolk of hard egg, crumb of bread, and ripe anagallis: their drink should be water, with a little liquorice in it.

Some young canaries, after feeding alone for a month, will fall into a languor, and require again to be fed by hand. This ought to be done as a sure means of getting them over the moulting, which is the cause of their languor and disgust to food.

The quality of the seeds given to canaries should be attended to. Under the name of rapeseed, many little grains are confounded, some of which are pernicious. The best rapeseed is of a violet colour, neither too young nor too old, and without any bitterness. As for millet, the whitest is the best, but it should not be given too abundantly, as it may make the birds too fat, and overheat them. Hempseed should be of a silver grey, and the smallest is the best. This also should be given sparingly to canaries, except in the depth of winter. Grass-seed, or what is called *canary*-seed, is the natural food of these birds in their own country. It should not, however, be given alone, or in large quantities, but mixed in small proportions with rapeseed. Lettuce-seed is occasionally wholesome, and causes evacuation. Plantain is also nutritive and stimulant, but should only be given from time to time. Oats are sometimes used, but it should be seldom; they are heating, and may even suffocate.

A word on the artificial education of canaries. The bird intended to be taught by the flageolet, or bird organ, should be placed apart by himself eight or ten days after he begins to eat alone and warble, which is a certain sign that it is a male. He must also be in good health. For the first eight days he should be kept in a cage, covered with muslin; he is then placed in a chamber, where he cannot possibly hear any other bird. If a flageolet be used, the tones should not be too high. Fifteen days after this, the clear muslin is changed for green or red serge, very thick, and he is left in this situation, until he learns his lesson perfectly. He should be supplied with two days' provisions at a time, and only at night, to prevent his attention from being distracted. A prelude, and a single well-selected air, are sufficient. A greater number, or too long a time, fatigue the memory. All these birds are not equally apt: some learn in two, some in not less than six months. Too many lessons do not ensure a more rapid progress, but fatigue and often disgust the pupil. Five or six a day are quite sufficient—two in the morning, two in the middle of the day, and two at night. Those in the morning and evening are best, as there is less then to distract attention. The entire air should be repeated nine or ten times in succession, with any repetition of the commencement or conclusion. Two birds should never be taught at once in the same apartment.

These birds are subject to various diseases, on which, or their remedies, it cannot be expected that we should dilate largely. The most usual causes of disease to these captive birds, are a too rich or abundant nutriment, excess in love, desires not satisfied, or the cares of their little family; prevention, in all such cases, as in every other case, is better than cure. It is after hatching that such maladies generally declare themselves, and the moulting is always apt to increase them. A bit of steel put into their water is then very useful, and it should be changed three times a week.

It may be thought, perhaps, that we have been too extended in our details respecting this bird; but, independently of its beauty, docility, and talents, and the interest taken in it by the fairer portion of our readers, there is much in its peculiarities to engage the attention of the philosopher; and the facility of observation which we possess respecting it, renders our researches the more valuable and satisfactory.

Canaries are in most general request. They are to be seen in the north and south of the old continent, and in America, from Canada to Cayenne, and most probably in every quarter inhabited by Europeans. It is said, that in some of the Antilles, they have escaped, multiplied in a state of freedom, and produced a new race.

The opposite is the figure of a finch, brought to this country in the curious collection of the Rev. Mr. Hennah, from Mexico, which does not appear to have been hitherto described. The general colour of the bird is blue slate, but on the top of the head this colour becomes nearly black; on the back are several oval patches, and on the throat and breast are waved spots of the like colour; the wing feathers, in general, are black, or dusky, with a dusky yellow margin, and the tail is nearly black.

A specimen of this species has been lately set up in the British Museum, and a drawing of it is understood to be in the collection of the Prince Musignano, under the name here adopted.

As for the other finches enumerated in the text, and additions, we could say little on them here calculated to interest the reader. Little, indeed, has been ascertained by naturalists respecting them, and to notice that little would only involve us in tedious and uninstructive repetitions. We therefore pass at once to the

Widow-birds.—Such is the appellation of a handsome family of birds, found not only in Africa, but also in Asia, as far as the Philippine islands. This name, which seems to suit them well enough, whether by reason of the black which predominates in their plumage, or their long sweeping tail, has, however, been given them through mistake. The Portuguese gave them the name of birds of *Whidha*, from a kingdom of Africa, where they are very common; and the resemblance of this word to that signifying widow in the Portuguese language, proved a source of deception to foreigners, more especially as the latter name agreed so well with many characters of the birds. The females are never adorned with the long tail, and the males have it only during six months, which are not the same for all. With the young it appears to depend on the day of their birth; with the adult, on the climate which they inhabit. The first moulting in which the

C. Hamilton Smith Esq.re del.

Rev. Mr Hennah's Collection

THE FIELD FINCH.

FRINGILLA CAMPESTRIS. Pr. Musignano?

London. Published by G.B. Whittaker. July. 1828.

males assume their bridal habits, and begin to sing, takes place in spring, and the second in autumn, or, to speak more correctly, at the epochs which correspond to those seasons in intertropical countries. After the last moulting, the males resemble the females so nearly, that they may be very easily confounded together, without that very accurate knowledge of them, which can only be obtained by frequent and almost habitual comparison. The females also undergo two moultings, but suffer no changes, except that in growing old; some are observed to assume colours almost similar to those of the male, during the season of reproduction. This observation was made by Mauduyt, on an individual of the species *Paradisea-Whidha-Bunting*, which he preserved alive for a considerable time. " In proportion," he says, " as she advanced in age, she became less like the male in her winter plumage, and more like him in her summer dress, still she was not so handsome, and wanted the long tail feathers."

It was the opinion of Montbeillard, and it has in a measure been adopted by the Baron, who has at least expressed himself as if such were the case in some species, that these long feathers were what is called a *false tail*, or a mere elongation of some of the upper coverts. This observation according to M. Vieillot holds true, only of one species, (*Longicauda*). It applies very well to some long feathers of this widow-bird, but, as it would appear, not to the others, in which the lengthened tail is not a developement of some of the upper coverts. These long feathers in the *paradisea*, *regia*, *serena*, and *panayensis*, are the four intermediate quills of the tail, which, with the other eight, for these birds have no more, constitute the full number of twelve, which the males, females, and young possess at all times. If these four quills did not form a part of the tail, it would only be composed of eight; from which it would result that the males in perfect plumage, would have a fourth less of tail feathers than in their autumnal habits. But this is inadmissible. Brisson and Latham concur as to their descriptions, in this point, with M. Vieillot. The latter gentleman assures us that his observation is founded on repeated examination of males, both dead and living, and that there can be no sort of doubt that the four large feathers are the intermediate quills of the tail, and not some of the upper coverts, and that these same quills do not differ from the other eight, when the males are clothed in the same livery as the females. A remark confirmatory of this opinion is, that if the long feathers are only two in number in one species, (*Superciliosa*), they are accompanied by ten lateral quills.

The widow-birds, according to travellers, employ nothing but cotton in the construction of their nest, and this nest has two stages or stories. The male inhabits the upper, and the female hatches in the lower story; but it does not appear, from any account, whether all the widow-birds construct nests of this description, or whether they are confined to a peculiar species; and if so, what is that peculiar species? On this point, travellers, naturalists, and even certain Dutch amateurs, who, it is said, have succeeded in getting these birds to hatch, in a state of captivity, are all silent.

Brisson, Montbeillard, and other French ornithologists, have placed the widow-birds with the sparrows and chaffinches, Latham and Gmelin with the buntings; but the conformation of the bill shows that their proper place is that assigned them by the Baron. The male of the *paradisea (Whidah bunting)* has rather an agreeable and varied song, though somewhat sharp. When in his summer plumage, he sings with more force, and if kept in a large aviary, will sing in flying These birds are found on the western coast of Africa, in Senegal, and the kingdom of Angola; their disposition is

SUPERCILIOUS WIDOW-BIRD.

FRINGILLA SUPERCILIOSA. Vieill.

London. Published by Whittaker & Co. Ave Maria Lane. Octr. 1829.

gay and familiar, and they are easily satisfied as to food. By being kept sufficiently warm, they might be brought to hatch in these countries.

Levaillant observes, that this species, and also the *serena*, in a certain season, furnish a conductor to the little troops of Senegalis and Bengalis. This conductor remains on a bush while the flocks are seeking their food on the ground, and when it flies away, they all follow; they also form flocks of their own species.

Levaillant tells us, that the female of *longicauda* enjoys a privilege which Nature has refused to the females of the other species, although, at a certain age, she has invested them with the colours of the male; this is the having of the long-tail feathers which, though always short before, after these females have lost the faculty of reproduction, becomes considerably lengthened, and assumes a vertical instead of a horizontal direction. He does not inform us, however, whether the feathers augment in number or not. Another attribute of this female, according to this writer, is to wear at all times, after the prescribed age, the uniform which is peculiar to the male during the season of reproduction only. From this it follows, that the individuals met with in this splendid livery, during the six months in which the male is in his winter costume, are certainly old females disguised in the male habit; while the males, on the contrary, are in the dress of the females. This is not the only peculiarity remarkable in this species; it also differs from all the rest of its order by being polygamous. Mr. Barrow tells us, that these handsome birds live in a sort of republic, where two males suffice for at least thirty females. This fact is confirmed by M. Levaillant, who assures us that they live in societies, and construct nests very near each other. A society is composed of about four-and-twenty females, and either from some law of nature, or some other unknown cause, the females are always nearly double the number of males.

The widow-bird, of the species *regia*, is the most remarkable for its elegance of form and charms of voice. These birds may be kept in properly constructed aviaries, where due attention is paid to warmth of situation and shelter. They require room for the full developement of their naturally graceful and supple motions, and gay and lively disposition. This widow-bird delights in bathing, and at the sight of fresh and limpid water, testified its joy by singing. It is difficult, if not impossible, to make them hatch in such climates as England and France; but by keeping them in a very warm situation, they will live a good while, with proper care. They are not unfrequently seen at Lisbon; but their native country is the African coast, generally, though it does not appear that they are found in Senegal. There are no more particulars, respecting any of the species, worth laying before our readers.

The primary character of the GROSBEAKS is that from which they are named. Their beak, in general, is extremely solid and powerful, and, except in the group of bulfinches, is conical and pointed. The tongue is also strong, and has a longitudinal furrow; the head is larger, and more fleshy than in the insectivorous birds in general; the internal toe is free, but the three exterior are connected at their base.

Notwithstanding these distinctive peculiarities, there exists considerable difficulty in separating this group of birds from many others; a difficulty, indeed, not peculiar to them, but prevalent in every branch of zoology, whenever the natural method is attempted to be made the sole basis of arrangement. The Baron, it has been observed, states, that there is a gradual passage, without assignable interval, from the linnets to the grosbeaks. The latter, moreover, differ among themselves in the relative character of the grossness of the beak, so that where the linnets shall end and the grosbeaks begin, must be, to a certain extent, arbitrary, as must be also the introduction of subgenera, or minor groups. In many instances, it is true, generic characters are sufficiently marked by the hand of Nature; but in many more, the passage from one genus to another or others, is so nicely graduated, as to be imperceptible.

The numerous species of the grosbeaks differ among themselves widely in habits and locality. Particular species are confined to particular countries; but the genus is spread over almost all moderate climates. The majority of them live in pairs only, solitary and silent; but others associate in flocks, and have a pleasing song. Some resort to the interior of woods, while others are found in the open country, in coppices, or in low and marshy situations; these construct their nests on the branches of elevated trees, or in the midst of thick bushes, while those commit their young to the shelter of some hole. In the nature of their aliment they seem more consistent. This, as is sufficiently indicated by the character of the bill, is composed principally of kernels and hard grains, from their facility in breaking which the word *coccothraustes* has been applied by Brisson, generally, to them all, though Gesner first used it, to distinguish the common species.

The *Green Grosbeak (Loxia Chloris)* is frequently confounded with the buntings, though destitute of that osseus tubercle in the bill which characterizes that genus. It is found in all Europe, and in northern Asia as far as Kamtschatka; its favourite resorts are orchards, gardens, and woods, and particularly the evergreens. Some of them are, it seems, found all the year in France; but many migrate, during winter, to the south. They are here, also, during all the year; pair in May, and build their nest of moss, lined with hair and feathers, generally in hedges. The eggs, five in number, are of a bluish white, speckled with brown; during the incubation of the female, her mate is busily employed in providing food for her, which he disgorges from the crop, in the manner of pigeons; at intervals he may be frequently seen flying in a circle, round above the nest, falling suddenly, and displaying many signs of vivacity and pleasure. In winter these birds associate with linnets, chaffinches, and others of the order.

They are not observed to feed at all on insects, but to confine themselves to berries, grain, the buds of trees, and the like. They are easily reconciled to captivity, may be made very familiar; and, as it is said, may be taught to pronounce words.

The *Ring Finch*, *Fringilla Petronia*, though very like the sparrow, is distinguishable as indicated in the text. It is found principally in Germany and in other parts of Europe, but not in this country. It is migratory, and associates in flocks during winter. It builds in the holes of trees, and is mostly attached to forests, living both on seeds and insects. This, like the last, is easily tamed, and reconciled to confinement.

The *Common, or Haw Grosbeak*, (*Loxia Coccothraustes*). This bird is about twelve inches long. It is by no means a common bird even in those parts of Europe which it never quits: with us it is migratory, arriving here in small parties in the autumn, and quitting again in April. In France it is found all the year, in the woods during summer, but near houses in winter. The male has a weak, unpleasant cry, which, when the bird is hurt or angry, is not unlike the noise made by a file. The haw-grosbeak builds its nest in a tree, ten or twelve feet from the ground, in the angle

H Kearsley del^t

Mus Lin. Soc

BLACK LINED GROSBEAK.

FRINGILLA BELLA. Vig. c Hors.

London Published by G.B.Whittaker Oct 1828

C. Hamilton Smith, Esq.^e del.

Life Jamaica.

VIOLACEOUS GROSBEAK.

LOXIA VIOLACEA.

London. Published by G.B. Whittaker, July, 1828.

formed by a large branch from the trunk. It lays four bluish-green eggs, spotted with olive brown, and irregular blackish bars. The young come from the egg covered with down, and are fed by their parents on insects. The old and young birds continue to form one family for the first season, feeding on nuts and hard seeds, which they easily break by means of their strong bills.

These birds afford little to interest in a state of captivity; and it seems necessary to keep them apart from other less powerful birds, whom they will kill, not with the bill, but by pinching out the flesh with their talons.

The foreign coccothraustes of Cuvier afford little matter of interest beyond their specific characters, at least so far as observations have been hitherto made and published upon them. The opposite figure (*F. Bella*) of one of these, remarkable for its beauty, is from the Museum of the Linnæan Society. It is the weebong of Port Jackson, and is little more than three inches long.

The bill and rump are bright red; round the eyes, and also at the edge of the forehead, is a black line. The rest of the bird is deep ashy-grey, rather lighter on the under parts, traversed every where by small transverse black bars.

A number of real or factitious species have been appropriated to this division; but there is great uncertainty about many of them; and as the primary character of the genus, the grossness of the bill, varies in degree in the several species, a perfect monagraph of this group in the present state of knowledge, is impracticable, if not altogether hopeless.

The foreign subgenus Pitylus, is distinguished, as we have seen, from the grosbeaks proper by a slight compression in the bill, with a convexity in the upper mandible, and occasionally an angle in the lower jaw.

Of the *Violaceous Grosbeak*, belonging to this subgenus, the opposite engraving is taken from a drawing from life, made in Jamaica, by Major Charles Hamilton Smith. This bird is the *Purple Grosbeak* of Dr. Latham. The general colour of of the plumage is violet-black, except the irides, a streak over the eye, chin, and vent, which are orange-red; the wings, a deeper violet. This bird inhabits the Bahama Islands, Jamaica, and the warmest parts of America, and feeds on the mucilage of the poison-wood berries.

We may now turn our attention to the BULFINCHES, (PYRRHULA).

Here again, unfortunately, we are not less involved in difficulties as to the species, different authors appropriating them to different divisions of the grosbeaks, and distinguishing the bullfinches from the rest by different characters. Thus Daudin, who makes of them the fourth section of his *Loxia*, distinguishes them by the short mandible, extremely convex, and forming almost a spherical cone. Temminck, who makes them the first division of his twenty-fourth genus, *Fringilla*, gives for their characters convex mandibles, the upper of which is bent at its point, and nostrils generally hidden by the feathers of the forehead. Vieillot treats them as a distinct genus, employing a union of characters drawn from the different parts in which generic characters are generally found; and Cuvier, as we have seen, distinguishes them only by a round and gibbous beak.

The *Bulfinch* is found in most parts of Europe, frequenting woods and gardens; builds its nest either in the fork of a tree, not very high, or in a bush, generally the white-thorn. About the end of April, or in May, they begin the affair of nidification. The nest is composed of small branches, interlaced on the outside, and the fibres of roots within. The female lays from four to six bluish white eggs, with red or brownish spots, particularly toward the gross end. Besides the buds of trees, on which these birds feed, they take also in summer grain and berries; and as has been said, sometimes insects. Some of them migrate, but others remain during the winter, and then approach nearer to human habitations. They live five or six years.

The natural note of this species is by no means interesting; but when caged, its powers of acquiring distinct tunes is very surprising. It may be brought also to articulate words, and the female is equally capable of these acquirements with the male. They shew also more attachment than other small birds in general, and can distinguish strangers from those who take care of them.

This species, which possesses many pleasing qualities in a confined state, is very destructive in a state of nature, by feeding on the buds of fruit-trees, especially pears, apples, and plums. They appear to associate in families of the parents and their young of the same season, an association not determined by the approach of winter, but which continues until the ensuing spring, when the young pair and breed. A woody country, in the vicinity of hills, is their favourite resort. They are most usually seen on the upper branches of trees; but should a hawk or any thing else alarm them, they descend rapidly into the middle of the thickest bush at hand, and remain there without uttering the slightest noise. In spring, on the contrary, when the family disperses, and the young males select their mates, they are no longer to be found on the tops of trees, but concealed in the thickest bushes, where they would escape all observation, but for the continual call they make use of to one another. The lateness in the season, compared with other birds, of the breeding time of the bulfinch is remarkable; but even this circumstance, trifling and unimportant as it may seem, of their economy is not without a substantial cause. The young are fed, in all probability, on grain, to the exclusion of insect and chrysalids, the usual food of other young birds of this order; and if they were hatched early in the season, it is obvious that grain would neither be so plentiful, or so fit as at a later period.

Bulfinches that are to be taught to whistle, should be taken with the nest when nearly feathered; they must then be kept on moss, and perfectly clean. The best food for them is a paste made of sopped bread, hempseed, and bruised rapeseed, the whole mixed up with yolk of egg; the hempseed, however, should be used sparingly, as otherwise it is pernicious, although these birds are extremely fond of it. As the young males, before their first molt, have nothing by which they may be distinguished from the females, a few feathers may be plucked from the breast, which will soon be replaced, and will sufficiently indicate the sex. The male bulfinch and the canary will breed together; but there seem to be considerable patience, care, and attention, necessary on the part of their keeper, to effect a union. Independently of the large variety of the bulfinch mentioned in the text, individuals, black, white, and spotted, are sometimes seen, particularly among those bred in confinement, under which circumstances the deviations of nature seem most prone to display themselves.

Bulfinches are taught to pipe with a bird-organ, or german flute. It is said that they are capable of improving an air they may have been taught, by adopting tunes and graces of their own; it is much less doubtful, however, that they will acquire an incorrect melody, if the teacher be inefficient or careless. Many of these educated birds are imported into this country, annually, from Germany, and afford great gratification by their pleasing imitative faculty.

We insert from the pencil of Major Smith, a figure of *Pyrrhula Sanguinirostris*, which is the *Red-billed Grosbeak* (*Loxia*) of Latham. The bill is thick, passing far back at the base, and deep blood-red. Forehead and chin, black.

C. Hamilton Smith Esq. delt.

RED-BILLED GROSBEAK.

PYRRHULA SANGUINIROSTRIS.

London, Published by Whittaker & Co. Ave Maria Lane, June 1829.

Rest of the head, and all above, rufous grey. Beneath, the bird is pale rufous, inclining to white on the breast and belly We also give a variety, whose plumage is very pale ashen above, the black extending a little above the base of the bill, over the eye, and a little irregularly behind. It barely trenches on the chin. From the projection of the breast to the abdomen is also black; bill, not quite so deep a red. This bird inhabits Africa.

The generic characters of the Cross-bills will be found in the text. Linnæus and Latham have placed them with the Grosbeaks, but the peculiarity of the bill indicates the propriety of a separation.

They are found in the Northern countries of Europe and America, and inhabit by preference the large pine forests, the fruit of which constitutes their principal food. They sometimes emigrate. These birds nestle in the most rigorous season of the year, and have a very variable plumage. There are three authenticated species, one of which inhabits North America.

The *Common Cross-bill* is extended from the North of Europe, as far as Greenland. Very numerous flocks of them appeared about sixteen or seventeen years ago in the neighbourhood of Havre de Grace. They did a vast deal of injury to the apples, which they tore in pieces to come at the pipps. The cross-bill is not a distrustful bird, and will allow itself to be approached pretty closely. It may even be taken by the hand when fatigued, and it exhibits no symptoms of impatience in captivity. It may then be fed on hempseed, but that of the pine is its aliment in a state of liberty. It makes its nest, in January, of moss and lichen, and fastens it to the branches with the resin of the pine, and covers it with this matter. It lays four or five whitish eggs, picked out, spotted, and striped towards the gross end with blood-red.

The *Loxia Pytiopsittacus*, given as a race in the "Règne Animal," seems to have some claims to distinction of species. Its bill is stronger, more curved, and less long than that of the preceding, and the point of the lower mandible does not pass beyond the edge of the upper. This species is found in North America, and particularly delights in the colder regions. It nestles on the branches of the fir-tree; and lays four or five ash-coloured eggs, marked irregularly with blood-red.

The *Leucoptera* is also a native of North America, from New York to Hudson's Bay. Nothing of its habits is known.

There are two species of the HARD-BILLS known, one inhabiting the North of Europe, Asia, and America, the other the Sandwich Islands. The first called the *Pine Grosbeak*, is a handsome bird, and sometimes migrates from the North in winter. In America it proceeds, but rarely, as far southward as New York. It is more common in Canada. As it loves cortical seeds, it delights in forests of coniferous trees. Its song in spring is agreeable, and often heard at night. The nest is placed at no great distance from the ground, where the female deposits four white eggs. The *Flamengo* is a variety of this. Of the habits of the *Psittacea* inhabiting the Sandwich Islands, we know nothing.

The COLIES live in families, and each family nestles in the same bush. They sleep suspended to the branches with the head down, and close pressed against each other. They walk like the martins, leaning on the length of the tarsus, and climb like parrots, using the bill to assist them. They live on fruits, grains, the buds of trees, and the tender sprouts of pottage plants. These birds belong to the old continent, and are found in the warm countries of Asia and Africa. Eight species are enumerated, of which nothing but external description can be adduced.

The BEEFEATER is but one species, not much larger than the crested lark, in bulk, but longer. It is found in Senegal, and lives on insects, and particularly on the worms, or larvæ, found under the skin of oxen. These birds are often seen perched on the backs of these animals, and other large quadrupeds, wounding the skin with their bills to get at the worms. From this their name is derived.

The CASSIQUES are birds which delight in woods, and do not frequent the open country. They seek their food on trees, bushes, and the ground. They walk with facility, and do not migrate like some of the following divisions. Their food is composed of worms, insects, berries, and grains, which they swallow entire. In captivity every aliment suits them, and they show much docility. They have an aptitude for articulating words, imitating the cry of animals, and learning tunes. Their nest is composed with art, and suspended to the extremities of branches. Some prefer trees on the water's edge, others deserted places covered with sallies. They lay three or four eggs, and have several broods in the year, in all seasons, like most birds which are sedentary in the torrid zone.

The *Crested Oriole*, is a native of Cayenne, Brazil, and Paraguay, where it generally receives the name of *yapu*. When it is perched, it is in the habit of crying, with its body stretched out, its head low, the wings open, and strongly agitated. Its cry is very singular, and agreeably varied. These birds nestle in common, on trees which are on the borders of woods, suspending their nests to the extremity of horizontal branches, at some considerable distance from the trunk. The male and female both employ themselves in its construction, interlacing it with strips of the bark of the *caraguata*, with little rushes, and many black filaments resembling the hairs of a horse's mane. They give it the form of a purse, or long pouch, about thirty six inches in length, and ten in width at its lower part, which is hemispherical. The entrance is towards the top, and the bottom is furnished with a thick bed of dry leaves. In this species some individuals are larger than others, which has given rise to an opinion, that there are two races differing in size.

Of the *Black* and *Yellow Oriole*, called *Yapou* in Brazil, Sonnini gives the following details:—

" It is a bird that can be educated with equal ease and pleasure. Its natural disposition, which leads it to seek the society of its consimilars, gives it an equal inclination to that of man. Its voice is strong, clear, and sonorous; and its aptitude to imitate the song of other birds, and even the cries of different animals, renders it susceptible of being taught to repeat airs and a variety of sounds. It can counterfeit the laugh of a man, the barking of a dog, &c. It is not difficult in the choice of its food, and will eat almost every thing that is presented to it. This bird exhales a kind of odour, which renders its flesh uneatable. This odour the people of Cayenne resemble to musk; but it is more like that of *castoreum*.

" In the wild state, the *yapous* remain in flocks, and when perched on trees, they appear, from the variety of their native and imitative sounds, to be mocking the passers-by. The Brazilian name is expressive of their natural cry. They have several other names, in those countries, formed by an *onomatopœia* of the same description.

" They live on insects and various kinds of grain. They suspend their nests to the extremities of the branches of the most lofty trees, almost always in open places, and near the water-side. The form of these nests is that of a narrow cucurbite, surmounted by its alembic. They are simply composed of dried plants, without horsehair or any similar substance, which some describers have taken for the small, dry filaments used by these birds. Many hundreds of these same

BLOXAM'S PLANT-CUTTER,

PHYTOTOMA BLOXAMI.

London Published by G.B. Whittaker; Nov. 1. 1827.

nests may be seen suspended to the same tree, and agitated by the wind."

The genus PHYTOTOMA is omitted by the Baron in the "*Règne Animal*," as he had not seen any of the species, and did not consider them sufficiently authenticated. It is necessary, however, for us to notice it, as there is in the British Museum an undoubted specimen of a new species of this genus, and of which the opposite is a figure.

The generic characters of phytotoma are, a strait, conical, robust, and pointed bill, the entire edges of which are finely serrated, and well adapted for cutting plants. The tongue is very short and obtuse. There are four toes on the feet, three of which are before, and a smaller one behind.

Of the species already admitted, which are called *P. rara*, and *P. tredactyla*, it is not necessary for us to enter into any description here. The bird before us has the character of bill which we have described above, except that it appears to be slightly arched. The crown and occiput are deep moronne-red; chin, breast, and almost to the vent, orange-rufous, plumage above, brownish-black, with a white spot on the carpus, and a small white streak a little lower down on the wing. The tail is square, and full at the end. The wings extend to something less than one-half its length. This bird was brought from South America, and is named after the discoverer.

The TROUPIALES and CAROUGES.—"M. Cuvier (*Règne Animal*) has divided the *troupiales* and *carouges*. But he gives to the first the principal characters of the bill which I have applied to the second, so that his *troupiales* are *my carouges*; and his carouges, *my* troupiales."

There are certain kinds of egotism and absurdity which are best exposed by allowing them to speak for themselves. When will naturalists learn, that the proper objects of writing are, to instruct, not to confuse;

to enlighten, not to obscure; to remove difficulties, not to create impediments; to describe their subjects, not to display themselves? When will they learn, that a jargon of names is not science; that ostentatious and trifling egotism is not fame?

We shall consider these two divisions together in a general way, and notice what may be remarkable in any of the species. For the preservation of the French names of the division, to which we were unavoidably compelled, we refer to the text, and to the passage just translated, for an apology.

The Troupiales are confined to the new continent. They usually live in pairs, and some exhibit the social instinct of the xanthorni, with which they often mingle in their emigrations. They do not frequent plains, but delight in coppices, woods, &c. and usually select such spots as are most thickly wooded. They seek their food on trees and on the ground, but seldom eating any grains of the corn kind. Their food is usually composed of berries and insects. They show as much art and ingenuity as the cassiques, in the construction of their nests, if not more.

The Carouges belong to the same continent. In their habits they have some analogy with the stares, with which they have been sometimes confounded. They fly, at certain seasons, in numerous bands, and withdraw during a portion of the day into reeds and rushes, where they also pass the night. They are numerous in Paraguay, and remarkably social in their manners. Even the season of nestling does not divide their union, and it is common enough to see many species of this family assemble and labour in concert, and even join with very different species. Their physiognomy is animated, and their motions lively and indicative of distrust. They fly with moderate rapidity, but for a good while together, and at a considerable height. Their song is a sort of whistling. They are very vigorous, walk rather with precipitation, and with the body nearly upright. Sometimes they are seen on the ground, and sometimes perched on trees. They do not seek concealment, and never enter the woods, or feed on fruits. Insects, grains, and small seeds, constitute the staple of their subsistence. They are easily brought up in cages. They take abundance of care to withdraw their nests from all eyes. These details, however, cannot be generalized for all the species, for among those who inhabit the Antilles, there are some that usually go in pairs, and sometimes in families; but they apply to the great majority. The carouges of North America quit it at the approach of winter, and are the first emigrating birds which return thither in spring.

We are obliged, for the sake of brevity, to pass over the Oxyrhinci and Pit-pits, to each of which our author allows but one species, and of which the details could produce nothing but tedium. We proceed to the

Stares.—More birds have been placed in this division by ornithologists than by Nature, whose several modes of arrangement are occasionally found to differ. Our observations here must be chiefly confined to the *common stare*, whose habits are best known.

The time of reproduction with the stares commences in the early days of spring. Then each pair forms its selection and isolates; but this union does not take place very peaceably. The males dispute with excessive vehemence, and the female is the lot of the conqueror. It is at this time that their song is heard, which is almost a continual chirping. They have also another cry, which is a very long and sharp whistle; this is usually a cry of uneasiness. Once paired, they seek out a convenient place for the cradle of their young. They sometimes possess themselves of the nest of another bird, sometimes build under the roofs of houses, churches, and even in the crevices of rocks. It is not certain that they construct their nests on trees. The materials which they employ are straw externally, coarse hay for the centre, and fine plants for the inside, and some feathers. In this nest, inartificially constructed, the female deposits four eggs of a greenish blue, about the bulk of those of the thrush. The male takes a share in the incubation, and the young do not leave the nest until they are well plumed. This bird is not in request for its natural song, but for its plumage, and its aptitude in learning all that it is taught. Its voice becomes clear and sonorous, and its whistle very agreeable. It pronounces words with facility, and sometimes entire phrases, and repeats airs with great perfection. To have a perfect singer, it must be taken in the nest, three or fcur days after its birth. If it be left longer, it will always remember its natural song and disagreeable cry. At this tender age, it must be kept in a small box with moss, and great cleanliness is necessary. It should eat but little and often, and the airs, &c. intended to be taught should be frequently repeated.

In our temperate climates the stares have but two broods in the year, and the second is not numerous. To procure young ones with greater facility, when the old ones have established themselves under the roofs of churches, or dove-cotes, earthen pots are attached to the walls, as for sparrows. The stares never fail to possess themselves of these, especially if they are disturbed in their usual haunts. The young may be fed with the same paste we have mentioned for the nightingales; but their food should be varied, for they will accommodate themselves to most food. In the wild state they live on slugs, small worms, scarabæi, various grains, berries, olives, cherries, and grapes. It is pretended that the last correct the bitterness of their flesh. It is generally dry, hard, and ill-flavoured. These bad qualities may be removed, according to some, by plucking out the tongue the instant the birds are killed, or bleeding them in the neck.

Others say, that it is only necessary to decapitate them; while some connoisseurs, remarkable for humanity, insist, that flaying them alive is the most sovereign remedy for their unsavoury peculiarities.

As the stares confined in cages are, like many other captived birds, subject to epilepsy, their flesh has been absurdly considered a specific remedy for men attacked with the same malady.

These birds do great mischief in the vine countries. On the other hand, they are very useful in corn-lands; for they destroy an immense number of those pernicious insects which would otherwise ruin the hopes of the agriculturist.

The stares generally live seven or eight years, and in a state of domestication, some have been known to arrive at twenty: they are very fond of society, and the moment their hatching is over, they assemble in numerous flocks, and do not quit each other, night or day. They retire, at sun-set, into marshes, covered with reeds, which they always chuse for their retreat. From the earliest dawn they commence chattering all together, quit their nocturnal asylum, and spread themselves throughout the country, where they often mix with the crows, jackdaws, fieldfares, song-thrushes, and even pigeons, but with these more rarely. They also like to mix with the cattle pasturing in the fields, and are often seen amidst a flock of sheep; it is not rare to see them perched upon their backs. They are attracted by the insects which are hovering about them and swarming in their dung, and by the worms which they expose in pasturing.

The stares have a mode of flying peculiar to themselves; their flight is circular and crowded. The circular flight enables the fowler to destroy many of them with fire-arms; for when one falls, the others return and circle round him. But the crowded flight is advantageous for an escape from birds of prey. When one of the latter attacks the stares, that instant they close their ranks; and whether the assailant finds himself embarrassed by their number, whether he is astounded by their cries and the noise of their wings, or whether he cannnot pierce them, or chuse his prey, he is almost invariably obliged to quit them. Noseman says that it is proved that the stares, when hard pressed by a bird or prey, shoot their dung with such force, that the assailant is obliged to quit his pursuit. This remark, however, requires further verification, though, certainly, in the matter of excrement, the authority of Noseman ought to have some weight.

Montbeillard, and several other writers, assert that the stares are not voyaging birds, but that they remain constantly during the winter in the countries where they were born. This is contradicted by other naturalists, who say that some of them change climate, and others do not. "Those," says Sonnini, "seen in Malta, are passengers, as are also those which appear in the southern islands of the Grecian Archipelago, in Candia, in Egypt, and, most likely, in Barbary, where, Porrit tells us, they are common in the autumnal season. It is certain, that in the country adjacent to Rome they disappear after the hatching time. At this epoch they are more numerous in the neighbourhood of Bordeaux, &c. in the south of France, than in the northern provinces of the same country. It is probable that they are erratic birds, drawn from place to place by the proportionate supply of food."

The stares are plentiful through the old continent. They are found in Sweden, Germany, Italy, the north of Asia, the Cape of Good Hope, according to Kolben; but Levaillant denies this. They are plentiful in England at all times; and large flocks of them in the winter, seem to countenance the idea of emigration from a colder climate.

They are hunted in various ways. The ancients, notwithstanding the badness of their flesh, hold them in some request, as an article of food. In Holland they are considered as a delicacy, and various means used to procure them. In the vast marshes there, frequented by the stares, it is customary, when night closes in, to attach to poles and spread several nets, each furnished with a lighted lantern. The rushes and reeds are then beat up, and the birds, attacked with blows of switches, and stunned by the noise, fly towards the light, and are entangled in the nets. Many hundreds are thus caught at a time.

An ingenious mode of destroying them is, by inclosing one's self in an artificial cow, made of osier, covered with a hide, and in which the imitation of nature is so perfect, that the birds mistake it. It is placed in the centre of a herd, and there the fowler may shoot hundreds of stares at his ease; for they flock abundantly about the cattle, without any distrust, and when one falls, the others, as we have said before, continue flying in circles around him.

The *Louisiana Stare* has no habits in common with ours, except that of frequenting meadows, which, with it, is a permanent one. It passes its life constantly in such situations, where it finds both food and lodging. It never perches on trees, except when pursued, and never rests there long, or passes the night there. Its slender head, rather compressed at the sides, and its elongated neck, feet, and claws, mark it as a native of the humid meadow. It is sedentary in Pennsylvania and the neighbouring countries; many individuals, however, emigrate on the approach of frost. They are more numerous in that state in autumn; for then those that nestle in Nova Scotia, Canada, and the more northern regions, quit their native country, and sojourn in the centre of the United States.

These stares run swiftly, and have a lively flight. If

pursued, they hover, and sail off, like our grey partridge; and as soon as they reach the ground, they squat themselves down at the foot of a bush, or in a high tuft of grass, always on the opposite side to the object which affrights them; they move the tail up and down, if undisturbed, and horizontally, if surprized. They remain in families during the winter, and go in pairs in the spring. Each couple appropriates a district, where they do not suffer others to enter. The male is much attached to the female, and both to the young. Their song is not disagreeable, and both male and female utter a sort of sharp whistle when annoyed. They place their nest on the ground, in the midst of brambles, or tufted plants, and construct it with dry grass, &c. They have but one brood annually, composed of five or seven eggs; the young do not separate from the parent until spring. Their food is worms, insects, and various grains.

The *Magellanic Stare (Sturnus Militaris)*, of which we give a figure, is found in the Malouine Islands, on the coast of the strait of Magellan, and it advances into America as far as Monte Video, or beyond. It forms flocks, more or less numerous, which seek their food on the ground, descending on the corn-fields, and eating up the grain.

The two last birds, with another, called *loyca*, are formed by M. Vicillot into a separate genus, which he names *sturnella.*

A general analogy between the Crows, properly so called, the pies, the jays, the nutcracker, the temia, and the glaucopis, leads us naturally to consider these birds as constituting a group; and whether we view them as forming so many genera of that group, or treat the crows as a typical genus, and the rest as subgenera, is perhaps indifferent. The advocates for simplicity will adopt the one, while the stricter systematists will probably prefer the other. Cuvier, as we

H. Kearsley del.t
Mus. Lin. Soc.

MAGELLANIC STARE.

STURNUS MILITARIS.

London, Published, by G. B. Whittaker, March 1829.

have seen in the text, makes three genera of his corvine family, including 1. The Crow, with their several subgenera, above-named; 2. The Rollers; and 3. The Birds of Paradise.

In addition to the generic characters applied in the text to the first of these, we shall merely add here that they have three toes before and one behind.

The crows proper differ from the pies and jays in their mode of locomotion on the ground. The former advance by walking, and that rather deliberately; but the latter move by vigorous leaps. In the disposition to steal, and hide any thing they can carry off; in the habit of laying up a store of provisions; and in the imitative faculty of the voice, they all agree; but in the nature of the food of each, in their gregarious or solitary habits, and in the shelter they select to build or breed in, they differ from each other materially.

The *Raven*, which is the largest species of its order, is black, but there are purplish reflections on the upper part of the body, and greenish tints underneath. The female is distinguished by a colour less deep, by a weaker bill, and by being rather smaller than the male. The plumage of the young also, is not so decided a black, and it is without reflections. The tongue is black, cylindrical at the base, flatted and forked at the extremity; the œsophagus is dilated at the point of its junction with the stomach, and forms a sort of gizzard, of which this bird is otherwise destitute. The stomach of the raven is neither muscular, like that of the gallinaceous and other birds, nor is it like that of the birds of prey, and of quadrupeds; but, in point of solidity of its coats is intermediate between the two. Thus a small tinned tube will not be altered in shape in the stomach of this bird, though it will in that of a pigeon; but, a tube of lead, being a softer metal, will be flatted in the stomach of the raven, though an ordinary membranaceous stomach will have no such effect upon it. In accordance with this comparative weakness of stomach in these birds, we find that when they take grain, they break it, by means of the bill or feet, before they swallow it.

Carrion and putrid animal matter, which they can smell at great distance, form the basis of the food of these birds; when such, however, is not forthcoming, they live on fruits, grains, insects, dead fish, and molluscous animals, whose shells they break against stones. It is said that they will at times attack living animals, as rats, partridges, frogs, &c., and that falling on the back of the larger sort, as asses, buffaloes, &c., will seriously injure them by repeated strokes with the bill. The fetid nature of their food renders these birds unfit for the table. They were unclean to the Jews, and are generally considered in a similar light by most savage nations.

Ravens, when threatened or attacked, fear neither cats nor dogs; and render themselves formidable, not merely to children, but even to men, whose legs they will peck at, and wound with some effect. Notwithstanding their courage, however, they may be brought to associate with man, and they have been employed for purposes of falconry.

They are also greatly attached to one another, and live in general in pairs, each pair remaining connected for several years, probably for the whole period of their lives. They make their nests in the crevices of rocks, in holes towards the tops of deserted towers, and sometimes on the summit of an isolated tree. This nest, which is very large, is composed exteriorly of branches and roots; bones of quadrupeds, or fragments of hard substances, form the second coat, and the interior is lined with moss, &c. The female lays, about the month of March, five or six pale greenish and bluish eggs, lined and spotted with a neutral tint; both the male and female sit, and the incubation lasts about twenty days. There may be generally found in the vicinity of these nests a considerable accumulation of grains, nuts, fruits, and other things, though it appears these hoards are made rather by the blind instinctive impulse of the bird, than for the use of the young. These at their birth are whitish.

It is said that the mother leaves the young for some time after their birth without nourishment, and that she prepares the first food they have for them in her stomach, in the manner of pigeons. As is generally the case with animals of a sort of domestic inclination, the male defends his young family with great courage and address, and will succeed in repelling the attack of the kite.

The young are ready about the month of May to quit the nest. So long as they are but partially able to provide for themselves, the parents bring them food during the day, and every evening the family reassemble in the nest, and this practice continues the whole summer, which has led to the presumption that they breed more than once in the year. Though not quick breeders, however, these birds are very long lived. When the young have attained strength enough to provide altogether for themselves, the old birds drive them from their own adopted vicinity to seek an asylum elsewhere.

The raven is capable of a very elevated flight, and accommodate sitself with ease to various temperatures; hence almost the whole earth is open to it, and we find it accordingly from the arctic circle to its opposite. Its selecting any particular country for its residence seems to be rather the result of the quantity of food to be found there, or of accident, than of high or low temperature, or of climate. The specific characters, however, of these birds are influenced by locality, and a disposition in the plumage to become partially white has been observed in such of them as inhabit high latitudes. Hence the varied raven, or *cacolotl* of Hernandez, found in Mexico, and the white raven of the North of Europe, are now considered as mere varieties of the common species. The *corvus crucirostra* of Maugé, and the ravens with one and two horns, which were presented to the Duke of Saxony and the Duchess Christine, were probably artificial impositions; the latter done in the manner in which the spur of a cock is sometimes affixed on the head.

The *White-necked Raven of Levaillant* (the *ruperine raven* of Daudin), deserves a notice. In size it is intermediate, between the raven and the hooded crow. The bill is peculiar, being surrounded at the base with feathers, which are directed forward, as in the genus generally, but it is convex above and compressed laterally. This bird is extremely ferocious, feeds on carrion, but it is apparently equally fond of fresh animal food, for it is said that it will not only attack and kill lambs, and small antelopes, first picking out the eyes and the tongue, but it even pursues the buffalo, the rhinoceros, and elephant; and perching on the back of these animals, will succeed in getting out the larvæ of such insects as are deposited under the skin.

The *Carrion Crow*, which is spread over both the old and new continents, feeds, like the ravens, on putrid flesh; to which, however, it adds insects, worms, fine grain, fruit, and eggs. In winter it associates with the rooks and hooded crows, and may be seen with them on fresh turned earth, searching for worms, insects, and their larvæ. At the approach of night they assemble in considerable flocks, and retire to the highest trees of the forest. Early in the spring, when the rooks quit the south of Europe in flocks, to build their elevated cities in the high trees in the north, the carrion crows separate into pairs, and proceed alone in the great work of nidification. The nest, like that of the rook, is constructed at the summit of a tree of slender branches, matted with clay and horse-dung, and lined within with fibres. The female lays four or five eggs, of a paleish green, with obscure spots and bars.

The old birds have great affection for their young, which remain a long time under their fostering care, and to which the parents have the address to carry the eggs of partridges in their bill. They lay but once in the year, unless the eggs or young be destroyed by any accident. They will engage with the accipitres in defence of their young and eggs. It is said that the conjugal union of these birds continues during life.

The carrion crow will learn when domesticated to speak like the raven, and displays in its general habits no small sagacity. It will sometimes seize chickens from the poultry-yard, and will also destroy such small birds as it can find in the fowler's snare, which it escapes falling into itself, probably by the perfection of its sense of smell; birdlime, or glue; and meat, infused with nux-vomica, are used therefore to take them instead of the net.

White and varied varieties are sometimes met with, especially in high latitudes, as in the ravens; indeed, a great general analogy prevails between these species, in nearly all their characters, essential and indifferent.

A third species, the *Rook, C. Frugilegus,* has also great analogy with the last—indeed is not to be distinguished from it without considerable difficulty. The principal distinctive character consists in the nudity of the base of the bill, and of the forehead, and upper part of the throat in the rook, which parts are covered with feathers in the carrion crow; but as even this is not observable in the young birds, it seems likely to be rather an artificial than a natural distinction, arising from the rook thrusting its beak into and raking the ground, and thus causing a trituration on the parts in question, which may efface the feathers. To recognize the rook, therefore, before its first moult, we must observe that its bill is longer than the head, and is entirely straight; while in the crow it is not longer than the head, and the upper mandible is bent at the point, and jagged towards the end on both sides; the feathers, moreover, of the front part of the neck are silky, and rounded at their end in the rook; but they are stiff and pointed in the crow. The iris also is hazel in the crow, but bluish in the present species. Rooks, however, do not feed on carrion, or the flesh of large animals, but confine themselves to grain, and the larvæ of insects, especially the chafer. It is therefore not easy to determine whether they are a greater good or evil to the agriculturist.

These birds seem partial to our island, as they remain here the whole year; but in France, and most parts of Europe, especially to the south, they are birds of passage; and in Spain, as it is said, are not known. Their gregarious disposition, particularly during incubation, on the tops of lofty trees, is well known. After the breeding season they disperse, and in a great measure abandon the trees in which they bred their young. The eggs, five in number, a little smaller than those of the crow, are bluish green, with dark blotches. They begin to build early in March, and the male and female sit by turns. White and pied varieties are sometimes seen.

The *Hooded Crow*, *C. Cornix*, continues the whole year in Scotland and in parts of Ireland; in England it is seen only during winter, in general in the open country, or on the banks of rivers, in small companies of ten or twelve. They are migratory also on the continent; are said to be common in northern Asia, and to be found also in India and in Africa. They nestle in high trees or rocks, and are then separated into pairs, though at other times they seem to mix promiscuously. Their eggs are like those of the carrion crow. The female and the young are without the grey, which predominates on the body only of the male.

The *Daw*, or *Jackdaw*, *C. Monedula*, is permanent in this country, and breeds principally in old towers; the nest is made of sticks, and lined with grass and wool. The eggs are five, greenish blue, with dark brown spots. These nests are sometimes found also in hollow trees; and it has been said by Pennant and others, that the daw will, in some parts of England, build also in rabbit-holes; this, however, seems doubtful, especially as it does not appear to have been observed on the continent. In other parts of Europe the daws are generally migratory.

They feed on fruit, worms, larvæ, and insects themselves, and do not touch carrion unless impelled by hunger. Like the raven and the pie, they have a strong disposition to hide whatever they can get possession of, when tamed, to which they submit easily, and may be taught to articulate words.

This species has a considerable tendency to vary, and the white-collared daw, the white daw with a yellow cere, the black daw with a white head, and daws with white wings and shoulders, have been noticed either as distinct species or varieties.

The *Common Magpie* is the type of Cuvier's subgenus of Pica, distinguished by the convex upper mandible, and the long and cuneiform tail. This, in the magpie, has the two middle feathers of the same length, but the side feathers decrease rapidly in succession; about the neck the feathers are loose. It is one of the most beautiful species of this country, but loses much of its lustre in a state of captivity. The magpie is very common in England, and feeds both on animal and vegetable substances, frequently killing young ducks and chickens. It is said also occasionally to pick out the eyes of lambs, hares, &c., if weak; it also eats insects, fruits, and even grain.

No birds display more industry in the construction of their nests; they generally select the summit of the highest trees, especially if standing alone or in a row; but in forests, or very retired situations, they sometimes chuse a mere bush for the purpose. The male and female begin this work together in February, placing the nest, not, like the rook, in full sight of all from the ground, but so enveloped and surrounded with branches, that when the leaves appear, the nest is quite concealed. It is made of small branches, well interlaced together, leaving an aperture only in the side. The bottom of the nest is furnished with a matting of soft and flexible roots; and although the diameter of the inside of the nest does not much exceed six inches, it is upwards of two feet on the outside. It is said to occupy the birds two months to build this nest, and M. Vieillot has observed that if the nest is destroyed, or the birds are prevented from finishing it, they either content themselves with an old nest of their own species, or take to an old crow's nest, after repairing the outside. The same gentleman has also noticed that, at the early part of the breeding season, each pair of these birds begin more nests than one, though they finish that only in which the eggs are deposited.

Ordinarily they have but one brood in a year; but if their young be destroyed, they will sometimes have a second, and even a third. The eggs, seven or eight, are yellowish-white, spotted with brown and grey. The male and female sit alternately, and the incubation continues about fourteen days. The young are born blind, and remain so several days. The parents display great care of them, and continue their attentions a considerable time.

Bennet's Magpie, of which a fine specimen is in the British Museum, is so named by Mr. Gray, from the donor of it to that establishment. It is a very splendid bird, about the size of the common magpie. The general colour of the upper part is blue; but round the eyes, on the throat, and beneath it, is white. From behind the eyes descends a black, irregular stripe, which, passing downwards, unite and enlarge into a patch; the cheek and throat are whitish. The two middle tail-feathers are nearly double the length of any of the others; the lateral feathers are graduated, and tipt with white. From

H. Kearsley del.t
Brit Mus.

BERNET'S MAGPIE.

PICA BERNETTII

C. Hamilton Smith Esqr del.^t

RED-BILLED JAY.

PICA ERYTHRORYNCHOS.

London. Published by G.B. Whittaker March 1828.

behind the ear there arises a long crest of erect feathers, which gives an additional beauty to this otherwise beautiful bird.

The *Jays* differ from the pies principally in the bill, which is more hooked, and in having some long loose feathers on the crown of the head, which are erected when the birds are excited; the tail, moreover, in these birds, is longer and more graduated. They may almost be said to be omnivorous, living in general in the woods, but occasionally resorting to gardens and cultivated lands, to both of which they are injurious and destructive, as well by what they eat at the time, as by what they carry off to increase their hidden stores. In summer they live in pairs, but in the opposite season assemble in small groups. They advance on the ground always by leaps, and seldom or never walk. In disposition they are very irascible, petulant, and inquisitive, and take their scientific generic name, *garrulus*, from their constant loquacity. The nest is built in trees, generally at about half-way from the bottom, of sticks, interlaced together on the outside, cased within with mud, and lined with dry grass and fibres: the entrance to it is at the side. The eggs are white, spotted with brown and grey, and are from six to eight in number.

The common jay does not seem to be very generally or exclusively located, and is partially migratory from the west and northern parts of Europe to the south east, as the islands of the Grecian Archipelago, and also Egypt, Syria, &c. Though many are thus said to migrate, it is nevertheless clear that some continue in our own country and in France the whole year.

The *Red-billed Jay* is a very splendid bird. The bill and feet are red; the neck and breast are black; the crown of the head dotted black and white; body, above and beneath, ashen; of the tail-feathers, the two intermediate are much the longest, and the lateral feathers are graduated; they are blue, tipt with white, and a black bar between that colour and the blue. Inhabits China, and is frequently rendered very tame and amusing.

The *Nutcracker*, though a single species, has been separated by Gesner, generically, under the name *caryocatactes*, a separation which has been adopted by other ornithologists. Linnæus included it in his genus *corvus*, nor does it seem to differ from the preceding subdivisions of that genus, the jays and the pies, in any thing but the straightness of the bill.

The nutcracker, which is about the size of the jay, is found principally in Germany, especially in the thick and elevated forests of that country. Occasionally however they visit France in flocks, when they are observed to be extremely eager for food, and to fall easily into snares of all sorts. A few stragglers have been met with in this country, though rarely. They build in holes in trees, and lay five or six pale yellow eggs, marked with black spots. In habits and manners they are said to resemble the jay.

The *Corvus varians*, changeable crow of Latham, which is perhaps more peculiar for the velvet-like black feathers which surround the base of the bill and the eye, than for the thickness of the bill, seems to differ in those respects only from the great genus *corvus*, and its subdivision of pies, jays, and nutcrackers. It inhabits Java, but is not a familiar bird, and is found for the most part in lands recently cleared for cultivation. Of its nidification we know nothing.

The *Wattle-bird* constitutes a genus, Glaucopis, with but this single species, which, as to the bill and the velvety feathers which surround its base, is allied to the changeable crow. But, in addition to these characters, the tongue is singularly shaped, being indented into three or four angles, and furnished with short bristles. At the base of the under

C. Hamilton Smith Esqre del.

Drew's Coll. Plymouth.

MADAGASCAR ROLLER. Var?

CORACIAS MADAGASCARIENSIS. Lath.

London. Published by G. B. Whittaker, July 1828.

mandible, moreover, on each side, is a round, flat, blue substance, not unlike the wattle of a cock, changing by degrees from the base to a fine orange. Dr. Foster first indicated this species, on whose habits we have little information.

The ROLLERS, in certain characters and colours, have some affinity with the jays, but are clearly distinguishable from them by the attributes noticed in the text. This family is considerably extended over the ancient continent. We possess but one species in Europe, and it is very uncertain if there are any in America. American birds have indeed been indicated as Rollers; but among such are many which have been very clearly recognized as forming no part of this genus.

The rollers live on berries and insects, nestle usually in trees, and lay four or five eggs.

The *Garrulous Rollers*, in countries where birch-trees abound, prefer them to every other tree for nestling in; but it is said, that where such trees are rare, as in Malta, that they nestle on the ground. They lay about five eggs, of a clear green, covered with innumerable little spots of a sombre colour. But M. Meyer tells us, that the roller makes its nest in the hollow of a tree, and that the eggs are of a lustrous white. Such contradictions, added to others that we have not mentioned, respecting the colours of the female, sufficiently prove how little naturalists are acquainted with this *European* bird.

The rollers appear but seldom in the northern provinces of France. They are sometimes seen in the environs of Strasbourg, from which circumstance some have chosen to call it the Strasbourg Jay, when in fact it is neither a jay nor a native of Strasbourg. Others have called it the German Parrot, with equal propriety, for it has as little analogy with the parrot as with the cock-pheasant. These rollers come to Malta twice a year, in spring and autumn. They seem to migrate from Africa, but they advance pretty far north, for they are found in Sweden, Denmark, and the southern provinces of Russia. It does not appear that in their passage they ever fix or stop in the intermediate temperate climates, for they are quite unknown in many considerable districts of Germany and France. Montbeillard has traced their route from Smaland and Scania into Africa, through Saxony, Franconia, Suabia, Bavaria, the Tyrol, Italy, Sicily, and Malta.

The rollers, wilder than the jays and pies, remain in the thickest and least frequented woods. Their mode of life, however, is not unlike that of those birds. Like them, and often in their company, they may be seen in the cultivated fields in the neighbourhood of their retreat, seeking the same food. They are said sometimes to attack carrion. In general, however, they are not considered as flesh-eaters. They grow fat in autumn, and are said to be then good eating, which could not be the case if they were feeders on carrion. The voice of the roller is sonorous, and, as its Latin name indicates, it is a garrulous bird, like the jays and pies.

The Rolles differ from the Rollers in the bill, wings, and feet, which has caused our author to separate them, though in other points the analogy is close between the two subdivisions. We have no certain information respecting their mode of life. It is thought, from the size of their mouth, by some naturalists, that their principal food is berries and insects, which they swallow on the wing.

There are few or no birds respecting which such absurdities have been put forth, as the Birds of Paradise, the very names of which are traceable to the marvellous attributes with which they have been endowed. These tales have equally arisen from the wanderings of an unregulated imagination, and from facts which were not understood. A moment's reflection on the ordinary mode in which the operations of Nature are performed, would have checked the propagation of such folly. Nature, always wise and regular in her productions, could not have formed any species destitute of the means of living and perpetuating itself. Monsters are only accidental deviations from the prescribed and customary succession of phenomena.

The first birds of Paradise transported from the Australasian regions had no feet, because the natives of New Guinea and the neighbouring islands, where these birds appeared exclusively to exist, used to make certain ornaments of them, and deprive them of those limbs which could not answer that especial purpose. The quantity of superabundant feathers with which the sides of these birds are covered, must have necessarily concealed over the dried skin, the places whose parts had been mutilated. A little attention, however, might have discovered the traces of them, in raising the sub-alar feathers; but the love of the marvellous, and the fondness for conjecture prevailed, and absurd theories were formed, to explain how these birds could live and propagate in the air. As they were rarely seen at the epochs of incubation, they were dismissed to nestle in the terrestrial Paradise, from which, without doubt, their name was derived; as in the same way, from certain virtues attributed to them by the soothsayers and priests of the country, was derived their native name of *Manucode*, which signifies *Bird of God*.

Pigafetta, who embarked in the fleet of Magellan in 1825, was the first navigator who ascertained that these birds had wings and feet like all other birds. When the islanders were informed that the Europeans were so tasteless as to prefer them in a perfect state, they began to preserve them accordingly. But, as for want of other means, they have continued to dry them, either in ovens, or hot sand, it is always difficult to restore to them their original forms. This is the less surprising, as we are told by Helbigius, that after having removed the entrails, they pass a red-hot iron through the body, which must injure the form.

Levaillant has remarked on this subject, that as the savages remove all the bones of the cranium from the birds of Paradise, and dry the skin, run through a reed, these operations considerably contract the head, thus deprived of its support, and draw out the eyelids. From this, the characters of small head, and eyes in the bill, scarcely visible, have been deduced. Again, from the inevitable approximation of the feathers, on a very small extent of the hardened skin, results that appearance of natural velvet which, according to this writer, is very improperly attributed to them.

The birds of Paradise, which had been imagined to live only on dew, are rapacious, according to Bontius, and pursue small birds. Helbigius tells us that they live on various berries, and Linnæus that their food is insects, and especially the larger butterflies. But it appears that spices are their favourite aliment, and that they never remove from the countries where they grow. In the nutmeg season, the Emerald-Paradise birds are seen flying in numerous flocks, like the thrushes in vintage time in France.

Some species frequent bushes and thickets—others prefer woods and lofty trees; but they never perch on the summit, where the strong winds, ruffling their luxuriant plumage, might overturn them. To the branches of these trees the Indians attach light huts, from which they shoot these birds with blunted arrows.

As naturalists have generally been so perplexed and discordant concerning the classification of even some of the best known species, it is not surprising that a more than ordinary degree of confusion should have prevailed respecting birds so little known as those of Paradise. Accordingly we find that most authors have varied in the number of species which they have thought proper to admit into this subdivision; others have divided into many genera those species of which so little is known; and all, according to their general and laudable custom, have introduced an interesting variety and discrepancy into their nomenclature. To discuss their ingenious systems, and to reconcile their inconsistencies, we must leave to abler hands, contenting ourselves with the humbler occupation of gleaning such information for, our readers as may not be utterly unworthy of occupying a portion of their time and attention.

Valentyn, in his voyage to the Indies, says that in the Papuan Islands and New Guinea, there are six species of the birds of Paradise, and that the most common is *The Great Bird of Paradise (P. Apoda).* It inhabits the *Aroo Islands* in the dry monsoon, and returns to New Guinea in the rainy season. It arrives there in flocks of from thirty to forty, under the guidance of one bird, whose colour is black, with red spots, which the islanders of Aroo call the king, and which always flies at the head of the troop. They never abandon him, and only reŝt when he gives the signal. This rest is sometimes fatal to many individuals; for, in consequence of the structure and disposition of their feathers, they rise again with considerable difficulty.

To this recital Helbigius adds, that the subjects of this pretended king, (whose size does not exceed that of the common sparrow, and who has two long caudal feathers, adorned with eyes at their extremity,) remain immoveable on the tree, where they are assembled together in the evening, until he passes and brings with him the entire troop; and that if this chief be once pierced with an arrow, those which remain are usually all killed, when day-light continues sufficiently long for the purpose.

Thus far these two worthy Belgians. It is evident that

the bird to which they allude is the *King Paradise bird, (P. Regia,)* which is sufficiently designated by the size, and the appearance of eyes upon the tail feathers. But even supposing that these birds had really been seen among the individuals of the species now under consideration, there was no foundation for drawing the very extraordinary conclusions we have just cited. Levaillant, who regards nature with the sober views of a philosopher, and with all the accurate observation of a practical naturalist, observes, in an article on this very bird *(P. Regia)*, that it frequently happens with birds living in flocks, that one of them having strayed from his associates, and being unable to find them again, will unite himself to a troop of another species, and remain attached to it for a whole season, particularly in those places usually inhabited by his own consimilars. These new comers into a country, with a species not their own, naturally exhibit habits different from their companions. They preserve in the midst of them a foreign air, and always remain a little apart, which gives them the appearance of commanding the troop, and directing its operations.

The extent and suppleness of the feathers of the Great Bird of Paradise, gives it a facility of rising to a very considerable elevation, and cutting the air with a lightness which has doubtless caused it to be sometimes named the *Ternate Swallow*. But when the wind becomes too strong, these birds are obliged to rise perpendicularly, until they arrive at a less agitated region of the atmosphere. Notwithstanding this facility of escape, sudden hurricanes will arise, and greatly discompose their feathers, and they are then heard to utter cries resembling those of the raven or crow. The islanders who hear these cries rush upon t e individuals who fall, and who can only escape death by alighting on some eminence sufficiently high to enable them to resume their flight. The natives also take them in other ways, by snares, bird-lime, &c. When taken alive they defend themselves courageously, giving violent blows with their bill. The Moors with these birds make plumes for their helmets, and sometimes suspend them to their sabres. The islanders of Aroo say that the tails of these birds, that is, their subalar and accessory feathers, fall during the eastern monsoon, and are not seen for four months.

The *King Paradise Bird* is solitary ; does not perch, it is said, on large trees, but hovers from bush to bush, feeding on the red berries which certain shrubs produce. All this very ill accords with the quality of chief, or king, of the birds of Paradise, which nestle on the lofty trees of mountains. The islanders take him with a bird-lime or glue drawn from the *autocorpus communis.* His general habitat is New Guinea, except during the western monsoon, when he remains in the Aroo islands.

Of the habits of the other species of the birds of Paradise, nothing is known.

We now resume the text of Cuvier.

The Fourth Family of the Passeres, or that of the Tenuirostres,

Comprehends the rest of the birds belonging to the first division; those whose bill is slender, elongated, sometimes straight, sometimes more or less bent, without any neck. They are to the Conirostres what the slender-beaks are to the other Dentirostres.

The Nuthatches. (Sitta. Lin.)

Which we shall name first, have a straight, prismatic pointed bill, compressed towards the end, which they make use of, like the woodpeckers, to open the bark, and draw out the insects; but their tongue is not extensible; and although they climb trees in every way, they have only one posterior toe, which, indeed, is very strong. They do not use their tail to support them, like the woodpeckers and true climbers. We have but one in France.

The *Common Nuthatch.* (*Sitta Europea, L.*) Enl. 623. I. Naum. 139.

Bluish-ash above; reddish underneath; a blackish band passes behind the eye. It is above the size of the redthroat.

Sitta Canadensis. Briss. Enl. 623. 2.

Plumbeous; head and neck above, black; beneath,

rusty; side tail feathers, black and white. North America.

Sitta Melanocephala. Gm. Catesb. 1. xxii. Vieil. Gal. 171.

Summit of head and neck, above, black; body, above, plumbeous; beneath, white; lateral tail feathers, black and white. North America.

Sitta Frontalis. Swain. Zool. Ill. 2. or *S. Velata.* Temm. Coll. 72, 3. or *Orthorhyncus Frontalis.* Horsf. Java.

Head, nape, and upper parts of body, rich azure; wings and tail, less pure; chin, white; all rest beneath, light purplish ash. Sumatra and Java.

Sitta Chrysoptera. Lath. 3. Suppl. 327.

Ashen above; croup, upper tail-coverts, and all below, clear blue; wings, orange at the edge. New Holland.

Sitta Pusilla. Lath.

Top of the head and neck, brown; white band on nape; wings, blackish; underneath, whitish. North America.*

The Sittines. (Xenops. Ileger.)

Which differ only in having the bill a little more compressed, of which the under ridge is more curved.

* To these are added—
Sitta Fusca. Vieil.
Sitta Chloris. Sparman.
Sitta Caffra. Sparman.

(*X. Rutilus.* Lich.) Col. 72. 2 *An. Neops. Ruficauda.* Vieil. Gal. 170.

Above, ferrugineous; head, brown, pale streaked; throat, white; belly, olive; tail, cinnamon; inner barbs, of four outer feathers, blackish. Brazil. Length, five inches.

(*X. Hoffmanseggii.*) Col. 150, 1.

Back, dead leaf colour; throat and chest, reddish white spotted; rest below, reddish white. Brazil.

(*X. Anabatoides.*) Col. 150, 2.

Above, brown red; underneath, clear reddish; a broad stripe of pure white over the sides of the occiput. Amazon river.*

The ANABATES. Tem.

Whose bill, on the other hand, has the upper ridge a little convex, almost like the bill of the blackbird, not notched. Some have a long cuneiform tail, and even worn, which proves that it supports them in creeping.

(*Ph. Superciliaris.* Spix. 73.) Probably the same as *Anabates Amaurotis.* Tem. col. 238, f. 3.

A band of brown feathers from the lower angle of the eyes, covering the auditory passage. Brazil.

* Add from Ill. *X. Genibarbis.* pl. col. 150. fig. 1. Vaill. t. 31. f. 2.

(*Sphenura Striolata.* Spix. 83, 2.) or *Anabates Striolatus.* Tem. col. 238, f. 1.

A long tail, much graduated, distinguishes this species from its congeners. Brazil*

SYNALLAXIS. Vieil.

Whose bill is straight, but little elongated, much compressed, slender, and pointed; their tail is generally long, and ending in a point.

There are some of them which have the stalks of the tail feathers strong, and extending beyond the barbs.

Synallaxis Ruficapilla. Vieil. Gal. 174; or, *Parulus Ruficeps.* Spix. 86, from which *Synal. Albescens.* Temm. Col. 227, 2, et *Cinerascens.* Ib. 3. do not appear to me to differ specifically.

Top of head red; back greenish-olive; tail long and graduated. Brazil.

* Others enumerate,—

Anabates Cristatus. Spix. 84.
An. Rufifrons. Spix. 85, 1.
Philydor Ruficollis. Spix. 75.
Ph. Albogularis. Spix. 74.
Anabates Erythropthalmus. Newied. H. iii. 43.
Anabates Leucophthalmus. Newied. iii. 22.
Anabates Atricapillus. Newied, iii. 43.

Here should be added the genus ORTHONYX, of Temminck, which has the bill short, and nearly straight, not notched: nostrils lateral; Tarsus lengthened, &c. See Temminck.

Orthonyx Spinicaudatus. Temm.

Also,

OXYRHYNCUS. Temminck.
Oxyrhyncus Flammiceps. Temm.
O. Cristatus. Swainson.

Syn. Rutilans. Col. 227. f. 1.

Forehead, eyebrows, cheeks, sides of neck, red; throat black; body above, and abdomen, olive. South America.

Syn. Tessalata. Col. 311. f. 1.

Top of head and carpus of the wing, moronne; above, ochre-brown; middle of the belly, white.

Syn. Setaria. Col. 311, f. 2.

A tuft of loose black feathers; two caudal quills much longer than the rest. Brazil.

Prinia Familiaris. Horsf. Java.

Bill, strong, short, and broad at base; long tarsi, and wedge-shaped tail; type of a genus. Java.

Le Flûteur. Vaill. Af. 112. or *Malurus Africanus.* Swain. Ill. 170, has only the bill a little higher. South Africa.*

Dendrocolaptes Sylviellus. Tem. Col. 72, 1.

Upper part of head and neck, olive-green; wings and tail, moronne-red. Brazil.

The birds to which have been applied the name of

CREEPERS. (CERTHIA. Lin.)

Have the bill bent, but that is almost the only character they have in common.

* *Syn. Ruficauda.* Vieil.

We distinguish, first,

The CREEPERS, properly speaking. (CERTHIA. Cuv.)

Thus named from their habit of climbing trees, like woodpeckers, making use of their tails as a prop. They may be recognized by the quills of the tail being worn, and finishing in a stiff point, like that of the woodpeckers.

We have one of them,

The *European Creeper*. *Certh. Familiaris*. Enl. 681.

A small bird, with whitish plumage, spotted with brown above, and tinted with red on the rump and tail. It nestles in the hollows of trees, and climbs with rapidity, seeking insects and larvæ in the clefts of the bark, under the moss, &c.

Cinnamon Creeper (*Certhia Cinnamomea*. Vieil. 62).

Plumage above, wings and tail, cinnamon-colour; underneath, white.

Thorn-tailed Creeper (*Motacilla Spinicauda*). Syn.II. Pl. 52.

Dusky reddish-brown, above; mottled-yellow on the crown; shoulders and underparts, white; tail cuneiform, with the shafts projecting into long points; size of a sparrow. South America.

America produces some true creepers, of a tolerably large size, which have been named

PICUCULI. (DENDROCOLAPTES. Herm.) *Grimpars.* Vieil.

Their tail is the same, but the bill is much stronger, and more broad transversally.*

Picucule Creeper. Climbing Grackle. Lath. (*Gracula Cayanensis.* Gm. *Gracula Scandens.* Lath.) Enl. 621, and Vieil. Ois. dorés, 76.

Rufous above, rufous-yellow underneath; every where marked with narrow transverse dusky streaks. Ten inches. Guiana.

The *Dendroc. Decumanus.* Spix. 87, and *Falcirostris,* 88, are at least very near to this.

Le Grand Grimpart. Vail. 42.

Distinguished from the last by the continued curve of the upper mandible only. Brazil.

Le Gr. Maillé. Vaill. 29. 2.

Top of head reddish brown; throat white; tail red; beneath, white in the middle, and edged with black. Brazil.

Le Grimpart Flambré. Vail. Promer. 30, or *Dend. Platyrostris.* Spix.

Head and neck dull brown; back and under part earthy brown; feathers of yellowish red, like scales, cover the throat and front of the neck. Cayenne.

Le Gr. Enfumé. Vaill. 28.

Plumage soot-colour, except two marks of a clear red on both sides of the head. Cayenne.

* Vieillot has called this genus DENDROCOPUS.

There is, indeed, one, which, by its straight and compressed bill, is allied to the nuthatches. It may, indeed, be considered as a nuthatch, with a worn tail.

Le Galapiot. Buff. *Oriolus.Picus.* Gm. et Lath. *Gracula Picoides.* Sh. Enl. 605, or *Dendrocolaptes Guttatus.* Spix. 91. 1.

Straight bill; top of head and neck behind, brown-red; feathers of the sides, front of neck and chest like white scales, edged with reddish-brown; upper parts of body lively red, lower parts clear red. Cayenne.*

Another, whose bill, twice the length of the head, is arched only at the end.

Le Nasican. Vaill. Promer. &c. 24.

Bill very long; head clear brown; two bands of dirty white on the back of the neck; throat and cheeks white; lower parts red. Brazil and Cayenne.

And there is one with the bill long, slender, and bent as much as in the *Heorotarii.*

Le Grimpart Promerops (Dendrocolaptes Procur-

* Others add,—

Dendroc. Tenuirostris. Spix. 91. 2.
Dendroc. Bivittatus. Spix. 90. 1.
D. Wagleri. 90. 2.
Dendroc. Fuscus. Vieil.
Dendroc. Rufus. Vieil.

vus). Temm. Col. 28, or *Dendrocopus Falcularius.* Vieil. Gal. 175.*

Olive-brown above; alar and caudal quills deep lively red. Brazil.

The Scalers, or Creepers of the Wall. (Tichodroma. Illiger.)

Have not the tail worn, although they climb along walls and rocks, as the common creepers do on trees; but they fasten themselves by their very large claws. Their bill is triangular, and depressed at the base, very long, and very slender.

But one species is known, which inhabits the south of Europe. *(Certhia Muraria. L.)* Enl. 372. It is a handsome bird, of a clear ash colour, with a lively red on the wing-coverts, and on the edges of a portion of the quills of the wing. The throat of the male is black.

This is the genus Petrodroma, of Vieil.

Brown Scaler (Certhia Fusca, Lath.*)*

Brown above; throat and breast barred brown

* Add from others,—

Dendrocopus Rubricaudatus. Vieil.
Dendrocopus Amgustirostris. Vieil.
Dendrocopus Guttatus. Licht.
Den. Pyrrhosophius. Vieil.
Dendrocolaptes Crassirostris. Vig.
Dendroc. Frontirostris. Vig.

and white. Six inches. South Sea Islands, appears to me to belong to this sub-genus.*

The SUGAR-EATERS. (NECTARINIA. Illiger.)

Whose tail, not worn, shews that they do not climb, but whose bill, of a moderate length, arched, pointed, and compressed, resembles that of the creepers. They are all foreign.

We more especially give the name of GUITGUITS to certain small species, the males of which have lively colours. Their tongue is bifid and filamentous.

Black and Blue Creeper (Certhia Cyanea). Enl. 83. Vieil. 41, 2, 3, and Gal. 176.

Black above; blue beneath. Four inches. Brazil. Young, green; wings and tail, blue.

(Cer. Cærulea. Edw.)

Head blue, with black spot round the eye; body violet blue; wings and tail black. Four inches. Berbice.

To these may probably be added:

Crimson Creeper (Cer. Sanguinea. Vieil. 66.*)*

Crimson; deeper above; quills black; vent white. Five inches. Sandwich Islands.

* Add,—

Petrodroma Bailloni. Viellot Dict. H. Nat. from New Holland.

Petrodroma Sanguinea of Vieillot, from *Certhia Sanguinea* of Linnæus, or *Meliphagu* of Temminck.

Cardinal Creeper. (*Cer. Cardinalis.* Id. 54, 58.)

Green gold above; carmine red underneath; two middle tail feathers elongated. Six inches. America.

Yellow-rumped, or Bourbon Creeper. (*C. Borbonica.*) Enl. 681.

Greenish brown above; rump yellow; beneath yellow grey. Five inches. Isle of Bourbon.

Vieil. Gal. 167, has named these birds *Coereba.*

N.B. *A. Armillata.* Sparm. 36. *C. Cayana.* 682. are mere varieties of *Cyanea,* or *Cærulea.*

We may separate from them certain species, larger and less handsome, whose tongue is short and cartilaginous.*

Baker-bird. (*Merops Rufus.*) Gm. Enl. 739. *Figulus Albogularis.* Spix. 78.

A bird of South America, as large as a red-breast; reddish above, whitish underneath; which constructs on the ground, under the shrubs, a nest, covered above like an oven.

This bird is the type of the genus Ophie, or *Opetiorhynchos* of M. Temminck; Furnarius of M. Vieil. Gal. 182. The genus Figulus, of Spix, is the same.

* From others—

Furnarius Fuliginosus. Zool. Coq. t. *Certhia Antartica.* Garnot. Ann. Sci. 1826. Isles Malomunes.

Furnarius Chilensis. Lesson. Zool. Coq. t. *F. Lessonii.* Dumont. Chil.

Furnarius Annumbi. Vieillot. *Annumbi.* Azara.

Picchion-baillon. Vieil. Gal. 172.

Greenish brown above; underneath, reddish white. New Holland.

Pomatorhinus Montanus. Horsf. Java.

Brown above, with olivaceous tint on upper parts of wings and tail; vivid chesnut on neck, back, and sides of breast and abdomen. The lowest part of the latter white. Java.

Pomat. Turdinus. Pl. Col. 441.

Grey; ashen brown, above; cheeks, chest, and sides clear ashen brown; rest beneath, dull white. New Holland.

Pom. Trivirgatus. Col. 443.

Top of head and neck marked by three broad bands; the middle ashen; the others white. New Holland.*

Climacteris Picumnus. Tem. Col. 281, 1.

Top of head, deep grey; nape and neck clear grey; a broad nankeen-coloured band over the middle of wing feathers. Timor.

Clim. Scandens. Tem. Col. 281, 2.

Head, neck, back, amber brown; throat and front of neck, pure white; chest and middle belly, is-abella. New Holland.*

Black and Yellow Creeper. Cer. Flaveola. Edw. I. 362. Vieill. 51.

Black above; stripe of white on each side of head;

* Dr. Horsfield makes a distinct genus of *Pomatorhinus.*

† M. Temminck makes this a distinct genus.

breast and underneath yellow. Jamaica and St. Domingo.

Cer. Varia (Mot. Varia. L.) Ed. 30. 2. Vieill. 74, which is the *Mniotille Varié.* Id. Gal. 169.

Head, bright red on top; hind head, blue; neck, back and rump, undulated with blue, black, yellow, and white. America.*

C. Semitorquata. Vieil. 56.

Streak of yellow on each side of neck, dividing it into two parts.

Prom. Olivâtre. Vail. Huppes and Prom. pl. 5 *(Mer. Olivaceus).* Sh.

Head and upper parts inclined to olive; paler beneath.

I suspect, also, that this is the place for *C. Virens.* Vieil. 57, and 58, and *Sannio.* Id. 64, which I have not seen, but which are distinguished by a slight furcation of the tail.

The Dicea. (Dicæum.† Cuv.)

Do not climb, and have not the tail worn. Their sharp, arched bill, not longer than the head, is depressed and widened at the base.

They come from the East Indies, are very small, and have generally some scarlet in their plumage.

* *Certhia Maculata* of Wilson, placed with *Sylvia* by Bonap. and others —type of the genus Mniotella of Vieil.

† The name of a small Indian bird, according to Ælian.

Red-backed Creeper. (*Cer. Erythronotos.*) Vieil. II. 35.

Top of head, neck, back and rump, crimson; throat, breast and belly, rufus white; wing-coverts, dark green. Three inches. China.*

The *C. Cruentata,* Edw. 81, is probably a variety of age.

Scarlet Creeper. (*C. Rubra.*) Vieil. 54.

Head, neck, breast, and streak down the middle of back, crimson; rest of body and wing-coverts, black. Three inches and a quarter. Island of Tarma.

The *C. Erythropygia.* Lath. is probably the female.

The *Nectarinia Rubricosa.* Tem. Col. 108.

Does not appear to me to differ from the above.

Orange-backed Creeper. *C. Cantillans.*

Blue grey above; breast and belly, and spot on back, orange yellow. Three inches. China.

The Heorotarii. Melithreptus. Vieil.

Have not the tail worn, and the bill is extremely

* Vieillot has referred to this genus—
Motacilla Hirundinacea. Sh. Nat. Mis. No. 114.
C. Tæniata. Son. 2d. Voy. pl. 107.
Dicæum Rufescens. Vieil. *Le Crombea.* Vail. O. D. iij.
Dicæum Rubescens. Vieil. O. D. t. 36.
Dicæum Choronotos. Vieil. O. D. t. 28, and perhaps Vl. Enl. t. 681. . 2.
Add,—
Dicæum Flavium. Horsf. Lin. Trans. Java.

elongated, and curved almost in a half circle. They come from the South Sea Islands.

One of them,

(*Certhia Vestiaria.* Sh.) Vieil. *Ois. dorés* II. pl. 52.

Is covered with scarlet plumes, which serve for the fabrication of the handsome mantles of this colour, by the inhabitants of the Sandwich Islands, which are held in such esteem.

Hook-billed green Creeper. (*Certh. Obscura.*) Vieil. *Ois. Dor.* 53.

Olive green; paler beneath; quills and tail dusky; seven inches. Honey-eater of Lath.

Great Hook-billed Creeper. (*C. Pacifica.* Id. Pl. 63).

Black above; rump and upper tail-coverts yellow; underneath dusky; eight inches. Friendly Islands. But the other *Melithrepti* of this Naturalist belong to very different genera, especially the Philedon, and to the Diceés.

* M. Vieillot has described the following species:

M. Vestiarius.
M. Obscurus.
M. Pacificus.
M. Pyrrhopterus.
M. Canescens.
M. Tenuirostris.
M. Cærulescens.
M. Fuscus.
M. Atricapillus.
M. Collaris.

The Soui-Mangas. (Cinnyris.* Cuv.)

Have not the tail worn; their very long and slender bill has the edge of the two mandibles finely denticulated, like a saw. Their tongue, capable of elongation beyond the bill, terminates in a furcation. They are small birds, and the males, during the season of love, shine with metallic colours approaching the brilliancy of the humming-birds, which they may be said to represent in the Old World, being principally natives of Africa. They live on flowers, of which they suck the juice. Their disposition is gay, and song agreeable. Their beauty has made them much in request in our collections; but the plumage of males and females, during the bad season, being altogether

M. Erythropigius.
M. Albicollis.
M. Flavicollis.
M. Flavicans.
M. Cardinalis.
M. Melanops.
M. Guttatus.
M. Sannio.
M. Ater.
M. Cucullatus.
M. Melanoleucus.
M. Novæ Hollandiæ.
M. Dibaphus.
M. Sanguinolentus.
M. Albicapillus
M. Agilis.
M. Virescens, &c.

Many of which have been transferred to other genera.

* The Greek name of an unknown bird.

different from their brilliant plumage, the species are characterized with difficulty.

Mellisuga of M. Vieillot.

The greatest number have the tail equal.

C. Splendida. Sh. Vieill. 82.

Head and throat, violet blue; upper parts, brilliant golden green; bright red bar across the breast. Africa.

C. Caffra. Edw. 347.

Shining green, above, glossed with green and copper; red bar across the breast. Cape.

(*Cert. Superba.* Vieill. 22.)

Bright green gold, above; throat, violet-blue and gold; bar of yellow across breast.

Loten's Creeper. (*Cert. Lotenia.*) Enl. 575. Vieil. 34.

Green-gold, above; beneath, black; black line between the bill and eye; violet band on sides. Five and a quarter inches. Madagacar.

(*C. Amethystina.*) Vieil. 5 and 6.

Velvet-black, with a gloss of violet; throat, amethyst-colour; rump, same. Cape G. Hope.

Collared Creeper. (*C. Chalybea.* Enl. 246. Vieil. 10, 13, 18, 24, 34, 80.

Bronzed gold, above; breast, red; collar, of steel blue. Four and-a-half inches. South Africa.

Green-gold Creeper. (*Cert. Omnicolor.*) Seba. 1. 69, 5.

Green, with a shade of all colours. Eight inches. Seba. Ceylon.

Copper Creeper. (*Cert. cuprea.*) Vieil. 23.

Reddish copper colour; under parts, black. Five inches. Malimba.

(*Cert. Purpurata.*) Edw. 265. Vieil. 11.

Deep violet-purple, glossed with green; tuft of yellow plumes on each side of the body. India.

Blue-headed Creeper. (*Cert. Cyanocephala.* Vieil. 7

Head, neck, and throat, violet blue; the remainder of the bird olive, above; ash colour, beneath. Five inches. Malimba.

Ceylonese Creeper. (*Cert. Zeilonica.*) Enl. 576.

Neck and back, black; crown, glossy-green; rump, purple; beneath, yellow. Size of a wren. Hindostan.

C. Dubia is treated by Latham as a variety of this.

Senegal Creeper. (*Cert. Senegalensis.*) Vieil. 8.

Top of head and throat, green-gold; rest of body, violet black; breast bright red, in some lights. Five inches. Senegal.

Red-breasted Creeper. (*Cert. Sperata.*) Enl. 246. Vieil. 16, 32.

Neck, back, and scapulars, purple, chesnut; rump

and tail-coverts, violet; breast, bright red: vent, yellow-olive. Four inches. The Philippine Islands.

C. Lepida. Sparman 35, is the female.

C. Madagascariensis. Vieil. 18.

Head, throat, and scapulars, brilliant green; rest, above, obscure olive; under each shoulder, a spot of fine yellow. Madagascar.

Grey Creeper. (Cert. Currucaria.) Vieil. 31.

Grey brown, above; spotted, on the lower part of underneath; wings, violet. Five inches. Philippine Islands.

Red-brown Creeper. (Cert. Rubrofusca.) Vieill. 20.

Dull gilded red above; wings, violet; belly, black. Four inches.

(Cert. Fuliginosa.) Vieil. 20.

Forehead, throat, and small wing-coverts, shining violet; the rest dusky. Five inches.

Pied Creeper. (Cert. Maculata.) Vieil. 21.

Spotted, black and white. Nearly five inches. Pennsylvania.

(Cert. Venusta.) Vieil. 79.

Forehead and chest, bright violet; throat, black purple; front of neck, and rump, azure; back of head, and back, and tail, emerald-green; wing, brown. Africa.

Green-faced Creeper. (Cert. Gutturalis. Enl. 578.*)*

Forehead and throat, gold green; head and body, blackish brown; fore part of neck, bright red. Size of a Linnet. Brazil.

Nectarinia Solaris. Tem. col. 341.

Throat, forehead, deep metallic green; body, beneath, orange; flank, pure yellow; above, olive green; wing, black, olive-edged; tail-feathers, black; two outer, white-tipt.

N. Eximia. Tem. col. 138.

Plumage, above, olive green; top of head and tail, deep emerald; rump, yellow; on the throat, a band of purple. Java.

N. Pectoralis. Id. col. 138.

Golden green, above, with brilliant reflexions; rump, blue; blue band on throat. Africa.

N. Lepida. Lath. col. 126. and Vieil. Gal. 177.

Forehead, deep green; throat, brown red; back, rump, and tail, fine violet; yellow, underneath. Malacca.

N. Hasselti. Tem. col. 376.

Top of head and occiput, green, glossed with yellow; top of neck, velvet black; rest, above, purple; belly, carmelite, very deep. Java.

N. Coccinogaster. Id. col. 388.

Top of head and nape, metallic green; velvety reddish brown on top of back, lower neck, and middle wing-coverts; belly, scarlet. Manilla.

Cinn. Eques. Less. and Garn. Voy. de la Coquille, pl. 31.

All the body, above and below, fuliginous brown; a narrow band, fiery-red, from the base of the throat to top of breast. Waigiou.

Of these birds, some are probably varieties of others.

Among some, the two middle quills are more elongated in the males.

Famous or Shining Creeper. (*Cert. Famosa.*) En. 83.

Green gold, glossed with copper ; wing-coverts, blackish ; under shoulders, a yellow spot ; between the bill and eye, a black stripe ; size of a linnet. Africa.

Beautiful Creeper. (*C. Pulchella.*) Enl. 670.

Green gold ; breast, red ; wing and quills, brown ; tail, blackish, and dyed with green gold ; seven inches. Africa.

Violet-headed Creeper. (*C. Violacea.*) Enl. 670.

Head, back, and lesser wing-coverts, violet ; lower part of back, rump, and upper tail-coverts, olive brown ; orange underneath ; six inches. Cape of Good Hope.

Saccharine Creeper. (*Sucrier Cardinal.*) Vail. Ap. 291.

Green gold above ; rump and tail-coverts, violet ; quills and tail, brownish black ; two middle tail feathers, longest ; six inches. South Africa.

Sucrier Figuier. Id. 293.

Head, neck, mantle, and upper wing-coverts, fine changeable green ; two long threads from tail : jonquil below. South Africa.

Nec. Metallica. Lich. Ruppel. pl. 7. and coll. 347.

Head, neck, back, and small wing-coverts, metallic green ; half-collar of lively metallic blue on the thoracic region : jonquil below. Nubia.

Nec. Mysticalis. Coll. 126.

Small moustache of metallic violet on each side of the bill ; calotte on the head, of same colour. Java.

N. Kuhlii. Col. 376.

Head and tail-coverts, metallic and bottle-green; belly, blackish green; sides, and under the wings, shining white. Java.*

We may still distinguish those whose bill is straight, or nearly so.

Cinnyris Elegans. Vieill. Gal. 177. or *Certhia Rectirostris.* Id. Ois. dorés II. pl. 75.

Head, back, rump, wing-coverts, and throat, brassy green; alar and caudal quills, clear green; under the neck, yellow; belly, dirty yellow. India.†

* Lichtenstein has added here—

Certhia Aurifrontalis.
Nect. Aurifrons.
Nect. Melanopogon.

† Lichtenstein has added here—

Merops. Cafer. and *Upupa Promerops* of Linnæus. *Nect. Cafer.* Licht.
Upupa Erythrorhyncus. Lath.
Nectarinia Melanorhyncus. Licht.

Lesson has given a monograph of this genus, in which he describes the following additional species: —

Cinnyris. Œneus. Vieil.
Cin. Nigralbus. Id.
Cin. Cinereus. Id.
Cin. Politus. Id.
Cin. Cirrhatus. Id.
Cin. Indicus. Seba.
Cin. Erythrorhynchos. Lath.
Cin. Solaris. Temm. *Nectarinia.*
Cin. Chrysogenis. Temm.
Cin. Congirostratus. Temm.
Cin. Mornatus. Affinis. Horsf.
Cin. Manillensis. Vieil.
Cin. Amboinensis. Lath. *Certhia.*

The ARCHNOTHERES. Tem.

Have the long bent bill of the *sonimanga*, but it is stronger, and without indentation; their tongue is

Cin. Chrysoptera. Vieil. after Latham.
Cin. Subflavus. Vieil.
Cin. Asiaticus. Lath. *Certhia.*
Cin. Ornatus. Temm.
Cin. Phœnicotis. Temm.
Cin. Rubrocana. Temm.
Cin. Clementiæ. Lesson.
Cin. Splendens. Vieil.
Cin. Splendida. Certhia. Lath.
Cin. Cinereicollis. Vieil.
Cin. Collaris. Vieil.
Cin. Aurifrons. Vieil.
Cin. Frontalis. Lath. *Certhia.*
Cin. Sericeus. Lesson.
Cin. Novæ Guineæ. Less.
Cin. Discolor. Vieil. *Cer. Senegalensis.* Lath.
Souimanga rayé. Vieil.
Cin. Iodeus. Vieil.
Cin. Nitens. Vieil.
Cin. Sugnimbindus. Vieil.
Cin. Smaragdinus. Vieil.
Cin. Sola. Vieil.
Cin. Melanurus. Vieil.
Cin. Erythrothorax. Vieil.
Cin. Perreini. Vieil.
Cin. Aurantia. Lath.
Cin. Bombicinus. Vieil.
Cin. Caudatus. Vieil.
Cin. Rubrater. Lesson.
Cin. Leucogaster Vieil.
Cin. Longirostris. Lath. *Certh.*
Sonimanga a gorge violette, et à poitrine rouge. Vieil.
Cin. Fuscus. Vieil.
Cin. Rubescens. Vieil.

short and cartilaginous; they are known only in the Indian Archipelago, and live on spiders.

Arachnothera Longirostra. Tem. Col. 84.

Ashen olive above; alar quills brown, edged with olive; caudal blackish, white tipped; throat and front of neck, white; rest clear yellow. Java.

Arach. Inornata. Id. ib. 2.

Small scaly plumes on the forehead; all above, green, bordering on olive; ashy whitish underneath. Java. *Cin. Affinis* of Dr. Horsfield.

After all these distinctions, we must still separate from the great genus *Certhia*, the *C. Lunata.* Vieil. *C. Novæ Hollandiæ.* White. New South Wales. *C. Australasiana.* Vieil. *C. Caruncalata.* Vieil. *C. Auriculata.* Vieil. *C. Cocincinica.* Vieil. *C. Spiza.* Enl. 578. Edw. 25. *C. Seniculus.* Vieil. *C. Graculina.* Vieil. *C. Goruck.* Vieil. *C. Cærulea.* Vieil. *C. Xanthotis.* Vieil. *C. Mellisora.* Vieil. All which are Philedons, by their indented bill, and their pencillated tongue.

THE HUMMING-BIRDS. (*Trochilus L.*)

These little birds, so celebrated for the metallic brilliancy of their plumage, and, especially by plates, which form on their throat or head, scaly feathers, of a peculiar structure, have a long and slender beak, enclosing a tongue, which is almost as much elongated as that of the pies, and is divided into two threads, which the bird employs to suck the nectar of the flowers. Nevertheless, they also live on insects. Their very small feet; their wide tail; their wings excessively long and narrow, in consequence of the rapid shortening of the quills; the short humerus; the sternum without slope, constitute a system of flight like that of the martins. Accordingly, we find the humming-birds balance themselves in the air as well as some flies. Thus they hover round plants, or flowery shrubs, and they fly more rapidly in proportion than any other bird. They live isolated, defend their nests with courage, and fight bitterly amongst themselves. We reserve the name of TROCHILUS to those which have the bill arched. Some are distinguished by the prolongation of the intermediate feathers of the tail.

We shall cite but one of the largest and most handsome.

The Topaz Humming-bird. Troch. Pella. Enl. 592.
Moronne-purple; head black; throat of the most brilliant topaz yellow, changing into green, enamelled with black.

Supercilious Humming-bird. (Trochilus Superciliosus.) Enl. 600, Vieil. 17, 18, 19.

Brown above, with a gilded gloss; over each eye, a white stripe; quills violet-brown; under part, rufous white. Near six inches.

Tr. Leucurus. Enl. 600, 3.

Green-gold above; beneath, greyish-white; crescent of red on the breast. Surinam.

Tr. Squalidus. Natt. Col. 120.

Plumage in general dull; two bands of reddish white on the sides of the head. Brazil.

Tr. Brasiliensis. Lath. Col. 120.

Green-gold above, glossed with copper; beneath, rufous-white. Brazil.

Others have the lateral quills of the tail very much elongated.

Long-tailed Humming-bird. (*Trochilus Forficatus.*) Edw. 33. Vieil. 30.

Top of head, blue; rest of plumage splendid green. About 7 inches.

(*T. Polythmus.*) Edw. 34. Vieil. 67.

Top of head and nape, black; body yellowish-green above; bluish-green underneath; edge of wing white. Nine and a half inches.

And especially the magnificent species of Peu, with the tail of shining gold, which is *T. Chrysurus.*

Many have the tail moderately furcated.

(*Tr. Elegans.*) Vieil. 14.

Green above; a spot of velvet-black on the breast, extending under the belly; tail forked. St. Domingo.

The greatest number have it round or square.

Tr. Mango. Enl. 680. Vieil. 7.

Golden-green above; throat, chest, and belly, velvet-black; wings violet-black. Great Antilles.

Tr. Nævius. Dumont. Col. 120.

Throat and front of neck, lively red; body above, scapulars, and wing-coverts, green, with slight metallic reflexions. Brazil.

Tr. Gutturalis. Enl. 671.

Upper parts and tail, deep gilded-green; dash of bright emerald green on throat. Cayenne.

Tr. Taumautias. Enl. 600.

All the plumage gold-green, except the wing, which is violet or brown. Brazil.

Tr. Violaceus. Enl. 600.

Head, neck, back, breast and belly, violet-purple; wings and tail gold-green. Cayenne.

Tr. Cinereus. Vieil. 5.

Fine glossy-green, above; beneath, ash-colour. South America.

Tr. Melanogaster. Vieil. 75.

Above head, and top of throat, brilliant green, with metallic reflexions; underneath black, feebly purpled; abdomen, white. Mexico.

Tr. Jugularis. Sh. Edw. 266. Vieil. 4.

Sides of head, throat, and breast, fine red; wings, fine dark green, with a polished gold gloss. Surinam.

Tr. Holo-sericeus. Sh. Vieil. 6, and 65.

Golden-green, above; tail-coverts and rump, brilliant blue; wings, blue. Mexico, Guiana.

Tr. Punctatus. Sh. Vieil. 8.

Plumage in general green-gold; throat, neck, and wing-coverts marked with small white spots. Mexico.

Tr. Pectoralis. Sh. Vieil. 9, and 70.

Above, obscure green, a little gilt; tail, violet; throat, and sides of neck, brilliant deep green. St. Domingo.

Tr. Aurulentus. Sh. Vieil. 12.

Top of head and neck, back and rump, obscure gold-green; chest, black, growing brownish under the belly. Porto Rico.

Tr. Aureo-viridis. Sh. Vieil. 15.

Colour, entirely green-gold, except quill-feathers, which are blackish, and tail steel-blue. West Indian Islands.

Tr. Hirsustus. Gm. or *Brasiliensis.* Sh. Vieil. 20.

Top of head, brown; neck, back, wing-coverts, gold-green; yellowish red beneath.

Tr. Albus. Vieil. 11.

Breast and belly, white, and sides of throat, but brown in the middle; above, variable gilded green. Cayenne.

Tr. Viridis. Vieil. 15.

Gold-green above, very brilliant; abdomen, bright grey, mixed with green. Guiana.

Tr. Margaritaceus. Enl. 680. Vieil. 16.

Body above, green-gold; beneath, pearly-grey. St. Domingo.

Tr. Multicolor. Gm. or, *Harlequin Humming Bird of Lath.* Vieil. 79.

Head, throat, front of neck, breast, and upper wing-coverts, green; a broad blue band between the eye and nape; black between this and top of back; belly, carmine, &c.*

We give the name of ORTHORHYNCUS to those whose bill is straight.

Among which are some with tufted heads; others have tufts or feathers prolonged on the sides of the head.

Crested Green Humming Bird. (*Tr. Cristatus.*) Edw 37 Enl. 227. Vieil. 47.

Top of head green; crest blue, upper part of body dark gilded green; under parts grey; outer tail feathers black; three inches.

* There have also been described,—
Trochilus Ater. Wied. A. y. 183.
Trochilus Pygmæus. Spix. t. 80. f. 1.
Trochilus Brevicauda. Spix t. 80. f. 2.
Trochilus Fulgens.
Trochilus Thalassinus.
Trochilus Melanotus, and
Trochilus Platycerius of Mr. Swainson, all from Mexico.

Crested Brown Humming Bird. (*Tr. Pileatus.* Vieil. 63.)

Like the last; but the crest wholly bright glossy blue; more elongated and pointed than it.

Orthor. Stephanoides. Less. and Gam. Voy. de la Coquelle, Pl. 31, No. 2.

Tuft, golden-green, with two longer blue feathers; above, gold green; below, blue; abdomen, pearl grey. Brazil.

Tufted-neck Humming Bird *Tr. Ornatus.* Enl. 640. Vieil. 49, 50.

Head and upper parts of body, green-gold; underneath, gilded greenish brown; rufous crest on head; tuft of long feathers on each side the head. Length about three inches.

Tr. Chalybeus. Vieil. Pl. Col. t. 66. f. 1.

Deep green plumes on sides of neck; bronzed green in general above; gilt tail; a little rounded. Brazil.

Tr. Petasophorus. Pr. Max. Col. 203.

Large tuft of violet plumes on head; green of varied shades all over; tail apparently forked.

Tr. Scutatus. Natter. Col. 299.

Throat and face golden-sapphire; black velvety band from one eye to the other; body above, sides, and small wing-coverts gilded-green; abdomen and tail-corvets white. Brazil.

Tr. Magnificus. Ill. Col. 299.

Tuft of orange-red on head; on both sides of neck unequal plumes white as snow, terminated by a small band of gold-green. Brazil.

Tr. Mesolencos. Tem. Col. 317.

Forehead bright sapphire; a long white band seems to divide the body in the middle. Brazil.

Others have the stalks of their primaries singularly widened.

Tr. Bilophus. Tem. Pl. Col. 18, 3.

Long tail, much graduated; top of head, bright emerald; two tufts of very bright gold colour above the eyes. Brazil.

Broad-shafted Humming Bird. (*Tr. Latipennis.*) Enl. 672. Vieil. 21.

Glossy green above; beneath, pale ash; quills, dusky; the three outer broad in the shaft, and bent.*

Among those which have no ornaments, we may distinguish the species with forked tail.

Tr. Ensipennis. Swainson, Zool. Ill. 107.

White-bellied Humming Bird. (*Tr. Mellivorus.*)
Enl. 640. Edw. 35. Vieil. 23, 24.

Head, throat, and neck, blue, glossed, with goldback; scapulars, and tail-feathers, green-gold. Four inches.

Sapphire Humming Bird (*Tr. Smaragdo-Saphyrinus.*) Vieil. 36, 40.

Body, green-gold; throat, rufous; lower belly, white. Four inches.

Red-throated Humming Bird. (*Tr. Colubris.*)
Edw. 38. Catesb. 65. Vieil. 31, 32, and 33

Green-gold above; grey beneath; throat, ruby-colour. About three inches.

Tobago Humming Bird. (*Tr. Maugeanus.*) Vieil. 37, 38.

Gilded-green above; brighter underneath; lower belly, white. About three inches.

Tr. Amethystinus
Gm. Enl. 672.

Throat and front of neck, brilliant amethyst, changing into purpled brown. Cayenne.

Tr. Furcatus. Vieil. 60.

Crown of head, blue; plumage in general, green; tail-feathers, stiff, forming a fork.

Tr. Langsdorfii. Vieil.
Pl. Col. t. 66. f. 1.

Tail-feathers, much graduated, and finishing in a point; brownish-green above; rose-coloured demi-collar on the chest. Rio Janeiro.

Tr. Enicurus. Vieil.
Pl. Col. 66. f. 3.

Gold-green above; a white gorget on sides of neck and top of breast; belly, white. Brazil.*

Among these there are some whose lateral quills, very much prolonged, are widened at the end.

Racket-tailed Humming Bird. (*Tr. Platurus.*)
Vieil. 52.

Green-gold, above; throat, emerald green; wings and tail, brown; shafts, very broad, and rufous-white. Two and a half inches.

Finally, we must remark, in consequence of its extreme smallness,

The smallest of the Humming-birds. (*Troch. Minimus.*) Enl. 276. 1 Edw. 105. Vieil. 64.

Of a grey violet, and as large as a bee.

And one other, on the contrary, because its size exceeds that of the rest of the genus.

* *Orthor. Cora.* Less. and Garn.

Tr. Gigas. Vieil.
Gal. 180.

which is nearly as large as our Martin.

Other species, with the tail square, slightly indented, are,

Tr. Moschitus. L.
Enl. 227.

Small feathers of bill and head, in one, light ruby; in another, purple; plumage, above, black, with green reflexions. Cayenne and Brazil.

Tr. Carbunculus.
Vieil. 54.

Throat, fore-part of neck, and breast, colour of deep carbuncle. Cayenne. A variety of the last, according to M. Vieil.

Tr. Ourissia. Enl. 227.

Coppery green-gold; part of back, breast, and belly fine deep blue.

Tr. Mellisugus. L.
Enl. 640.

Front, and upper parts of body, green-gold, changing into brown, according to the light. Porto Rico.

Tr. Rubineus. Gm.
Enl. 276. Vieil. 27.

Part of throat, very lively red; green-gold above. Cayenne.

Tr. Auritus. Sh. Vieil. 25.

Bright-gold-green above; white underneath. Cayenne.

Tr. Collaris. Vieil. 61 and 62.

Head, rich variable green, and gold; ruby-coloured ruff round the neck. Nootka Sound.

Tr. Superbus. Sh. *Longirostris.* Vieil. 59. Col. 299.

Crown of head, sky-blue; throat, brilliant scarlet, with full feathers; rest, gold-green, paler beneath. S. America.

Tr. Mellivorus. Enl. 640.

Head and under neck, deep blue; neck, above, green-gold, crossed by a milk-white bar. Cayenne.

Tr. Leucogaster. Gm. Vieil. 43.

Crown of head, green-brown, with golden reflexions; white, underneath, especially on belly. Cayenne.

Tr. Maculatus. Vieil. 44.

Crown of head, brown-gilt; plumage, in general, green, above; a white spot on lower part of breast enlarging on belly.

Tr. Saphyrinus. Sh.
Vieil. 35, 57.

Above, brilliant gold-green; front of neck and breast, rich sapphire, with violet reflexions.

Tr. Squamosus. Tem.
Col. 203.

Feathers on throat and neck black, in the middle, and white on the edges and end. Brazil.

Tr. Albicollis. Id. Col. 203.

Throat, part of front, neck, middle of belly, and lower tail-coverts, snow-white. Brazil.

Among the HOOPOES (UPUPA. Lin.) we place at first,*

The CRAVES. (FREGILLUS.† Cuv.)

Whose nostrils are covered by feathers directed forwards, which has occasioned them to be united, by many authors, with the crows, to which, in manners, they present some resemblance; their bill is a little longer than the head.

* There have also been described,—
Orthor. Amazilia. Less. and Garn. Voy. de la Coq. 31.
Colibri Crispus. Spix. t. 81. f. 1.
Colibri Leucopygus. Spix. t. 81. f. 3.
Colibri Albogularis. Spix. t. 82. f. 1.
Colibri Helios. Spix. t. 82. f. 2.

† M. Vieillot has changed this name into CORACIAS, which in Linnæus is applied to the Rollers.

The Crave of Europe. (*Corvus Gracula.* Lin.) Enl. 255.

Is of the size of a crow, black, with the bill and feet red; its wings reach or exceed the end of the tail.

It lives on the highest mountains of the Alps and Pyrenees, and builds there, in the clefts of rocks, like the chocard; but it is less common, and does not unite in flocks. It eats both fruits and insects. When it descends into the valleys, it is a sign of snow and bad weather.*

Corvus Affinis. Lath.

Above, dusky greenish-black: breast and belly, cinereous. Cayenne.

New Holland Crave.

Entirely black.

The HOOPOES proper. (UPUPA.)

Have on the head an ornament formed of a double range of long feathers, which can be erected according to the inclination of the bird.

We have one in Europe,

Upupa Epops. Lin. Enl. 52. Naum. 142.

Of a vinous red; wings and tail black; two white

* We know not what combination of the history of this bird with bad figures, has given birth to the imaginary species *Corvus Eremita*, Lin., a pretended bird of Switzerland, which nobody has seen since Gesner.

bands across the wing-coverts, and four across the wings of the quill. They seek out insects in humid ground, lay in the holes of trees or walls, and quit us in the winter.

Cape Hoopoe. (Upupa Capensis.) Enl. 697.

Is more analogous to the cranes, because the anterior feathers of the tuft, short and fixed, are directed forward, and cover the nostrils.

Upupa Minor. Vieil. Prom. pl. ii, and Gal. pl. 184. Vaill. Prom. 23

Body, above and beneath, deep brown-black. Length, nine inches,

The PROMEROPS.* Poriss.

Have no tuft on the head, and have a very long tail. The tongue, extensive and furcated, permits them to live on the juice of flowers, like the sonimangas and calibris.

We are well acquainted with the

Cape Promerops. (*U. Promerops*, or *Merops Capd.* Enl. 637.)

Brown above; rump, and under tail-coverts, olive-green; throat and belly, white. Seventeen inches; body, thin.

* M. Vieillot, in his Galerie, pl. 185, has changed this name into *Falconellas.*

M. Vaill. thinks that *Up. Fusca*, Gm., or *Papuensis*, Lath. Enl. 638, is the female of Enl. 639. *Upupa Paradisæa*, Seb. I. pl. 30, is the *Muscicapa Paradisi*, with the bill badly drawn. *Up. Aurantia*, Seb. I. 46, is, according to all appearance, a Cassique. *U. Mexicana*, Seba I. 45, is not from Mexico, as Seba states, in applying to it a passage from Nieremberg, lib. 10, in which a duck only is spoken of.

I doubt whether we ought not to place here *Promerops Cæruleus*, Shaw; *Prom. bleu*. Vieil.; *Upupa Indica*, Lath.; or whether we ought not to put it near *Up. Erythorhynchos*.

The EPIMACHI. (EPIMACHUS.* Cuv.)

Have, with the bill of the hoopoes and promerops, scaly feathers, which cover the nostrils, as in the birds of Paradise. They come from the same country, and are equally brilliant in their plumage. The feathers of the flanks are also more or less prolonged in the males.

The Epimachus, with frizzled ornaments. (*Upupa Magna*. Gm. *Up. Superba*. Lath.)

Black; wedged-tail three times longer than the body; feathers of the flanks elongated, reined, frizzled, brilliant at the edge, and of a polished-steel blue, which also shines upon the head and belly.

I know not whether we ought to place here, or near to *Merops*, the

* The name in Greek of a beautiful bird of India, of an indetermined species.

Promerar. Vaill. 8 and 9.

Lustrous-black above, with a cast of sombre-green; beneath, brownish black. Africa.

Promerupe. Vaill. 11 and 12.

Wings and tail, bay-red; tuft, black. East-Indies.

Prom. Siffleur. Vaill. 10.

Above, clear brown, shaded with olive; white below. Africa.

These beautiful birds of New Guinea, rare in the cabinets, are there often found deprived of their legs, which prevents the possibility of determining their place.

Ep. Albus. Paradisæa Alba. Blumenb. Abb. 96. Vaill. *Ois de Par.* pl. 16. and 17. and better Praner. 17. Vieil. pl. 13. and better Gal. 185.

Placed for a long time with the birds of Paradise, on account of its long fascet of white feathers on the flanks, the elongated plumes of which have six threads on every side. Its body is commonly of a violet-black, with a border of emerald-green to the feathers on the lower part of the breast; but there appears to exist some varieties altogether white. The primary quills of the wings are short, and much less in number than in other birds.

The Epimachus Promefil. Epimachus Magnificus.
Cuv. Vieil. Prom. 16.

Of a velvet-black, tail moderate, and a little forked; head and chest, shining with the most beautiful-polished-steel blue; the feathers of the flanks are elongated, slender, and black.

Epimachus Regius.
Less. and Garn. Voy. de Duperrey. pl. 28.
Ptiloris Paradisæus. Swainson.

Of a purple black; the upper part of the head, and top of chest of a brilliant green; the feathers of the flanks are rounded, edged with green.*

* *Certhia Antarctica,* Garnot. Ann. Sci. Nat. 1826.
Beak and feet black, throat variegated, yellow and ferrugineous; head and body, entirely reddish-brown; ferrugineous under the wings.

Generic Characters of Birds.

ORDER PASSERES.

Fam.
Tenuirostres
and
Syndactyles.

1. *Sitta*
2. *Xenops*
3. *Synallaxis*
4. *Certhia*
5. *Colibri*
6. *Orthorhynchus*
7. *Upupa*

Fam. Syndactyles.

8. *Merops*
9. *Alcedo*
10. *Prionites*

SUPPLEMENT ON THE TENUIROSTRES.

To the general characters of this family, as described in the text, we have nothing to add. The first genus is SITTA, which we have chosen to translate by our English name of NUTHATCH, which, however, is much more extensively applied by Dr. Latham. The habits of this genus resemble those of the woodpeckers, the creepers, and titmice; and many names have been imposed upon them in consequence of this analogy, which, however, convey only a partial idea of their peculiarities. Their relations with the woodpecker and titmouse, consist in their striking the bark of trees with their bill, and climbing along the trunk like the former, and they have much of the air and countenance of the latter; but they differ from the woodpecker in the form of the feet, tongue, and tail, and from the titmouse in the bill. They also possess, in their manner of climbing on the trunks and thick branches of trees, some analogy with the birds to which usage has consecrated the name of creepers; but they differ again from these in the conformation of the bill and tail. It is certain that all the species whose mode of life is known, nestle in the hollows of trees, and have at least one brood annually.

Many birds have been classed in this genus, whose right to that distinction is more than doubtful. Some of them have the lower mandible a little turned up.

The *Common Nuthatch* is rather sedentary in the countries which it inhabits; it approaches inhabited places in the winter, and sometimes shews itself in orchards and gardens. But the woods are its habitual dwelling, and the trunk of the tree which furnished its cradle usually constitutes its place of nocturnal retirement. Here also is its little magazine, where it hoards its provisions for the winter; for this bird is provident against the want which is consequent on the rigour of that season. It is observed in autumn to be always busy in making its provision of nuts and different grains, such as hempseed, &c. It is not by breaking them, like the small granivorous birds, that the nuthatch extracts their substance. It pierces them with powerful strokes of the bill, after having fixed them firmly in some cleft or hollow. Its mode of perching is peculiar to itself; for it has been remarked that it often suspends itself by the feet, or reposes itself on the side, and never in the manner of other birds. The nuthatch runs up and down trees in all directions, to chase the insects, on which it also feeds when grain is deficient. It runs up and down with equal facility, differing in this respect from the woodpecker, which rarely descends, and never except in an oblique line.

The character of this bird is very solitary. Its flight is gentle, and its motions are neat and graceful. Its ordinary cry is *ti, ti, ti, ti, ti, ti,* which it repeats, with increased precipitation, in climbing about the trees. Besides this cry, and the noise which it makes in striking the bark, it produces a very singular sound in putting its bill into a cleft, or rubbing it against dry and hollow branches. This noise is so loud, that it may be heard at a very considerable distance. In spring the male has a sort of song, like *guiric, guiric,* which it frequently repeats. He and the female labour conjointly in the arrangement of the nest, which they fix in the hollow of a tree, and often in a hollow which has been abandoned by the woodpecker. They will even make a hollow themselves in the tree with their bills, if the wood be worm-eaten. If the external aperture be too large, they contract it with unctuous earth. From this circumstance they have received in French the somewhat ludicrous denomination of *torchepots,* (*torchis,* yellow clay,) and also that of *pic-macon.* In England they are known by such names as *nut-jobber*, *wood-cracker*, *twit*, *nut-cracker*, *blue woodpecker*, *loggerhead*, *and jacbird.*

The female lays five, six, or seven eggs, of a dirty white, pointed with reddish, and deposits them on the dust of wood and moss. She evinces so much attachment in incubation, that she will suffer herself to be taken, rather than abandon the eggs. On thrusting a stick into the hole, she will hiss like the titmouse. It is even said that she never quits the eggs to look for food, subsisting only on what is brought her by the male, who, however, is singularly attentive in the performance of this duty. The little ones come out in May; and as soon as they can do without the assistance of the parents, all the family separate, and each lives alone during the rest of the year. These birds rarely have two broods. Though of a solitary character, and shunning the society of its consimilars, the nuthatch appears to take some pleasure in the company of birds of other species, for it is sometimes seen to associate with titmice and creepers.

The nuthatch inhabits the woods of this country throughout the year, but especially in the southern parts. It is scarce in France, though found pretty far north on the continent of Europe, in Russia, Sweden, and Norway, and even in Kamtschatka and Siberia.

Belon describes a variety under the name of *sitta minor*, which is altogether like this bird, but smaller and more noisy. It is seldom seen but with its female, is very quarrelsome, and fights bitterly. Dr. Latham thinks it a young bird of the common sort.

The *Sitta Pusilla* is a bird of the United States of America, and also found in Jamaica. It is a lively, alert bird, and difficult to approach. It is fond of the company of the sklit woodpecker, with whom it is often found in the pine forests. Latham also notices a variety found at Hudson's Bay, though perhaps it is a different species. The natives give it an unutterable name, which in their language signifies the voracity of this bird for berries, which it eats to excess, and also its pugnacious propensities, which it exercises with great effect on all small birds who come to dispute its favourite nourishment. It makes its nest in the oziers, and emigrates during winter.

The XENOPS have great affinity with the last; but of their habits nothing of the least interest is known. The same may be said of ANABATES and SYNALLAXIS. We insert a figure of the *xenops anabatoides* of Temminck, as illustrative of this genus and its subdivisions.

The CREEPERS, as will be seen in the text, have been divided into several sub-genera by our author. The first, or TRUE CREEPERS, has but few species.

The *European Creeper* is a little bird, found in the different countries of Europe, and is very common in England. It is to be met with as far as Siberia, and the north of Asia. Catesby has seen it in Carolina, and M. Vieillot in another part of North America. It is continually occupied in climbing along trees, in search of insects, and their larvæ, which constitute its principal food. It is often seen to pass from one tree to another, and its voice consists in a feeble, but sharp cry. It remains during night in the hollows of the same trees, and makes its nest there, composed of fine plants and moss, connected with spiders' web. The female lays five, six, seven, and sometimes nine eggs, which are white, with small red spots, as may be seen in the work of Lewin, though other writers describe them as ashen with points of a deeper colour.

The *Certhia Major*, of Brisson, is only a variety of the common creeper, distinguished by superior size. Its plumage and habits are similar. Scopoli has described a bird found in Carniola, which he regards as another variety, or merely a difference of sex. But the great difference of

NATTERERS NUTHATCH.

XENOPS ANABATIODES.

London, Published by Whittaker & C° Ave Maria Lane, June 1829.

colours, as Dr. Latham observes in his synopsis, would lead us to consider it a distinct species.

The Picuculi are small birds of this family, all of the New Continent. They all alike inhabit the forests, and climb against trees, leaning on their tail. They live on worms, which they extract from the bark; lay their eggs in the hollows of trees; do not walk on the ground, and they fly in the manner of the creepers and woodpeckers. They remain single or in pairs, and never in families; begin to climb against the trees, about three feet from the ground; and do not draw the worms from the bark with their tongue like the woodpeckers, but introduce their bill to seize them. If the worms and insects are too much concealed, they tap with their bills against the trees like the woodpeckers. They also sometimes use them as a lever to raise the bark.

Of Tichodroma there is but one species in Europe, the *Certhia Muraria* of Linnæus. This inhabits divers countries, but is not found in England: nor would it appear to be a native of Sweden, as Linnæus does not class it among the birds of that country. It is seen on the range of the Caucasus, and is supposed with some reason to be also a native of China. Its usual haunts are peaked and rugged rocks, and the walls of ruined towers, and such like ancient structures. It does not climb on trees like the common creeper, and chooses for its nestling place the clefts and crevices of solitary rocks. It voyages alone, and retires southwards about autumn to pass the winter. Its disposition is gay, and its voice agreeable. Of the habits of the other species, nothing is known.

Of the subdivision Nectarinia, those birds called Guit-guits are many of them found in South America. They live on insects, to which some of them unite the sweet and viscous juice of the sugar-cane, which they extract by sinking their bills into the clefts or fissures in the stalks, from which a superabundance of this saccharine liquid trickles.

Some of these birds live in flocks, with their congeners, and with various other little birds; others remain in pairs, but none of them climb. The Creoles of Cayenne confound them with the humming-birds, because, like them, they flutter round flowers, to catch with their bills the insects there concealed. They make their nests, at least the species whose mode of life is known, with great art, suspending it by the base to the extremity of a weak and mobile branch, with its aperture turned towards the ground. This construction and position places the brood and the mother in a state of shelter from spiders, lizards, and other enemies. Four eggs is the usual number laid, and this is repeated many times in the course of the year. Of the habits of DICEUM and MELITHREPTUS, nothing is known.

Belonging to the latter genus is the bird here figured under the name of Byron's creeper, which was brought by Captain Lord Byron to England from the South Sea, and is in the British Museum under M. Temminck's generic name of Drepanis, the Melithreptus of Vieillot. It is yellowish-red, except the quills, which are deep chocolate; the edges of the outer wing-feathers are edged with white.

The SOUIMANGAS, (CINNYRIS,) are so called from the name given by the inhabitants of Madagascar to a bird of their country, and the word in their language signifies *sugar-eater*. Montbeillard has generalized it for all these birds. Linnæus, Latham, and other naturalists, have classed them in the division of creepers (*certhia*), with which, in fact, they have no relation, except in the curvature of the bill; and even this, in most of them, differs, by having the two mandibles toothed like a saw on their edges; but the dentelations are so fine that they cannot be perceived but by the aid of a convex lens. They correspond so closely that they catch in each other.

The term creepers cannot, with any sort of propriety, be applied to these birds, for they do not climb; and their

BYRON'S CREEPER.

DREPANIS BYRONENSIS.

Sowerby del & sc

Mus. Brit.

London, Published by G.B. Whittaker, Nov. 1. 1827.

habits and manners are altogether opposite to those of the true creepers.

They have also been confounded with the humming-birds, by travellers, and even by naturalists. But they have attributes which are foreign to the latter. They have twelve quills in the tail, and the bill is slender, and forms a more acute angle than in the humming-birds. They are also distinguished by the length and nakedness of their tarsi, the conformation of their toes, claws, and wings. Besides, it is now an ascertained fact, that the entire tribe of the humming-birds is confined to America. It therefore appears, that all the African and Asiatic birds, to which that name was given, belong to the family of the souimangas, which supply the place of the others in the old continent. Like the humming-birds, they are adorned with the richest and most brilliant colours. The males have been particularly decked in this way by the lavish hand of nature. This, however, is only the case during the season of love. At all other times they resemble the females so closely, that it is impossible to distinguish them by the mere guidance of the plumage.

The souimangas moult regularly twice a year, and change colours at each moulting. But this change occurs in the males only. The females preserve constantly the same tints in all seasons, when they are once clothed with the colours of perfect age. This does not, however, appear to be the case with the humming-birds, at least not invariably, though M. Levaillant has expressed a contrary opinion. It would rather seem, that when these latter have attained their complete perfection in this way, it remains for the rest of their lives.

It is only when the souimangas are thus decorated with their brightest plumage, that they employ themselves in the construction of their nests, and the education of the young. After which, they resume their winter plumage, or, to speak more correctly, that which characterizes them during the rainy season, the only winter of the sunny climates of the tropics. This they preserve until the reapproach of the season of reproduction, when they moult a second time, and resume their brilliant colours. During both moultings, however, they are found more or less variegated with the colours peculiar to themselves, according as they are more or less advanced in each moulting. From this change of colour, twice a year, from the variegation during the two moultings, from the difference of the sexes, and the young, it has happened that many more species have been described than in reality existed. For a proper determination they should be studied in their living state, as M. Levaillant has studied them, though unfortunately but a small number has come under the inspection of that eminent naturalist. It is, therefore, more than probable, that many will be found in our additions to the text, which are merely nominal.

The souimangas live on insects, to which they add the melliferous juice of flowers. This has been denied by some naturalists, who have declared that they could not do so, as their tongue is unfitted for the purpose: an assertion which only proves that such naturalists never beheld the tongue in question. It is conformed exactly like that of the humming-birds, who live precisely in the same manner. "The parietes of this tongue," says M. Levaillant, "are of a substance which is corneous, and hollowed into a gutter, forming a sort of proboscis, the extremity of which is provided with many nervous threads, which, by their nature, constitute the first seat of taste. These threads serve not only to taste the liquor, but also, as a kind of sieve, to prevent grosser materials from passing along with the saccharine juice through the tube of the tongue, which they would thus obstruct. The hinder part of the tongue, which corresponds to the œsophagus, is furnished with two elongations, which, passing on each side of the larynx, proceed behind the head to implant themselves in the forehead, and serve, as with the woodpeckers, to push the

LONG-BILLED SOUIMANGA. *Tem. Pl. col 84.*

London. Published by G.B. Whittaker & C.o Ave Maria Lane. June, 1829.

tongue out of the bill, according to the degree of depth of which the bird has need to attain, for the purpose of finding its favourite nutriment.

The souimangas have a cheerful song, and possess much vivacity. They are fond of the society of their consimilars. They all construct nests, and some of them in the hollows of trees. They are known at the Cape by the name of *blom-suyger* (flower-suckers).

One of them *(Cinnyris Lotenius)* makes its nest on trees, between the branches of which it fixes it horizontally. Its form is hemispherical and concave, and it is composed almost entirely of the down of plants. The female lays five or six eggs, but it is not unusual for her to be driven from the nest by a sort of spider, as large as herself, which sucks the blood of the young.

As for the rest of this tribe, there is nothing to be said of them, except in the way of description, from which our plan precludes us.

The figure of the *long-billed souimanga* is from M. Temminck's works. For the specific characters we refer to the table. It is the *Prit-andun* of Java.

Passing the ARACHNOTHERES, respecting which there are no details of interest, we proceed to the HUMMING-BIRDS.

These are generally natives of the hottest parts of South America, and mostly confined between the tropics. Such as remove from thence, only sojourn in the temperate zones during summer. They follow the sun—advancing and retiring with him. Of two species which are found in North America, one penetrates into Canada, and the other to the North-west, as far as 54° 12' n. lat., where it has been met by Mackenzie. Those of South America do not proceed so far from the tropics. M. D'Azara tells us that they do not pass 35° s. lat. and many of them are sedentary at Buenos Ayres.

It seems certain that none of these birds are found in any part of the Old Continent, nor yet in Australasia, or Polynesia.

Nature has confined to America this one of her *chefd'œuvres.* She has been unusually prodigal of her favours to this little race. The brilliancy of their colours, and the elegance of their forms, are but ill conveyed by description. The Indians, struck with the fire and splendour of their hues, which shine with the united radiance of gems and gold, have given them the expressive name of *hairs of the sun.* These colours have the property of presenting a variety of different shades, according to the direction of the light. They all employ the same materials in the construction of their nests; most of them place it in similar situations, and the eggs are never more than two in number: the male and female work at the nest, and partake the labour of incubation. The nest is composed of various sorts of cotton, or of a silken down, collected from flowers: its texture is so strong and close, that it has the consistence of a soft and thick skin; it is usually placed on a branch, or attached to a single twig.

Their flight is continuous, humming, and so rapid, that the motion of the wings is imperceptible; when the bird hovers in the air, it appears completely immovable; it is observed to rest some instants before a flower, and then dart like an arrow to another, plunging its tongue into the bosom of all which it visits.

These birds never walk, or place themselves on the ground. They pass the night, and the heat of the day, perched on a branch, and often on a thick one. In general, their cries are never heard, but when they quit one plant, or tree in flower, to seek another. Their cry resembles the syllables *te-re,* more or less strong and sharp. They are solitary; and a single one on a tree is never approached by the others; still they will sometimes assemble, hovering about in great numbers, and crossing each other with excessive rapidity, above plants and flowering shrubs. They fight desperately, but disappear before one can discover the result of the combat. They do not exhibit less courage in attacking other birds which approach their nest: sometimes they will even attack them without motive, put them to flight, and even pursue them; larger birds, are thus frequently obliged to give way.

The description given of the tongue of the souimangas, will serve for the humming-birds. Naturalists are at variance on the point whether the humming-birds feed on insects as well as the juice of flowers. Don Felix d'Azara suspects that they have other means of nourishment besides the latter; he tells us, that in the environs of the river Plata, where some of them remain all the year round, and where there are neither woods nor flowers during winter, he has seen them in that season visit spiders' webs, and he thinks it probable that they feed on those insects. He adds, in support of this suspicion, the authority of M. Fr. Isidore de Guerra, a gentleman worthy of the highest credit, who had reared many of them, and who informed him that they actually do eat spiders. But M. Badier, who observed these birds in Guadaloupe, denies that they live at all on the juice of flowers; he assures us that they use their tongue only to catch the little insects in the calix of flowers; and that such as he attempted to rear with syrup, invariably degenerated and died. He adds, that on opening them he found the sugar crystallized in the intestines, and that a part of their intestines had lost its flexibility and become hardened and brittle. Of all those persons who have studied these birds in their native country, there are but two, as above-mentioned, who make them feed on insects at all, and M. Badier alone excludes the juice of flowers. M. Vieillot has also observed them in a living state, and he grants the possibility of their drawing up very small insects, which may be at the bottom of the calices of flowers, and also the dust of the stamina; but though he has killed many of them immediately after eating, he was unable to find any insect in the œsophagus, or in the stomach, or any remains of such; which causes him to believe that the juice of flowers constitutes their food; he also believes, in spite of what M. Badier has said, that this may be replaced by syrup or melted sugar. M. Mondidier has preserved some of these birds from five to six months, by feeding them on a very fine mixture made with biscuit, Spanish wine, and sugar, the substance of which they took by passing their tongue over it. Honey would seem preferable to this aliment. Dr. Latham cites a fact, which, though very singular, rests on the best authority, that humming-birds have not only been brought alive to this country, but a female taken on the nest has hatched her eggs in captivity.—"A young man, a few days before his departure from Jamaica, surprised a female humming-bird, which was hatching; having caught it, and desiring to procure the nest without injuring it, he cut the branch on which it was, and carried the whole on board ship. The bird became sufficiently tame to suffer herself to be fed with honey and water during the passage, and hatched two young ones. The mother, however, did not long survive, but the young ones were brought to England, and continued for some time in possession of Lady Hamond. The late Sir H. Englefield and Hans Sloane Stanley, Esq. both witnesses of the fact, informed the doctor, that these little creatures readily took honey from the lips of Lady H. with their bills; one of them did not live long, but the other survived at least two months after their arrival."

The doctor also informs us, that General Davies preserved several ruby humming-birds alive, for several months, by feeding them on honey, or syrups, or a mixture of raw sugar and water, which he put in the bottom of the calix of artificial flowers, made in the form of a pipe, and whose colours and arrangement approximated to nature as nearly as possible.

In fine, according to M. d'Azara, Don Pedro de Melo, governor of Paraguay, preserved one of these birds, taken adult, for four months. This bird was so familiar that it knew its master exceedingly well, would bestow caresses upon him, and hover about him to demand food. Then Don

1, LANGSDORFF'S CRESTED HUMMING-BIRD WITH NUCHAL RUFF.

2, LANGSDORFF'S HUMMING BIRD WITH ULTRA-MARINE NUCHAL RUFF.

London Published by Whittaker & C.º Ave Maria Lane, June 1829.

Melo would take a vase of very clear syrup, and incline it so that the bird could dip its tongue into it. He would also occasionally give it flowers. With these precautions, the bird lived as well as in the fields, and only perished through the negligence of a domestic.

These facts seem very clearly to establish, that the nectar of flowers, which may be replaced by honey, syrup, or sugar in a state of fusion, is a fit nutriment for these birds, notwithstanding the contrary opinion of M. Badier.

It would be useless to repeat all the little marvels which imagination has added to the natural history of the humming birds, such as their metamorphoses, lethargy during the bad season, death, and resuscitation with the flowers; and this, too, in a country where there is no season without flowers. All these fictions have been rejected by naturalists of good sense.

These birds are not distrustful, and suffer themselves to be approached within five or six paces, and sometimes nearer: so that a person placing himself in a flowery bank may catch them with a wand, covered with a gluey gum, in his hand. It is sufficient to touch them when they are humming before a flower; but it requires a correct eye, as they are in perpetual motion. This mode of catching them is attended with the inconvenience of injuring their plumage; therefore glue should be avoided, as the only object of pursuing them is for their plumage. They may be killed by shooting small peas with a pop-gun, which requires very great address. They may also be brought down by inundating them with water from a syringe, or by using sand instead of shot in a pistol. Even an explosion from the powder alone, taking aim very near, will be sufficient to make them fall. A net of green gauze, such as is sometimes used for butterflies, may be employed, and is sure to preserve their rich plumage; but it requires great patience, and can only be used on plants and dwarf shrubs.

We have now given every thing that is known concerning these birds except specific descriptions.

We insert figures of two species of humming-birds, brought from Brazil by M. Langsdorff. The first is crested with straight simple reddish feathers; above, the bird is green, inclining to reddish on the rump and tail, the quills vinaceous; the front, throat, and breast of a deep green waved; the colour of the belly and vent like that of the back, but divided from the deep green of the breast by a white patch. The lateral crest is of broad feathers, inclining backward; green at the tips; whitish towards the insertion.

The other is deep-green, waved all over, except the quills of the wings, which are blackish; and the lateral crest, which is of ultra-marine blue, mixed with red.

The figure of the giant humming-bird is from M. Vieillot. It is green-brown above; deeper on the back; the rump is white, mixed with red; the upper wing coverts are white, bordered with red; the rectrices have a little white spot at their extremity, and there is a white triangular spot at the end of the remiges; the abdomen is white. It is about nine inches long, and is much the largest of all the known species.

The habits of the Craves are like those of the rollers, with which they are classed by most naturalists.

Of the Hoopoes proper, we shall only notice the *Upupa epops*. It arrives in Europe in spring, spreads itself as far as the most northern countries, and quits this quarter of the globe in autumn to pass the winter in Africa. The species is sedentary in Egypt, and almost domestic, for it lives in the most populous cities, and nestles in the terraces of the houses. In France it is solitary, few of them being ever seen together. This bird delights in humid places, where it finds a more abundant nutriment. It is seldom found on high mountains. When it perches, it is at a moderate height. It is also at a small elevation that the hoopoe chuses a hole in which to construct its nest. Sometimes it takes one in a wall; some-

GIANT HUMMING BIRD.

TROCHILUS GIGAS.

London. Published by Whittaker & Co. Ave Maria Lane, June 1829.

times in an old tree, as a willow, &c.; and sometimes it places it on the ground amid the roots. It is pretended that it invests it with potters' clay, and also with the most infectious substances, which produce a disagreeable exhalation from the young birds. This, however, appears to be more than doubtful; for the bird, in general, is known to make its nest of dried leaves and moss, without any fetid materials. It lays from four to seven eggs, ashen grey, of an elongated form, and a little more bulky than those of the blackbird. It has various cries, but no song.

In a state of liberty the hoopoe feeds on terrestrial insects, worms, berries, and vegetable substances. In captivity it is nourished with raw meat, in long cuts. It grows very fat in autumn, and its flesh is in great request in Italy, in the islands of the Archipelago, and in various districts of France.

The hoopoe, taken young or old, soon grows familiar, and will accommodate itself to various kinds of food which it would not take when free. It must never be kept in a cage, but suffered to run freely through gardens and houses. Like all insectivorous birds, it drinks little.

Of the Promerops and Epimachi we are unable to add any thing to the specific descriptions of the text and table. The Epimachi are classed with the Promerops by most naturalists.

We now proceed to the last family of the Passeres, resuming the text of M. Cuvier.

The second and smaller division of the Passeres comprehends those in which the external toe, almost as long as the middle, is united thereto as far as the last articulation; we make but one group of them.

The Syndactyli,

Long since divided into five genera, which we preserve.

The Bee-Eaters, (Merops,)

With elongated bill, triangular at base, slightly arched, and terminating in a sharp point. Their sternum has behind on each side a double slope. Their long and pointed wings and short feet cause them to fly like the swallows. They pursue insects in large bodies, especially bees, wasps, &c., and it is remarkable that they are not stung by them.

There is one species common in the south of Europe, but rare in our latitudes.

The Common Bee-Eater. (Merops Apiaster.) Enl. 938, Naum. 163. Vaill. Guep. 1 and 2.

A fine bird with a yellow back; forehead and belly blue; the throat yellow, surrounded with black; which builds in hollows; it digs in banks four or five feet deep. The young continue there a long time with their parents, which gave rise, among the ancients, to the idea that the young took care of their parents in their old age.

The two middle quills of the tail are a little elongated, the first indication of a considerably greater prolongation in the majority of the foreign species.

Indian Bee-Eater (*Merops Viridis*). Enl. 740. Vaill. 4. Green-gold above; beneath green; throat blue; band of black on side of head. Nearly nine inches. Bengal. *Apiastes, Bengalensis, Torquatus,* and *Merops Egyptius* are varieties of this.

Variegated Bee-Eater (*Me. Ornatus.*) Lath.

Back and wings green, varied red and yellow; head varied red, black, blue, and yellow; tail blue; outer feathers red. Two middle tail-feathers longest. New Holland.

Superb Bee-Eater (*Me. Superbus.*) Nat. mis. 18.

Plumage generally red; forehead, eye, throat, and rump blue. Middle tail-feathers longest. Nine inches.

Chesnut Bee-Eater (*Me. Senegalensis.* Enl. 314, *et Badius,* 252.) Vaill. 12, 13.

Upper part of head, neck, and scapulars chesnut; brown stripe on side of head; body blue green. Eleven inches. Isle of France.

Supercilious Bee-Eater (*Me. Superciliosus.*) 259. Vaill. 19.

Body and wings dull green above; throat yellowish; below chesnut; eye in a black patch, with a greenish white band over and under it. About 11 inches. Madagascar.

The *M. Savignii* of Vaill. 6, seems to be the same as *M. Superciliosus.* M. Levaillant speaks of three distinct varieties, proper respectively to Persia and Egypt, to Senegal, and to Malimba and Madagascar.

Cuvier's Bee-eater. M. Cuvieri. Vaill. 9, and Swainson's Illustrations 76, under the name of *Savigny.*

Bar across the eyes, and rump black; throat white; forehead and sides of head with a white band; reddish green on the back; bright green beneath.

Lamarck's Bee-eater. M. Viridis, Gm. *M. Lamarck.* Vaill. 10.

Green-gold above, inclining to blue; tail-coverts green below; throat blue.

Many species, nevertheless, have the tail nearly squared.

Philippine Bee-eater. (*Me. Philipinus.*) Enl. 57.

Dull-green, above; rump and tail, blue-green; stripe of black on side of head; beneath, pale-green. Nine inches. Philippine Islands.

Cayenne Bee-eater. (*Me. Cayanensis.* 454.)

Light brownish-green; quills and tail, rufous.

N.B. This bird does not belong to Cayenne.

Red-winged Bee-eater. (*Me. Erythropterus.*) 318.

Dull-green, above; pale rufous-chesnut, beneath. Senegal.

Malimbic Bee-eater. (*Me. Malimbicus.*) Shaw. Or *Bicolor.* Daud. Ann. du Mus. I. 62. Vaill. 5. Vieill. Gal. 186.

Head and neck, slate colour; body, wings, and tail, vinaceous; dark streak through the eyes. Ten inches. Malimba.

Red-throated Bee-eater. M. Gularis. Nat. Mis. 337.

Velvet-black, throat; blood-red, forehead; rump, scapulars, and two middle tail-feathers, edged with blue. Sierra Leone.

The Ruffed Bee-eater. (*M. Amictus.*) Col. 310.

Top of the head, green; blue and pink, throat; and fore-part of breast covered with large, loose, deep-red feathers. Sumatra.

The *M. Daudin.* Vaill. 14, seems allied to the Philippine Bee-eater.

Coromandel Bee-eater. (*M. Coromandus.*) Lath. Sonnerat's Second Voy. or *G. Cytrin.* Vaill. 11.

Head and neck, yellow; beneath, greenish-yellow; black streak through the eye.

Varied Bee-eater. M. Quinticolor. Vaill. 15.

Throat, orange-yellow, with a blackish-green collar; scapulars, and upper wing-coverts, green; rump, blue; quills of tail, blue; above, greenish; underneath, breast, and belly, yellowish-green. From Ceylon.

Red-winged Bee-eater. M. Minutus. Vaill. 17.

Above and middle tail-feathers, green; quills, brick-dust color, tipped with black.

Leschenault's Bee-eater. (M. Leschenault.) Vaill. 18.

Very like *M. Quinticolor*, but with a black band.

Bullock's Bee-eater. (M. Bullock.) Vaill. 20.

Feathers, shorter and more silky than in the other species; yellowish-green, above; bluish on the head; middle wing-feathers tipped black; red round patch on throat; breast and belly, yellow; vent, blue.

Sometimes the tail is a little forked, but that again depends often on the state in which they are killed.

Swallow-tailed Bee-eater. (M. Taiva.) Vaill. 8.

Yellowish shining green, above; black band through the eyes; black patch on throat; collar and rump, blue. South Africa and Egypt.

Pirik Bee-eater. (M. Urica.) Swainson. Ill. 8.

Glossy olive-green, above; and beneath crown and

neck, chesnut; chin and throat, sulphur coloured; band, black. Java.*

The *Me. Congener.* Aldr. I. 876. is not authenticated. The *Cafer*, Gm. is *Upupa Promerops.* The *Brasiliensis* of Seba, is probably a Troupiale. The *M. Monachus, Corniculatus*, and *Cyanops* are thrushes. *M. Phrygius Cincinnatus, Cuculatus, Cyanops, garulus, Fasciculatus Caruncidatus* of Latham, appear to me to be Philedons, and I have satisfied myself that most of them are so. The *M. Cinereus* is a Soui-Manga with a long tail.

We should place near the bee-eaters certain birds with a long tail and metallic plumage, placed hitherto among the promerops, but of which the two external toes are united nearly as much as in the bee-eaters.

Red Billed Promerops. Promerops Moqueur, Upupa Erythrorynchos, Lath. The young with a black bill.

Black with a red gloss in some lights; white spot on inner web of first six quills. Abyssinia.

Namaquois Promerops. Prom. Namaquois, Vaill.

* To the above may be added,—
Merops Javanicus. Horsf. Lin. Trans. XIII. *M. Melanurus.* Vig. and Horsf. *M. Orientalis, M. Cærulescens, M. Albifrous*, and *M. Chrysocephalus* of Latham, with other from the General History which are more doubtful. Also, *Guepier gris rose; G. Sonini; G. Adanson*, and *G. Rongegorge*, from Le Vaillant.

Falcin. Cyanomelas, Vieill. Bill thinner and more curved than in the last; tail not so long or so much cuneiformed. Shining black with bluish reflexions, and the white spots on the tail.

The bee-eaters appear to be wanting in America, where their place appears to be in some measure supplied by

The Motmots. (Prionites, Illiger.)

Which have similar feet and gait, but differ by a stronger bill, the edges of which are serrated in both mandibles, and by a bearded tongue like a feather, after the manner of that of the toucans. They are handsome birds, about the size of the pie, with the plumage of the head loose like the jay's; a long wedged tail, the two middle quills of which are barbed in the adult, for a small space, not far from the end, which gives their tail a very peculiar form. They fly badly, live in solitude, nestle in holes, feed on insects, and even pursue small birds.

Brasilian Motmot. (Ramphastos Momota.) Gm. Enl. 370. *P. Brasiliensis.* Ill. Vaill. Ois. de Par. pl. 37 and 38.

Green above; quills, blue; reddish-green beneath; black dot on chest; spot of velvet black on the head, with green before it, and blue behind. Size of a magpie. South America.

Red-headed Motmot. (Motmot Dombey.) Vaill. loc. cit. pl. 39, and Vieill. pl. 190.

Like the last, but the head is red; and it has but ten caudal feathers, of which the other has twelve.

Pr. Marcii, Spix. 9.

Head, green; cheeks, and spot on chest, black; quills bluish-chesnut.

The *Tutre* of Paraguay of Azara, No. 52, is at least nearly allied to these.

The KINGFISHERS. (ALCEDO L.)

Have shorter feet than the bee-eaters, and the bill much longer. It is straight, angular, and pointed. The tongue and tail are very short. They live on small fish, which they catch by precipitating themselves into the water from the top of some branches, where they remain perched to watch their prey. Their stomach is a membranous sack. They nestle like the bee-eaters in holes on the rivers' banks. They are found in both continents.

The European Species *(Alcedo Ispida.)* Enl. 77.

As large as a sparrow; is greenish above, waved with black; a broad band of the finest aquamarine blue, predominates along the back; the under part, and a band on each side of the neck, are reddish.

The foreign species, like our own, have almost all

a loose plumage, varied with different shades of blue and green.

They may be distinguished according to their bills, which are sometimes simply straight and pointed, as in the common species.

Great African Kingfisher, Alcedo Afra. Sh. Maxima. Lath. Enl. 679.

Deap lead-colour; ferrugineous beneath, covered with small white spots; crest on head and neck; size of a crow. Africa.

Belted Kingfisher. (*Al. Alcyon*, 715 and 593, and Wilson's Amer. III.)

Bluish-ash, above; white beneath; collar and spots white; crested. Size of a blackbird. North America.
The *Ispida Carolinensis* of Briss. and *Jaculott Cinereus* of Klein.

Cinereus Kingfisher. (*Al. Torquata.*) 284.

Bluish-ash, above; chesnut underneath; collar and spots, white. Size of a magpie. Mexico.

Black and White Kingfisher. (*Al. Rudis.*) 62 and 716.

Black and white above; white beneath. Eleven inches. Asia and Africa.

Rufous and Green Kingfisher, or Spotted K. of Shaw. (*Al. Bicolor.* 592.)

Green above, with a few white spots; under part gilded rufous. Eight inches. Cayenne.

White and Green Kingfisher. (*Al. Americana.* 591.)

Dark green above, marked with white; beneath, white Seven inches. Cayenne.

Indian Kingfisher. Al. Bengalensis. Ed. II.

Blue-green above; rufous beneath; head striped blue and rufous. Four and a half inches.

There is a smaller variety without the rufous on the sides of the head.

Blue-headed Kingfisher. (*Al. Cæruleocephala.*) Enl. 356.

Back and wings blue; throat white; beneath rufous. Four and a half inches. India and Madagascar.

Crested Kingfisher. Al. Cristata. 756.

Bright blue above; scapulars violet; beneath pale-rufous; stripe on neck, blue. Five inches. Philippine Islands.

Purple Kingfisher. Al. Purpurea. 778. 2.

Head, rump, and tail, golden rufous; back and wing-coverts, blue black; underneath, rufous white. Near seven inches. India.

Rufous Kingfisher. Al. Madagascariensis. 778. 1.

Rufous above; beneath white, and slightly rufous. Five inches. Madagascar. Much allied to the last.

Supercilious Kingfisher. Al. Superciliosa. 756.

Green above; throat orange; belly whitish; stripe over the eye, orange. Five inches. Cayenne.

Gray-fronted Kingfisher. A. Cinereifrons. Vieill. Gal. 187.

Blue above; wing-coverts and scapulars black; throat and belly whitish; upper mandible yellow, with red and black spots; forehead grey. Malimba.

Biru Kingfisher. A. Biru. Horsf. Java. Col. 239. Pale azure; quills, brown within, beneath white. Java.

Half-collared Kingfisher. A. Semitorquata. Swainson Ill. 154.

Bluish-green; beneath, buff; black bars on the head, and collar, and on the breast. Africa.

Asiatic Kingfisher. A. Asiatica. Ib. 50.

Shining-blue, above; beneath, rufous; head, black, transversely banded with blue. India.

And sometimes they have the lower mandible swelled out.

Cape Kingfisher. Al. Capensis. 599.

Blue green above, inclining to ash; underneath, fulvous; chin white. Fourteen inches. South Africa, and probably India and China.

Black-capped Kingfisher. Al. Atricapilla. 673.

Head and neck, above, black; upper parts deep blue; throat and belly white and rufous. Ten inches. China.

Smyrna Kingfisher. Al. Smyrnensis. 232, and 894.

Chesnut; throat white. Eight and a half inches. Smyrna.

Varieties of this seem to be found in Java and India.

This is one of the two species distinguished by Aristotle.

Ternate, Kingfisher. Al. Dea. 116.

Head, glossy blue; back brown, margined with blue; underneath, rose white. Thirteen and a quarter inches. Island of Ternate.

M. Vigors makes this the type of his genus TANYSIPTERA.

Green-headed Kingfisher. Al. Chlorocephala. 783.

Head, green; back, dusky black; beneath, white. Nine inches. The Molucca Islands. Probably allied to the collared species.

Coromandel Kingfisher. Al. Coromanda. Sonn. 218.

Reddish lilac above; throat and beneath, white and rufous; white spot on rump. Size of a blackbird. Coromandel Coast.

White-headed Kingfisher. Al. Leucocephala Javanica. Sh. 757.

White above; head, streaked with black; wings and tail, greenish blue. Twelve inches. Java.

Senegal Kingfisher. Al. Senegalensis. 594 and 356.

Back, blue; head, grey; black mark before the eye. Nine inches. Senegal.

Crab-eating Kingfisher. Al. Cancrophaga. Sh. 334.

Blue-green above; black patch on wings; pestocular

streak, black; fulvous-yellow beneath. Twelve inches. Senegal.

Black-billed Kingfisher. Al. Melanoryncha. Tem. Col. 391.

Bill large and black; head and under parts, buff waved; wings, rump and tail, green. Celebes.

Varied Kingfisher. A. Omnicolor. T. Col. 135.

Bill and feet, red; back and belly, deep blue; quills, green; coverts and head, black; neck, waved black and deep blue; throat, waved red and black. Java.

Double-eyed Kingfisher. A. Diops. Id. Col. 212.

Above, ultramarine, with the vent and collar of same; a white patch between the nostril and eye; and under parts, white. Amboina, Celebes, &c.

Stubbed Kingfisher. Dacelo Concreta. Id. Col. 346.

Short and thick; feathers of the head rather elongated; bill yellow, except the upper part of the upper mandible; the whole bird varied with deep and light blue, orange, green, and white. Sumatra.

Dacelo Cinnamoninus. Swain. Ill. 67.

Blue-green; beneath, cinnamon. New Zealand.

It is of this division that Dr. Leach has formed his genus Dacelo.

N.B. In several of the figures in the *Planches Enluminées*, the bill is not sufficiently swollen.

There are, however, some in New Holland and in the neighbouring countries, with the mandible crooked at the end. In many of them the plumage, being greyish, and not smooth, shows that they do not frequent the waters. They live on insects, and have been called in French *Martins-chasseurs.*

Great Brown Kingfisher. Al. Fusca. Gm. *Gigantea,* Sh. Enl. 633.

Olive brown above; pale blue green underneath. Eighteen inches being the largest species. New Guinea.

Little Kingfisher. Col. 277. *Dac. Pulchella.* Horsf. Tem.

Bill deep red; back and tail barred blue and black; crown of head blue, with whitish spots; throat white; belly orange; front, back and sides of the head dirty red; quills black, spotted with white. Java and Sumatra.

Blue-eared Kingfisher. D. Cyanotis. Col. 262.

Bill red; feathers of head and neck brown; top of head and tail orange; through the eyes, and round the head and coverts, deep blue; underneath white. Java.

D. Gaudichaud. Quoy and Gaim. Pl. 25.

Deep black on the head and shoulders; white patch on throat, extending to sides of neck, and forming a collar; back, rump, and upper wing-coverts bright blue. New Holland.*

* Other species have been mentioned, as—

Alcedo Melanoptera. Horsf. *A. Collaris, A. Sacra,* and *A. Maculata.* Lath. Also, the Guriac, Black-winged, Dun, Black-backed, Lybian, Nubian, Ferrugineous, Rose-cheeked, Blue-crested, and the Trinidad Kingfishers, by the same; and *A. Venerata, A. Tuta,* and *A. Crithaca,* by Gmelin. These seem to be more doubtful—*A. Ægyptia, A. Novæ Guinæ, A. Flavicans, A. Cærulea,* and *A. Cayanensis* of Gmelin, *A. Amazonia,* and *A. Surimensis* of Latham, and *A. Leuchoryncha* of Seba.

The Ceyx. Lacepede.

Are kingfishers, with the common bill, but are destitute of the internal toe. There are two species in the Indies.

Three-toed Kingfisher. Al. Tridactyla. Pal. and Gm.

Azure blue above; yellowish white underneath. Four inches. Java.

M. Temminck connects this with the purple Kingfisher, notwithstanding the difference in the toes. If he is correct the number of toes must vary.

Azure Kingfisher. Al. Tribrachys. Sh. Nat. Mis. Pl. 681.

Fine deep blue above; buff underneath; streaks buff and white. Seven inches. New Holland.

Meninting Kingfisher. Alc. Meninting. Hors. Col. 239.

Light-blue above, speckled with white on the head; beneath white, with a blue collar; tips of wings and tail black. Java.

The Todies. (Todeus. Lin.)

Are small American birds, very like the Kingfishers in general form, and which also have the feet and bill elongated; but the latter is flattened horizontally, and obtuse at its extremity. The tarsi are more elevated, and the tail less short. They live on flies, and nestle on the ground.

Green Tody. Todus Viridis. Enl. 585. Vieill. Gal. 124. Beautiful green above; yellowish-white beneath; throat spot red. Four inches. Jamaica and the Continent.

The *T. Ceruleus* appears to be a Kingfisher. Ed.

The *Todus fuscus*, *T. ferrugineous*, *T. plumbeus*, *T. sylvia*, *T. cinereus*, and *T. varius*, have also been named.

The following species have been improperly referred to this genus:—*Todus Regius*, Enl. 289; *Paradisæus*, ib. 234; *Leucocephalus*, Pallas, and the two Platyrhinci of Desmarest, which are *Todus Rostratus and Nasutus* of Shaw, or the *To. Platyrhynchos and Macrorhynchos* of Gm. Vieillot gives the first, Gal. 126.

We shall terminate the history of this order by the most extraordinary of its genera, which have not so much resemblance to the other Syndactyli as they have to one another, and which might very well make a particular family. These are

The Hornbills. (Buceros, L.)

Large birds of Africa and the Indies, which their enormous denticulated bill, surmounted by prominences sometimes as great as itself, or at least strongly swelled out above, renders so remarkable, and connects with the toucans; while their port and habits approximate them to the ravens, and their feet are those of Merops and Alcedo. The form of the excrescences of the bill varies with age. The interior is usually cellular. The tongue is small, at the bottom of the throat. They take all kinds of food, hunt mice, small birds, and reptiles, and do not disdain carcasses.

HORNBILLS with protuberance on the bill.

Rhinoceros Hornbill. (*Buc. Rhinoceros.*) Enl. 934. Vaill. Cal. 1 and 2. *B. Africanus*, Vail. pl. 17, may probably be a variety of age. *B. Niger*, according to M. Temminck, is the same, badly preserved.

Black; dirty white, beneath; protuberance on the bill as large as bill itself, and turning upwards. Bird, three feet and upwards long, from tip of bill to end of tail.

Unicorn Hornbill. (B. Monoceros.) Sh. Enl. 873. Vaill. 9, 10, 11, and 12.

Black, glossed with green and purple; three outer quills of wings and tail white; the prominence on the bill prolonged into a kind of horn.

Great-billed Hornbill. (B. Cassidix.) Tem. Col. 210.

Bill, very large, bright yellow; prominence on bill very large, and semi-circular; body, wings, and thighs greenish shining black; top of head marron. Celebes.

Malabar Hornbill. (P. Malabaricus, Lath. or *Albirostris.)* Sh. Col. 14.

Black above, with violet and green; beneath white; first quills of the wing, and three exterior tail feathers, white; helmet varying with age. Malabar, Coromandel Coast.

Trumpet Hornbill. (*B. Buccinator.*) Col. 284.

Green, above; belly and vent, white; tail tipped with white; prominence, or bill, trumpet-shaped. Cape.

Gingi Hornbill. (*B. Gingianus.*) Son. Second Voy. t. 121. Vaill. 15.

Cinereous grey, above; black band beneath the eye; breast and belly, white; tail, cuneiform, with two middle feathers rufous grey, tipped with black. Gingi.

Bifronted Hornbill. (*B. Bicornis.*) Vaill. 7. the adult female; *Cavatus,* Id. 4, is the middle-aged male. Plates 3 and 5 are altered individuals.

Black with white patch on second quills; protuberance forming a double horn. Philippine Islands.

Bontian or Indian Hornbill. (*B. Hydrocorax.*) Enl. 282, the young. Col. 283, adult. Vaill. t. 6.

Brown; cheeks and throat, black; greyish half-collar on neck. Molucca Islands.

Violaceous Hornbill. (*B. Violaceus.*) Id. 19. *Abyssinicus.* Enl. 779, in the middle age. Vaill. Ap. 230 and 31, adult birds. Vieill. Gal. 191.

Like *B. Monoceros,* but with the three last quills on each side the tail, white. Coromandel.

Channelled Hornbill. (*B. Sulcatus.*) Col. 69. (*B. Leucocephalus.*) Vaill.

Blackish-green, face; front of neck, whitish; beneath, reddish; prominence, channelled longitudinally; feathers of the head, long, red. Celebes.

Panayan or Furrowed Hornbill. (*B. Panayensis.*) Enl. 780, female; 781, the old male. Vaill. Col. 16,

17, 18. *B. Manillensis.* Enl. 891, may be the young, the *B. Enleirostris.* Wagler.

Greenish-black, above, changing into blue in different lights; under parts, more dusky; feathers of neck, elongated. Isle of Panay.

Striped-tailed or Angola Hornbill. (*B. Fasciatus.*) Vaill. Af. 233. *B. Melanoleucus.* Vaill.

Tail, with alternate black and white bands. Africa.

Plowed Hornbill. (*B. Exaratus.*) Col. 211.

Blackish green, all over; prominence channelled longitudinally; feathers of the head, long. Celebes.

Hornbills without a prominence on the bill.

B. Javanicus. Vaill. Cal. 22, the young male. Id. Ap. 230, old male, the same as the *Col. de Waidjiou.* Labill. Voy. *B. Undulatus.* Shaw. Vaill. Col. 20 and 21, are the females. *B. Erythrorynchos,* Enl. 260. Vaill. Af. 238, the young.

B. Hastatus. Cuv. Enl. 890. Vaill. 236 and 237.

Bill, much arched; greyish, above; each feather edged with white. Senegal.

Crowned Hornbill. (*B. Coronatus.*) Shaw. Vaill. Af. 234 and 5.

White band from the corner of the eye passing round the back of the head. South Africa.

(*B. Bengalensis.*) Vaill. Cal. 23.

Bill, smooth, large; black-brown; wing, blue-grey; coverts, black tipt. Ceylon.

N.B. The *B. Galeatus*, of which the head only is known, Enl. 933, and which Vaillant erroneously thinks an aquatic bird, is a true hornbill, in which the prominence is furnished with a horn excessively thick, especially at the anterior part. Gen. Hardwick has lately described and figured the examples. Third Lin. Trans. xiv. t. 23.

B. Ambigenas. Swainson. Ill. *B. Malayanus.* Raff. Lin. Trans. *B. Cornutus.* Id. ib. *B. Jubatus.* Vaill. Dict. *B. Plicatus.* *β.* Lath.

B. Griseus. and *B. Viridis.* may probably be Philedons. *B. Ruber.* Lath. and Sh. is *Coracia Militaris.* *B. Albus*, according to Wagler, is an aquatic bird. *B. Orientalis.* Sh. is *Scythrops Novæ Hollandie.*

The *Corbicalao* of Vaill. is *Philedon Corniculatus.*

SUPPLEMENT ON THE SYNDACTYLI.

The SYNDACTYLIS, or second division of the passerine birds of Cuvier, includes those only which have the middle and external toes united as far as their penultimate articulations. The word has been applied by other ornithologists, particularly M. Vieillot, to certain of the aquatic birds whose feet are more or less palmated, and affords, therefore, an example of the difficulties imposed on the uninitiated in zoology by the various terminology of the professors and masters of the science. Those who have bestowed a life of severe labour in the praiseworthy investigation of animated Nature, seem too often blind to their own legitimate merit, while they seek for reputation in the puerile invention and application of new names; but, not to digress, suffice it to say, that syndactyli, with our respected author and his followers, means a division of the present order; with other authors, it means something else.

This small division will be seen by the text to include only the bee-eaters, the *Prionites* of Illiger, which M. Vieillot calls the *Bariphonus*, i. e. the Motmots of Latham, &c.; the Kingfishers, with their sections; the *Ceyx* of Lacepède; the Todies; and the Horn-bills.

In the form of the body, mode of flying, locality, and even habits, there is some analogy between the BEE-EATERS and the swallows; so much so indeed that in the neighbourhood of the Cape, where they most abound, the Dutch colonists call them mountain swallows.

In addition to their generic characters, already indicated, we may add, that they have one toe turned backward and three forward; and, that in common with the present division, the outer toe and the middle are united. The tail varies in shape, being square and forked in some species, and having the two middle tail-feathers larger than the rest in others. This has given rise to three divisions of the genus. The males are rather larger than the other sex, and more brilliant in colours.

Like the swallows, the bee-eaters eat, drink, bathe, &c. in flying, and seem to perch, not, like most other birds, to search for food, but merely for the purpose of rest. The winged insects, and more especially bees, constitute the food of these birds; but although they are named in French guêpiers, or wasp-eaters, it seems probable that they do not attack these insects, but confine themselves, if possible, to the honey-bee. Our countryman Ray seems to conclude that they feed on small fish, probably from their habit of flying over the surface of water, and frequenting marsh lands. They nestle, in general, like the kingfisher and river-swallows, in holes on the banks of rivers; and although they are said to make these holes themselves, by means of the bill and claws, they are sometimes expelled from them by the swallows.

In their social habits the several species of bee-eaters seem to differ, some living in large societies, some in single pairs, and some in small families. M. Levaillant observes that in the parts of Africa he visited, the square or forked-tailed bee-eaters live in small families: while those which have the two intermediate tail-feathers longer than the rest, live in larger bands.

It has been said, absurdly enough, that these birds can fly backward. The fact is, that their rapid turns at times, while on the wing, in pursuit of insects, which endeavour to evade their destroyer by acute angles in flight, have induced this delusion.

They are, it seems, good eating, and are said to imbibe a sweet flavour from the honey of the bees which they feed on. The skin, however, is coriaceous.

Enough may have been said already of the specific character of the *Common Bee-eater* to distinguish it. It is ten or twelve inches long, and is found in various parts of Europe, but no where more plentifully than in the neighbourhood of Gibraltar, where they make their nests in holes in sand-banks. These holes penetrate the bank about three feet, and then take a rectangular direction for about three feet more, where a large cavity is formed for the nests, in which are deposited six or seven white eggs, rather less than those of the blackbird.

The bee-eater is not, however, confined to Europe; but is common in the southern latitudes of Russia, in India, and especially in Southern Africa, where it is said to guide the Hottentots to the wild honey in the woods.

It has been, though very rarely, seen in this country; a flock of them is recorded in the Linnæan Transactions, to have appeared in Norfolk, in 1793; and we have inserted the figure of one shot by Mr. Geslin, in Devonshire, in the year 1827.

We have also inserted a figure of a bee-eater, in the extensive and valuable collection of the Zoological Society. This species was brought from Sumatra by the late lamented Sir Stamford Raffles, and belongs, like the common species, to that division of the genus, distinguishable by the elongation of the middle tail-feathers.

The top of the head, and half down the back, are of a rich brown colour; the throat, rump, and tail-feathers, are ultramarine blue; the breast and wing-coverts, sea-green; the primary quills of the wings are nearly black, as are also the bill and tail.

In illustration of that division of the Bee-eaters, which is characterized by a forked tail, we have also inserted a figure drawn by Major Hamilton Smith, from a specimen which was in Mr. Bullock's late Museum, and named by him *Merops puella*. This bird is extremely delicate and pretty. The head

SMITH'S BEE EATER.

MEROPS PUELLA. Hamilton Smith.

London Pub.d by G.B. Whittaker March 1829.

SUMATRA BEE-EATER.

MERORS SUMATRANUS. Raffles.

London Pub.d by G.B.Whittaker, Nov. 1828.

THE BEE-EATER.

MEROPS APIASTER. L.

London, Published by Whitaker & C° Ave Maria Lane, June 1829.

and neck are white; the back, sides, and two intermediate tail-feathers, which extend a little beyond the other, are of light buff colour; the wings are white, but the humeral feathers and the quills for about half their length, are of a fine ultra-marine blue, tipt with black; the rump also is blue, which colour extends up the back to a point; the outermost or long tail-feathers are tipt with black, and the lateral tail-feathers are ash-coloured, for about half their length from the insertion, then white, and finally tipt with black.

Fernandez originally described under the name of MOTMOT the type species of the genus now so named. We have seen that it is not far removed from the bee-eaters, and differs, perhaps principally, in the indented bill and pencillated tongue. Brisson, by suppressing the *t*, latinized this word *momotus*, and applied it to a genus. Illiger, ever proud of his classical learning, however happy, in general, in the application of it to zoology, designated the genus by the name *prionites*, in reference to the indentations of the bill—a character, by the way, we must observe, by no means peculiar to the motmot, and, therefore, not with propriety one from which it should be named. M. Vieillot, having appropriated the Greek of Illiger to one of the families of his Sylvia, has coined a new name for the motmot—this name, *baryphonus*, has reference to the voice of the bird, said to be not unlike that of a man.

We are not as yet well-informed on the natural habits of the motmots, though D'Azara has written at some length on their manners in captivity. Piso states, that they build on the summits of high trees; but the better opinion appears to be, that, like the preceding genus, they betake themselves to holes in the ground for the purpose of nidification; D'Azara describes those he had, as heavy and stiff in their movements, which, on the ground, were by sudden and oblique leaps, with the legs very wide: their tail was in constant motion; they were fed with bread and raw meat, the latter of which they preferred. If a small bird, or mouse, were let loose in the chamber in which the motmots were kept, they would pursue it in a most determined manner, and when possessed of their prey, would strike it violently against the ground with their bill. This did not appear to be done merely for the purpose of killing it, but to break the bones, in order to swallow the whole more easily.

The two middle tail-feathers are more than three inches larger than the others, and these feathers are frequently found partially denuded of webs; this peculiarity is indeed so common, that it was for some time considered generic; but, as it is never found in the young birds, and not always in the old, there seems to be no doubt that it results from some hitherto unknown habit, and it is not improbable that if these birds breed under ground, the effect in question may result from some friction during the breeding season.

The KINGFISHERS, ALCEDO, have in general a body thick in proportion to its length; a large head; the bill long, angular, and thick at the base; the nostrils small, and in general covered with feathers; the tongue fleshy, short, and sharp; the tarsi very short; the toes, four; and, as in all the syndactyli, the outer toe and the next are united; the tail is in general very short. Some of these birds have a crest, but more of them are without. Their stomach is like that of the rapacious birds, and, like them, they return at the mouth, in small pellets, the indigestible parts of their food.

The species of this genus are very numerous, and are spread over almost all the world. In Europe, however, we have but one, which is as vivid and highly ornamented, as to colour, as any European bird, though much inferior in beauty to those species of the genus which are found within the tropics. The largest of the species are nearly as big as a crow; but the smallest do not much exceed the size of the nightingale.

Few birds have given rise to more superstition and folly in mankind than these. The Greeks called the common species *Alcyon*, from Alcyone, the daughter of Æolus, and wife of Ceyx, on whose death, by drowning, as the poets inform us, she threw herself into the sea, and both were metamorphosed into kingfishers. M. Lacépede has separated the kingfishers with three toes into a distinct genus under the name of Ceyx. The ancients had, moreover, many superstitious notions with regard to these birds; and even in the present day, and in very different parts, as in Siberia and the South Sea Islands, they are still looked upon with veneration and awe.

It was till lately considered that fish was the general if not the sole food of all the birds of this genus, though at times they are obliged to satisfy themselves with worms and insects. This is certainly true as to such of the species as are inhabitants of the banks of rivers.

The rage for improvement, and *assisting the memory* by the invention of new genera, has extended itself to the kingfishers, although the numerous species of these birds, widely spread over the surface of the earth, are as yet insufficiently known for the purposes of interrelative comparisons. Cuvier has distinguished, as subgenera, or groups, the common kingfisher, with its similars with a straight and pointed bill; others with the under mandible swelled out; and others, again, with the mandible bent at the end, and the feathers loose and unlike aquatic birds; to which he has added the genus Ceyx of Lacépede, or kingfisher with three toes.

M. Vieillot has also divided these birds according to the number of the toes; but he has subdivided those with four toes into three sections; the first of which is distinguished by a straight quadrangular bill; the second by a straight triangular bill, having the lower mandible convex; and the third by a triangular bill, with a furrow in the upper mandible. The last of which sections comprehends three species, one of which is *Alcedo Gigantea* of Lath., the *Fusca* of Gm. This species Dr. Leach has made a distinct genus, under the name Dacelo, distinguished by a conical quadrangular bill, opening under the eyes, the upper mandible larger than the under, and strongly furrowed toward the point.

M. Temminck admits this genus Dacelo of Leach, which he calls the martin-chasseur. With reference to the insectivorous regimen of the species, and in contradistinction to the martin-pêcheur, which feeds principally on fish; and in addition to the character designated by Dr. Leach as peculiar to his Alcedo, M. Temminck adds, among others of less importance, that the plumage is loose in the Alcedo, and not smooth and shining, and calculated, therefore, for aquatic habits, as it is in the common species. It would be a solecism in English to speak of the insectivorous and piscivorous kingfishers; and if the two genera are to be considered as established, we must have recourse, in conversation, to the scientific appellatives for one or both, to distinguish them from each other. The Baron, we may observe, in conclusion, has not deemed the insectivorous regimen a sufficient ground for a generic separation of the species to which it is more particularly applicable.

It might, perhaps, have been well if the genus Ceyx, which seems to possess (if any) more merit for its classical and pretty allusions than for its utility to science, had never emerged from the great genus Alcedo, to which it originally belonged: differing, as it appears to do, nothing from the common kingfishers but in the uninfluential character of being tridactylous.

The *Common Kingfisher*, which is found in Europe and in Asia, is six or seven inches long. This species has no crest on the head. The bill is black, but the inside of the mouth is yellow; but we shall not repeat here its specific characters.

The flight of the kingfisher is extremely rapid, a fact so much the more remarkable as the wings are very small in proportion to the body, and must, therefore, be furnished with very powerful muscles. Destined to live by the destruction of other beings, its leading habits are patience, perseverance, and ferocity. Perched on a slight branch overhanging a stream, the kingfisher will remain a great length of time waiting the uncertain passage of its prey beneath: or, moving rapidly along the bank from one little elevation to another, it is more actively, but not more abstractedly, occupied in the same predacious avocation. The eagerness and impetuosity with which it darts under the water in pursuit of its finny prey are excessive; and it is remarked, that, in order to give a greater impetus to its descent, the kingfisher first mounts a few feet immediately over the spot where it is about to dive; and that this preliminary ascent is always proportioned to the size of the fish about to be attacked. At other times, this bird will skim rapidly along the surface of the water, uttering a sharp cry, and seizing such small fish as may thus come within its power. The kingfisher, unlike some other ichthyophagous birds, does not swallow the fish whole, but carrying it on land, breaks and tears it by means of its strong bill. Severe winters are frequently very destructive to these birds, when the frozen surface of the waters shuts up the finny race from their attacks.

These birds begin the affair of propagation about the middle of March. The female deposits from six to eight white eggs, almost without a nest, in some rat-hole in a river bank. It does not appear to be ascertained whether she has more than one brood in a season.

It is extremely difficult, or, rather, it is impossible, to preserve the kingfisher in confinement; for though it should be fed daily with fresh fish, and attended to with the greatest care, it seems never to survive any length of time in captivity;

indeed, its natural localities are so opposite to any thing which can be afforded to it in an artificial condition, and its habits are so independent of mankind, that we need not be surprised at the difficulty we find in our attempts to domesticate this bird. Their flesh has a musky scent, and is not eatable, and their fat has a reddish tinge.

The *Sacred Kingfisher, Alcedo Sacra,* Gm. is nine inches long. The bill is lead-coloured, with a white spot at the base of the under mandible; the general colour is light blue above, and whitish underneath; over the eyes is an arch of pale red, extending to the neck, and under is a blue stripe of the same length as the other. There are, however, four varieties of this species known, one of which has the supercilious stripe, white; another has a black head; the third has a greenish head, and the fourth a white collar.

This species is found in the Society Islands, New Zealand, and the last mentioned, or white collared variety, is proper to the Philippine Islands. The epithet sacra is given to this species in consequence of the veneration paid to it by the inhabitants of Otaheite, who treat it even as a divinity under the name *Eatua.*

On the second division of the text, of kingfishers with the under mandible swollen, we find nothing connected with their habits hitherto recorded, and it is needless to repeat mere specific descriptions.

The third division, proper to the remote countries of New Holland, and the neighbouring islands, or some of the species of it at least, seem to possess legitimate grounds for separation from the rest, in the important character of their food, and the habits consequent thereupon. They seem rather to be inhabitants of the woods, and to feed on insects, than to frequent the rivers and lakes for the sake of the fish.

The opposite figure of Gaudichaud's kingfisher belongs to this division.

C. Hamilton Smith Esq. del.t Paris Mus.

GAUDICHAUD'S KINGFISHER.

DACELO GAUDICHAUD.

London, Published by Whittaker & C.o Ave Maria Lane, June 1829.

Mus. Lin. Soc.

LEACH'S KINGFISHER.

DACELO LEACHII.

London. Published by Whittaker & Co. Ave Maria Lane. June 1829.

It is from a fine specimen, brought by the circumnavigators M. M. Quoy and Gaimard, from the South Sea Islands, and deposited in the Museum at Paris. It is named by these gentlemen from M. Gaudischaud, the botanist, who accompanied them in their voyage. This species is found in the woods of the Islands of Rawak and Waigion Marian, and in New Holland. It is about a foot in length. The bill, which is thick and square, is horn-coloured, with a tinge of green on the sides, and is sharp pointed. The plumage is deep black on the head and back; the throat has a white patch, which passes in a narrow band round the neck, and a white streak also passes through the eye; the lower part of the back, and the upper wing-coverts are bright blue; the quills are deep blue, tipped with black; the under parts of the body and tail are red brown.

Leach's Kingfisher, from New Holland, is from a specimen in the Museum of the Linnean Society. It is whitish spotted, and streaked with dusky; rump, fine blue; quills, black edged with blue.

The genus Todus, which has been said erroneously to include many species, is very limited. The horizontal flatness of the bill forms the principal feature of distinction between it and the kingfisher. The generic characters are sufficiently though shortly stated before.

The *Green Tody, Todus Viridis*, like its congeners, is an inhabitant of America, and its islands. It is a small bird, about four inches long, and is called in St. Domingo the ground parrot, from its habit of being almost always on the ground. The female places her nest, generally, like the kingfisher, in a hole on a river's bank, but sometimes in a mere concavity, of her own making, on the ground, formed of straw, moss, cotton, and feathers. She lays four or five grey-blue eggs, spotted with deep yellow. It is said in Buffon's work, that this bird has, during the breeding season, an agreeable song, but M. Vieillot denies this. It feeds on insects, particularly the winged sort, which it catches with much address. Its flight is short, and when at rest, its appearance indicates stupidity.

The head and upper part of this bird are fine green; at the base of the lower mandible is a white border; the throat and fore part of the neck are red, and the wings are brown within; the belly and vent are pale yellow, mixed with a rosy tint; the bill is reddish above, and horn-coloured underneath.

There is nothing remarkable about the other species.

The last genus, BUCEROS, of this comprehensive and ill-defined order of birds, is so peculiar as almost to merit an order by itself.

There is, perhaps, no genus of birds which, taken altogether, presents so many diversities as that of the hornbills, by the varieties in the shape of the bill in each species, so that each species might appear to belong to a distinct genus, if we were to adopt in their classification the characters proper to this organ alone. The bill, in all of them, is not merely excessive in bulk, but may be said to be deformed by the protuberance, casque, or helmet on the upper mandible, and which is infinitely varied in form. The size of the bill would indicate, at first sight, that it was a very powerful weapon of offence, but its awkward malformation, if we may so say, at once corrects such an idea, and shews that, however bulky this monstrous bill, it is by no means formidable.

Buffon has well observed, that the great bill of these birds, as well as that of many of the toucans, cannot have much strength, not having sufficient support; he compares this bill to a lever placed too far from its fulcrum.

The form of the bill not only differs in each species, but it varies also with the different ages of the bird, especially in those which have the additional protuberance on the upper mandible; for these are born with the bill destitute, or nearly so, of this protuberance: when young, there is simply a small excrescence, which increases with the age of the bird,

C. Hamilton Smith Esq. del.t Jamaica.

THE GREEN TODY.

TODUS VIRIDIS.

London, Published by Whittaker & C.o Ave Maria Lane June 1829.

gradually changing its shape until it assumes the form proper to the adult bird. Hence species have been erroneously multiplied, when the differences, taken as specific, were merely those of non-age.

A general uniformity in the structure of the feet of these birds, however, prevails, notwithstanding their aberrations from any given type as to the bill. These, in all the species, are covered with large scales; the three toes which are directed forward, are nearly equal in length, and are nearly united together at the base, so as almost to form a sole; the hind toe is large and flat, and gives a powerful support to the bird, which, nevertheless, does not in general move by walking, but by leaping, with the feet together. They, however, seldom descend to the earth, and are generally seen perched on the largest trees, especially those that are dead, into the holes of which they retire for concealment, and, at the proper season, for nidification.

All the hornbills have, moreover, a few bristles, like eyelashes, above the eyes. The tongue, which is very small, is cartilaginous, and is attached to the bottom of the throat. They live gregariously, in large bands, and feed principally on insects, lizards, and frogs. They also pursue small mammalia, which they swallow whole, after having killed and comminuted them in the bill, and they will even attack and feed on carrion. They do not, at least in a natural state, as it is said, ever take fruit, though, when domesticated, they soon become accustomed to it, and to bread and vegetables.

The bony part of the bill of all birds, which is in fact an elongation of the jaws, is covered with a horny case, which is easily removed entire; but in this genus, the case appears to be more adherent than in others—so much so, that it can only be removed piecemeal. The protuberance on the upper mandible of some species of the hornbills is, however, entire; it has no bony core, and is consequently very light, and necessarily so, or the weight would be too much for the bird to carry. There is, however, within this protuberance, and on the upper mandible, the rudiment of a bone, but extremely porous and light; and this rudiment becomes graduated in proportion to the size of the protuberance in the different species, and in the same species at different periods of its age.

The cutting edge of these large bills is naturally indented; but in consequence of their brittleness, and thinness, they become much more and irregularly jagged by use; the horny case of the bill, however, like the nails and hoofs of quadrupeds, is constantly growing, and thus the accidental injuries done to the edges are as constantly repaired.

M. Levaillant found the dissected head of a large rhinoceros hornbill to weigh four ounces, and that of a raven, more than one ounce, although the size of the former was twenty times that of the latter, and hence he satisfactorily inferred that the bony substance of the hornbill must be much less compact than that of the raven, and in all probability of all other birds whose bill is of moderate dimensions.

The *Rhinoceros Hornbill* is the largest, and attains four feet in length, with an expanse of wings of about three feet. The bill is nearly a foot in length; and, stuck on it, as it were, at the base of the upper mandible, is a second bill, or prominence, which is turned upwards and backwards, like the horn of the rhinoceros.

Notwithstanding the formidable appearance given to this bird by its monstrous bill, it is utterly useless to it as a weapon; and the bird itself is of too cowardly a disposition to make use of it offensively, even if it were more effective. It advances by leaps, and displays in general every appearance of cowardice and stupidity, except indeed when food is offered, which causes it to assume a momentary air of confidence and vivacity by spreading the wings, opening the monstrous bill, and uttering a cry of satisfaction, feeble indeed for a bird of its dimensions. Levaillant had an opportunity of seeing one at the Cape in a vessel which touched there from Batavia. It was fed in general with biscuit moistened in water, and with meat, both cooked and raw, and ate also rice and vegetables. It appears therefore that these naturally very voracious birds easily accommodate their appetite to circumstances; though, in a natural state, insects, snakes, lizards, and even carrion, are their usual food. M. Le Vaillant on one occasion offered the hornbill in question some small birds he had shot, which were seized immediately; and after being for a short time pressed and rubbed in the bill, were swallowed whole. The sailors stated that he hunted the rats and mice whenever they came within sight, though he never had dexterity enough to catch them; he swallowed however very readily all that were offered to him. The habit of this bird appeared to be to seize every thing that was given, and afterwards to reject such as did not suit its taste, to which it never recurred.

The true bill of this bird is black at the base, then slightly reddish, and afterwards light-yellow to the point. The false bill or prominence is red or flesh-coloured on the upper side, and light yellow beneath. A black line marks the contact between it and the true bill; which, rising with the bend of the false bill on each side, gives it the appearance of a perfect beak, with the two mandibles closed.

The eyes are large, and are furnished above with black lashes. The feet are strong, and covered with large brown scales; and the nails, flatted laterally, have their points blunted and injured by friction on the ground.

The general colour of the bird is black; but the tail, which is slightly cuneiform, is tipt with dirty white.

Bontius relates of this bird, in a state of nature in Java, that it lives on carrion, and will follow the sportsman to feed on the entrails of the animals he may take, and that it hunts rats and mice and small quadrupeds.

The rhinoceros-hornbill is found in India, as well as in

Java and the Philippine islands; nor does it appear to be very rare, if we may judge from the number of dried heads in the European collections, though entire specimens of the bird are of very rare occurrence.

It is quite unnecessary here to recapitulate the species: for a list of them we therefore refer to the preceding part; to which we shall merely add here figures of two species of these remarkable birds from that highly elegant work the "Planches Coloriées" of M. Temminck.

The *Crested Hornbill, Buceros Cassidix*, Tem., has the bill of a bright yellow, furnished at the base of both mandibles with an additional horny substance, covered with transverse rays; that on the upper mandible is nearly semicircular, large at the base, forming a cutting edge in front; its colour is deep purple-red.

The specimen figured was brought from Celebes by M. Reinwardt, and was a male adult. In what, if any, particular the adult male differs from the female and the young is not known. The top of the head is marron-colour; the neck is bright golden-yellow; the body, wings, and thighs are shining metallic black-green, and the tail is perfectly white. The beak is of a brilliant golden-yellow, the base covered with a thick horny bed, transparent, and marked diagonally with three deep black-coloured furrows, the ridges between which are reddish-orange; round the eyes and the guttural skin are livid yellow, tinted with blue. The total length of this bird is near four feet, and the bill ten inches.

This species inhabits high woody mountains, and lives principally on fruit.

To this figure we have added a plate representing the principal differences to be found in the bills of several of the species of this very extraordinary genus.

Bills of Hornbills.

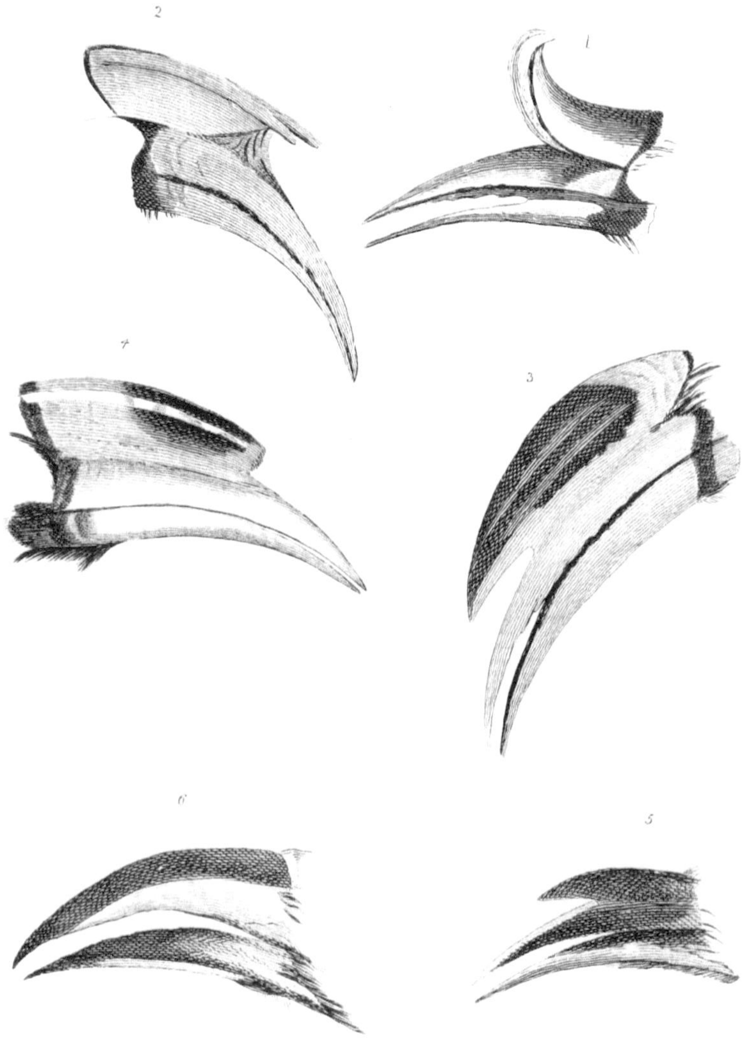

1 . B. Rhinoceros — 2 B. Bicornis.

3 . B. Monoceros. — 4 . B. Violaceus.

5 Gingianus. — 6 . B. Bengalensis.

Published by Whittaker & C°. Ave Maria Lane. Jany. 1830.

CRESTED HORNBILL.

B. CASSIDIX TEM.

London Published by Whittaker & C° Ave Maria Lane. June 1829.

The Third Order of Birds, Scansores, or Climbers,

Is composed of birds whose external toe is directed backwards, like the thumb, whereby they have a more solid support, of which some of the genera avail themselves, by hanging and climbing on the trunks of trees. Hence they have been named in common, Climbers, Scansores, although, strictly speaking, the term does not apply to all of them, and many birds climb without belonging to this order, by the arrangement of the toes, as we have seen already in the creepers and nuthatches.

The birds proper to this order build, in general, in holes of old trees; their flight is middling; their food, like that of the Passeres, consists of insects and fruits, accordingly as their bill is more or less strong; some, as the woodpeckers, have peculiar means of obtaining their food.

The sternum in most of the genera has two slopes behind; but in the parrots there is nothing but a hollow, and this is often filled up.

The Jacamars. (Galbula. Briss.)

Are nearly allied to the kingfishers, by their elongated, sharp, quadrangular bill, and by their short feet, whose anterior toes are, for the most part, united; nevertheless these toes have not the same formation exactly, as in the kingfishers; the plumage moreover, of the jacamars, is not so smooth as in the kingfishers, and has always a metallic lustre. They live solitarily in humid woods, feed on insects, and build their nests on low branches.

The American species have the bill larger than the others, and quite straight.

Paradise Jacamar. (*Galbula Paradisea.*) Lath. Enl. 271.

Deep glossy-green; head, violet-brown; throat, neck, before, and under wing-coverts, white. Size of a lark. Surinam.

Green Jacamar. *Alcedo Galbula.* Lin. (*Galbula Viridis.*) Lat. Enl. 238.

Brilliant glossy-green; belly and vent, rufous; chin, white. Size of a lark. Guiana.

Rufous-tailed Jacamar. (*Galb. Ruficauda.*) Cuv. Vaill. Ois. de Par. II. pl. 50, or *G. Macroura.* Vieill. Gal. 29.

Green; chin and throat white, beneath, and tail, rufous.

White-billed Jacamar. (*Gal. Albirostres.*) Lath. Vaill. pl. 51. Vieill. *Ois. Dorés*, I. pl. 4.

Rather less than *Gal. Viridis.* Glossy-green, above; head, brown; a triangular white spot on the chin; beneath, rufous.

White-bellied Jacamar. (*Gal. Albiventris.*) Vaill. 46.

Golden-green, above; pure white, beneath, with a patch of green on the breast. Brazil.

But there are some in the Indian Archipelago, whose bill, shorter, thicker, and a little bent, approximates them to the bee-eaters. The anterior toes are more separated. These are the JACAMEROPS of Le Vaillant.

Great Jacamar. (*Alcedo Grandis.*) Gm. *Galbula Grandis.* Lath. Vaill. pl. 54.

Copper-green, above; beneath, ferruginous. Eleven inches long. Surinam.

Le Vaillant describes one of these birds, whose bill has no angle above.

Vaillant's Great Jacamar. Vaill. I. pl. 53.

Head, cheeks, and rump, green-golden-blue; quills and throat, white; neck, breast, and belly, red.

Jacamaciri is the Brazilian name of this bird, according to Margrave. *Galbula* seems to have meant the oriole with the Latins; it is Meering who has transferred this name to the jacamar.

Lastly, there are some (the *Jacamar Alcyon*) which have only three toes; they live in Brazil.

Three-toed Jacamar. (*Alcyon Tridactyla.*) Vail. Jac. Sup. f. L. Lath. Sup. 50, and Spix. 57—2.

Black-green, above; reddish-white, beneath; vent, black. Brazil.

The Woodpeckers. (Picus.* Lin.)

Are well defined by their long bill, straight, angular, compressed in a corner at the extremity, and fitted for cleaving the bark of trees; by their slim tongue, armed towards the end with spears, inclining backwards, which, thrust forward by the long elastic horns of the os-hyoïde, can reach out much beyond the bill; and by their tail, composed of ten† quills, with the stems stiff and elastic, which sustains them by its support when they climb along the trunks of trees. These birds are eminent climbers; they support themselves in all directions on the bark of trees, which they strike with their bills, and in the clefts and holes of which they drive their long tongue in search of the larvæ of insects, on which they feed. Their tongue, independently of its bristles, is supplied with a viscous liquor, furnished by the large salivary glands; it is withdrawn by two muscles, rolled like ribbons round the trachea. In this retractile state the horns of the os-hyoide mount under the skin, and round the head, unto the upper base of the bill; and the sheath of the tongue is

* Picus is the name of this bird in Latin.

† There are properly twelve: but those on the side are extremely small.

folded at the bottom of the gullet. Their stomach is almost membranaceous; they are without cæcum; nevertheless, they also eat fruit. Fearful and wary, they pass the greater part of their time alone. In the season of love, the male calls the female by striking rapidly on a dry branch. They nestle once a year in the hollow of a tree, and the two sexes sit alternately.

There are six or seven species in Europe.

Great Black Woodpecker. (*Picus Martius.*) L. Enl. 596. Naum. 131.

Almost as big as a crow; entirely black; the male has a cap of bright red, but in the female it is a mere spot. This species prefers the fir-tree forests in the north.

Green Woodpecker. (*Picus Viridis.*) Enl. 371. Naum. 132.

As large as a dove; green above, whitish underneath; the top of the head red, the rump yellow; is one of our finest birds. The young is spotted with black underneath, and with a black streak on the chin. It prefers small woods, in level countries, the beech-tree and the elm, and seeks its food on the ground.

The *Picus Canus* of Gm. Edw. 65. Naum. 133.

Is of a more ashy tint, with the bill more slender, and with a black moustache. The male is red only on the top of the head, and the female is without that colour. This species goes less to the southward, and is more rare in France, than the preceding, with which it corresponds in habits. Ants are its favourite food.

Greater Spotted Woodpecker. (Picus Major.) Enl. 196 the male, 595 female. Naum. 134.

Of the size of a thrush; varied above with black and white; the back and the rump black; underneath white; about the vent red, with a red spot also on the occiput of the male. The young has nearly the whole cap red; prefers green trees; frequently approaches human habitations, but scarcely ever descends to the ground.

Middle Spotted Woodpecker. (Picus Medius.) Enl. 598. Naum. 136.

The size of a sparrow; varied with black and white above; greyish white underneath; red upon the head of the male only. Of the north and middle of Europe. It is said that they descend to the ground in search of ants; but Naumann asserts that this opinion is without foundation.

In the north-east of Europe there is a species rather larger, though similar to our first; but this has always the lower part of the back and the rump white, and the cap of the male red. It comes sometimes into Germany. It is the *Picus Leuconotos* of Bechstein. (Naum. 35. t. 65.)

The foreign woodpeckers are very numerous, and considerably resemble one another, even in certain distributions of colours; as, for example, the red on the head.

Species analogous to the black woodpecker.

Pileated Woodpecker. (*P. Pileatus.*) L. Enl. 718.

Black, with a red crest; sides of head and neck with white streaks; throat and spots on wings, white. North America.

Lineated Woodpecker. (*P. Lineatus.*) L. Enl. 717.

Above black; neck and breast same, edged with white; crest, head, and nape red. Cayenne.

White-billed Woodpecker. (*P. Principalis.*) L. Enl. 690.

Bill, white and channelled; crest, red; neck, black; white about rump. America.

Helmet Woodpecker. (*P. Galeatus.*) Natt. Col. 171.

Crest, red; neck and mantle, dusky; beneath, yellow, with black spots; half-collar on throat, black and spotted.

The above four species are nearly allied; to one of which probably belongs the

Buff-crested Woodpecker. (*P. Melanoleucos.*) Gm. Lath. Syn. I. 2, t. 25.

Crest, buff; blackish-brown, with a waved white band, from bill, joining on the back; belly same. Surinam.

Red-necked Woodpecker. (*P. Rubricollis.*) Gm. Enl. 612.

Head, crest, and neck, crimson; deep-brown, above; cream-colour, underneath. Brazil.

Robust Woodpecker. (*P. Robustus.*) Spix. t. 44.

Head and neck, deep-crimson; crested; ears, blackish.

Spix's White billed Woodpecker. (*P. Albirostris.*) Spix. t. 45.

Black; bill and side of neck, white; crown, red; belly, black; white banded.

Strong-billed Woodpecker. (*P. Validus.*) T. Col. 378, and the female, 402.

Bill, large and strong; body, red; cheeks, yellow; crest, red triangular; wings, black, with reddish yellow bars; tail, black. Indian Archipelago.

Red-headed Woodpecker. (*P. Erythrocephalus.*) L. Enl. 117.

Head and neck, crimson; back, rump, breast, and belly, white. Virginia.

Dusty Woodpecker. (*P. Pulverulentus.*) T. Col. 389.

Body, wings, and tail, dingy; head and neck, slate-coloured, with numerous little yellowish spots; under the eye, a red patch; throat, yellow. Molucca.

Concrete Woodpecker. (*P. Concretus.*) Reinw. Col. 90.

Form, thick and stunted; black; edges of feathers on the upper parts, yellow; crest, orange, long, triangular. Java.

Chili Woodpecker. (*P. Chilensis.*) Lesson Voy de la Coq. 32.

Brown, barred with whitish; lower part of the back and rump, pure white; bill, black; crown, ash; cheek, red; throat white; tail, brown. Chili.

Lewis's Woodpecker. (*P. Torquatus.*) Wilson, Amer. III. 20.

Head, back, wings, and tail, black; top of head, red; collar, white. America.

St. Domingo Woodpecker. (*P. Dominicanus.*) Spix.

Not crested; white; occipital spot and belly, yellow-green; back, scapulars, wings, and end of tail, black.

These species are analogous to the Green Woodpecker.

The Pierced Woodpecker. (*P. Percussus.*) Col. 390, and the female, 424.

Above, green; beneath, dirty-yellow, covered with black spots, like gashes; top of head and neck, and patch on breast, red; cheeks, slate, with a black patch behind the eye. Cuba.

Bengal Woodpecker. (*P. Bengalensis.*) L. Enl. 695, of which *P. Aurantius.* Gm. Briss. IV. is probably a variety.

Top of head, black with white spots; crimson, crest; above, black with white spots; beneath, white. Bengal.

Goa Woodpecker. (*P. Goensis.*) Gm. Enl. 696.

Crown and crest, crimson, edged with white; wings, yellow: above, black, with zigzag-white stripes; beneath white. Goa.

Azara's Woodpecker. (*P. Aurulentus.*) Illig. Col. 59, or *Macrocephalus.* Spix. 53.

Yellow-green, covered beneath with lunated darker spots; top of head and stripe from the gape, red; the eye in a black stripe. Paragua.

Red Woodpecker. (*P. Puniceus.*) Horsf. Col. 423. Like the *Gaudy Woodpecker*, but smaller, and bill not so robust.

Gaudy Woodpecker. (*P. Mentalis.*) Col. 384.

Body, dark-green; wings, red; tail, black; crest, triangular, green, yellow, and dingy-red; throat, black, with small white stripes. Java.

Ceylonese Woodpecker. (*P. Ceylonus.*) N. Nat. Forsch. 13, pl. 4.

Hind toe, very small; forehead, with long, sharp feathers, scarlet; chin and throat black; beneath, white.

Crimson-rumped Woodpecker. (*P. Goertan.*) Gm. Enl. 320.

Grey-brown; crown and rump, crimson. Senegal.

Manilla Woodpecker. (*P. Manillensis.*) Gm. Sonn. pl. 36.

Dirty-green; top of the head, spotted white; and red spot on wings. Luçonia.

Gold-back Woodpecker. (*P. Senegalensis*). Gm. Enl. 345.

Crown, red; back, golden-yellow; beneath, brown and white. Size of a sparrow. Senegal.

Passerine Woodpecker. P. Passerinus. Gm. Briss. IV. Spix. t. 56. f. 1. 2.

Olive-yellow, above; beneath, barred white and brown; spots on wings, white; crown, red; Size of last. St. Domingo.

Surinam Woodpecker. (P. Luzonicus.) Cuv. Sonn. pl. 37.

Greenish-black, crown; slightly crested; chin and beneath, white; feathers, black edged; tail, black-brown, with a central row of white spots. Surinam Islands.

Red Woodpecker. (P. Miniatus.) Gm. Ind. Zool vj.

Above, red; belly, white. Java.

Yellow-headed Woodpecker. P. Chlorocephalus. Gm. Enl. 784.

Head and neck, yellow; crest, crimson olive-brown, marked with white spots. Guiana.

Yellow Woodpecker. P. Exalbidus. Gm. Enl. 509.

Yellowish white; red stripe from gape to ears; tail black. Cayenne.

Ferruginous Woodpecker. P. Cinnamoneus. Gm. Enl. 524.

Reddish-cinnamon above, with whitish spots; crimson spot on each side of the throat. Cayenne.

P. Palalaca. Cuv. Enl. 691.

Forehead and crest, scarlet-reddish; golden-olive, cheeks; reddish-white, body; beneath, reddish; feathers, black edged; tail black.

P. Jumana. Spix. 47.

Deep cinnamon-red; crested; sides, back, rump, ochraceous; cheek-band, scarlet; tail-quills, black. Brazils.

Yellowish Woodpecker. P. Flavicans. Id. 51.

Deep-ochraceous; cheek-band, scarlet; wing-coverts, yellow-edged; tail, blackish-brown. Brazil.

The following species are analogous to the *Picus Medius.*

Red-billed Woodpecker. P. Rubriventris. Vieill. Gal. 27.

Black; not crested; forehead and chin, silky-yellow; crown, chest, and belly, scarlet; chest, olive; sides, legs, and vent, greenish-white; black banded.

Swallow Woodpecker. P. Hirundinaceus. L. Enl. 694.

Black; patch on the head, and down the middle of the breast, red; rump, white. Cayenne.

Yellow-bellied Woodpecker. P. Varius. Gm. Enl. 785.

Belly, pale yellow, mixed with black; above, black, with white spots on wings; head, banded with black and yellow. North America.

Canada Woodpecker. P. Canadensis. Gm. Enl. 345.

Above, black, with a little white on the middle of back; beneath, whitish; head, &c., banded black and white.

Hairy Woodpecker. P. Villosus. Gm. Enl. 754.

Black, above; white beneath, with hair-like feathers dividing those colours. Carolina.

Red-cheeked Woodpecker. P. Undatus. Enl. 553.

Lion-colour, marked with dusky bars; plat of red under the eyes. Guiana.

Little Woodpecker. P. Pubescens. Gm. Catesby, 31, 11. Wilson, I. 9.

Like *P. Villosus*, but smaller. Six inches long. America.

The following species are striped across.

Malacca Woodpecker. (*P. Malaccensis.*) Gm. Enl. 748.

Back, reddish-grey; belly, rufous-white; top of head, crimson; quills, red and brown, with white spots. Malacca.

Encenada Woodpecker. (*P. Bicolor.*) Pl. Enl. t. 748. f. 1.

Grey-brown and white, irregularly blended. America.

Rufous Woodpecker. (*P. Rufus.*) Gm. Enl. 694. f. 1.

Nearly allied to *P. Undatus.*

*Carolina Woodpecker.** (*P. Carolinus.*) Gm. Enl. 597 and 692.

Top of head and neck, red; back, black and white, banded; tail-feathers, black and white. Jamaica.

* This species has but three toes, but figured in the Pl. Enl. with four.

Cayenne Woodpecker (*P. Cayanensis.*) Gm. Enl. 613.

Olive striped, with black ; beneath, yellowish ; top of head, red. Cayenne.

Gold-crested Woodpecker. (*P. Melanochloris.*) Gm. Enl. 719.

Black and yellow striped, and spotted; crest long, gold colour. Cayenne.

Rayed Woodpecker. (*P. Striatus.*) Gm. Enl. 281 and 614.

Forehead, cheeks, and throat grey ; above, black, striped with olive ; beneath, olive. St. Domingo.

Supercilious Woodpecker. P. Superciliaris. Col. 433.

Black and white ; beneath, reddish ; eyebrows, black.

Yellow-crested Woodpecker. P. Flavescens. Gm. Brown II. pl. 12, and Spix 49.

Back, black with yellow bars ; belly and tail, black ; head and long crest, yellow. Brazil.

Cardinal Woodpecker. P. Cardinalis. Son. pl. 35.

Crown, red ; from the eye backward, a white stripe ; back, black, margined white ; wings and tail, spotted with white ; beneath, white.

Sklit Woodpecker. P. Querulus. Wilson Amer. II. 15.

Back and scapulars, black and white alternate. Georgia and Virginia.

The Field Woodpecker. P. Campestris. Spix 46.

Head, chin, and throat, black; front of neck, yellowish, gular streak black; body beneath, white. America.

Mace's Woodpecker. (P. Macei.) Col. 59.

Top of head, red; above, covered with black and white bars; cheeks and throat, white with triangular black patches; beneath, dirty-yellow, with darker spots; vent red. India.*

* There have been named in addition to the above—

P. Borei. P. Machloti. P Vieillottii. (the *P. Borealis* of Vieillot.) *P. Rupelii. P. Lichtenstenii. P. Punctiligerus. P. Erythraucen. P. Radiolatus. P. Spilogaster. P. Meropirostris. P. Ischnorynchos. P. Scutatus. P. Caniceps. P. Xanthotaenia,* (the *P. Auratus* of Vieillot) by Wagler.

Also. *P. Javanensis,* by Horsfield. *P. Leucogaster,* by Reinwardt. *P. Ruber. P Minor. P. Mahrattensis. P. Pectoralis. P. Icterocephalus,* (the *P. Chlorocephalus* of Gm.) *P. Fasciatus,* of Latham. *P. Rotatus,* (the *P. Rubicus* of Latham.) *P. Maculipennis,* of Lichtenstein. *P. Squamosus. P. Maculatus. P. Rubricollis. P. Chlorolophos. P. Brachyurus. P. Vittatus. P. Icteromelas. P. Mystaceus. P. Fuscescens. P. Erythrops. P. Flavicollis. P. Punctatus,* (not Cuvier,) and *P. Cristatus,* by Vieillot. *P. Affinis,* Swainson; (the young of which is *P. Ruficeps* of Spix.) *P. Rubiginosus,* Swainson, and some others.

It may be remarked, as more than probable, that—

Picus Strictus, Horsf. and *P. Peralaimus,* Wagler, are the same as *Picus Goensis.*

Picus Candidus. P. Melanopterus, Pr. Max., and *P. Bicolor,* Swainson, are all *P. Dominicanus* of Spix.

Picus Chrysochloros of Vieillot, and *P. Braziliensis* of Swainson, is *P. Aurulentus* of Licht.

P. Tristis, Horsf. *P. Poicirolophus,* Temm.

P. Olivaceus, Lin. is *P. Arator,* Cuv.

P. Leucotis, Illiger, is *B. Querulus.*

P. Variegatus, Lath. *P. Minor,* and *P. Moluccensis,* Latham, is the same as *P. Bicolor,* Gmel.

It should be remembered, that these distinctions of analogy, founded especially on colours, are of little importance; and that it may be that several of these species may be little else than varieties.

M. de Lacepede has named PICOIDES some species of woodpecker, which are destitute of the external toe, and have only, in consequence, two toes before and one behind; otherwise, they are in all respects assimilated to the ordinary species.

We have one in the north and east of Europe.

Northern Three-toed Woodpecker. (*P. Tridactylus.*) Edw. 114. Naum. 137.

Intermediate in size between *P. Martius* and *P. Minor;* is black, spotted with white above; white underneath; the rump of the male is orange, that of the female white.*

One might also make a sub-genus of those whose bill, slightly arched, begins to approach that of the Cuckow.

P. Icterocephalus, Lath. is *P. Chlorocephalus.*—See *Spix. t.* 54. f. 2. (not. f. 1.)

P. Maculifrons, Spix. t. 56. f. 1. is the young male of *P. Passerinus.*

P. Flavifrons, Vieil. Spix. t. 52. f. 2. and *P. Coronatus*, Lich. is the *P. Rubriventris.*

P. Flavicans, Spix. t. 31. f. 2. is the female of *P. Exalbidus.*

P. Lathami, Wagler, is *P. Cafer*, Lath.

P. Philophenarum, Lath. is *P. Palalaca*, Cuv. and *P. Bengalensis*, Gmel.

P. Malacunus, Lath. *P. Rubescens*, Vieil. is *P. Mineatus*, of Forster.

* *P. Tiga.* Horsf. Raffles. Java. *P. Hirsutus. P. Tridactylus.* Lath.

Such are the

Gold-winged Woodpecker. Picus Auratus. Buff. *Cuculus Auratus* of the tenth edition. Enl. 695. Wilson, Amer. L. 3.

Transversely striated, black and grey; sides of throat and middle of breast, black; hind part of head, red.

Yellow-shafted Woodpecker. P. Cafer. Lath. *Promépie.* Vaill. Prom. 32.

Brown, above; vinaceous beneath, with round black spots; whiskers, crimson. South Africa.

Banded Woodpecker. P. Poicilophos. Tem. Col. 197, f. 1.

Banded, black and yellow. Java. Six inches long.

One of these seeks its food only on the ground, although it has the same tail as the rest.

Digging Woodpecker. P. Arator. Cuv. Vaill. Afr. 255 and 6.

Olive-brown, spotted on all the upper parts with minute white spots; throat and breast whitish. Africa.

I have only taken from the genus Picus, the *Minute Woodpecker* of Lath. *Yunx Minutissimus.* Gm. Enl. 786. Vieill. Gal. 28, which is properly a Wryneck.

The Wrynecks. (Yunx.* Lin.)

Have the tongue extensible, like the woodpecker,

* The Greek name of the bird in question. Torquilla is its Latin name.

and by means of the same mechanism; but it is without spines; their bill, moreover, straight and pointed, is nearly round; their tail has the feathers only of an ordinary form. They live nearly like the woodpeckers, except that they climb but little.

We have one of them in Europe.

The *Wryneck*. *Yunx Torquilla*. Lin. Enl. 698. Naum. 381.

Of the size of a lark, brown above, and prettily marked with little blackish waves, and longitudinal yellow, and black reticulations; whitish striped across, with black underneath.

Its name is taken from the singular habit which it has, when surprised, of twisting the neck and head in various ways.

The PICUMNES of Temminck differ from the wry-necks only in having a very short tail. They are very small birds.

Least Wryneck. *P. Minule*. Tem. *Yunx Minutissima*. Gm. Enl. 786. f. 1.

Size of a wren; top of head, red; occiput, black, white-spotted; body, red; beneath, grey; hen, crown black. Cayenne.

Crested Wryneck. *P. Cirrhatus*. Tem. Col. 371. f. 1. Vieill. Gal. 28.

Above, brown; beneath, brownish-white, with transverse bars; crest, black and red, spotted white. South America.

Little Wryneck. P. Exilis. Tem. Col. 371. f. 2.

Like the last, but with the cheeks and neck, yellow. South America.

Some of these picumnes have only three toes, like the picoides.

Anomalous Wryneck. P. Abnormis. Tem. Col. 371.

Green, above; beneath, brightish-brown. Java.*

The CUCKOW. CUCULUS.† L.

Have the bill moderate, rather deeply cleft, compressed, and slightly arched; the tail long. They live on insects, and are migratory. We subdivide this numerous genus as follows.

The TRUE CUCKOWS.

Have the bill of moderate strength; the tarsi short; the tail with six quills. They are celebrated by the singular habit of laying their eggs in the nests of other insectivorous birds; and what is not less extraordinary, the foster parents, frequently belonging to much smaller species, take care of the young cuckow as of their own offspring, even when the introduction of the cuckow's eggs has been preceded, as is often the case by the destruction of their own. The

* Add *Yunx Minutus*, Vieillot, the *Carpentero Nano* of Azara.

† Κοκκυξ, *cuculus*; cuckow, expresses the cry of the European species.

cause of this phenomenon, unique in the history of birds, is still unknown. Herissant has attributed it to the position of the gizzard, which is deeper in the abdomen, and less protected by the sternum than in other birds. The cœcum of these cuckows are long, and their lower larynx has but one proper muscle. We have one cuckow, generally spread throughout Europe.

Common Cuckow. Cuculus Canorus. L. Enl. 811.

Of an ashen grey, with a white belly, striped across with black; the tail spotted with white on the sides; the young has red instead of grey.

But there also, sometimes, comes here a species spotted and crested, whose cry is more sonorous.

(*C. Glandarius.* Edw. 57.) Naum. 130, the male, Pl. Col. t. 414, the female. The *C. Pisanus* of Gm. is the young of this.

The warm countries of both continents produce many others.

Cape Cuckow. C. Capensis. Vaill. Af. pl. 200.

Which is probably only a variety of the common species. Pl. Enl. 390.

Solitary Cuckow. C. Solitarius. Cuv. Vaill. 206.

Obscure-brown, above; light-brown, underneath, with darker transverse bars; tail, black, white edged. Africa.

Panayan Cuckow. C. Radiatus. Son. Voy. pl. 76.

Black-brown, cheek and throat, vinaceous; chest and belly, yellow, black banded; tail, black, white tipt. Panay.

Noisy Cuckow. C. Clamosus. Cuv. Vaill. 204, 205.

Deep-blue; quills of tail, brown; female, barred, greenish, underneath. Africa.

Edolio Cuckow. C. Edolius. Cuv. Vaill. 207, 208.

Colour, dingy blue-black and green, mixed; crested; female, white, beneath. Africa.

N.B.—*Cuc. Serratus.* Sparm. Mus. Carl. III. is the male of this. *C. Melanoleucos.* Enl. 272, is the female.

Collared Cuckow. C. Coromandus. Lin. Enl. 274, and a variety. Vaill. 213.

Tail, wedge-shaped; head, crested; body, black beneath; collar, white.

Carolina Cuckow C. Americanus. Enl. 816, or *Carolinensis,* Wil. III. 28. f. 1. Catesby 1. t. 9.

Above, ash; beneath, white; tail, wedge-shaped; lower jaw, yellow.

Black-billed Cuckow. C. Erythropthalmus. Wilson III. 28. f. 2.

Bill, black; red round the eye; like the last, but without the cinnamon colour on the wings. America.

Yellow-bellied Cuckow. C. Flavus. Enl. 814.

Tail, wedge-shaped, pale-brown; beneath, yellowish-red; crown and throat, ash; tail, black, white banded. Panay.

There are moreover in Africa some pretty species, of a more or less golden green. Their bill is rather more depressed than in the common cuckow.

Gilded Cuckow. C. Auratus. Enl. 657. Vaill. 211.

Tail, wedge-shaped; above, golden-green; beneath, white; head with five white streaks; two outer, and tips of other tail-feathers, white. Cape of Good Hope.

Klaas Cuckow. C. Clasii. Vaill. 210.

Green above, the feathers edged with golden-yellow; beneath, white; middle tail-feathers, green; other quills, black and white. Africa.

Shining Cuckow. C. Lucidus. Lath. Syn. I. t. 23, and Col. 102. f. 1.

Fulgent; tail, nearly square; above, golden-green; beneath, white, waved with golden and brown. New Zealand.

Cupreous Cuckow. C. Cupreus. Lath. and Vieill. Gal. 42.

Bright copper-colour, above; beneath, yellow; legs, black. Africa.

Brazen Cuckow. C. Chalcites. Col. 102.

Reddish-brown, above; each feather with a greenish metallic line in the middle; beneath, whitish, waved with light-brown. South Sea Islands.

Other species, the greater part of which are spotted, have the bill higher vertically.

Rufous Spotted Cuckow. Cu. Punctatus. Enl. 771, and *Scolopaceus*, Enl. 586, perhaps also *Maculatus*, Enl. 764, may be mere varieties.

Tail, wedge-shaped; above, blackish, red spotted; beneath, red, black streaked; tail, red, banded. India.

Sacred Cuckow. C. Honoratus. Lin. Enl. 294. Vaill. 216.

Tail, wedge-shaped, blackish, white spotted; beneath, white and ash, banded. Malabar.

Society Cuckow. Cu. Taitensis. Sparm. Mus. Carl. 32.

Brown, ferrugineous, spotted; beneath, white, brown streaked; tail, wedge-shaped, with sublimate ferrugineous bands. Pacific Islands.

Mindanao Cuckow. C. Mindanensis. Enl. 277.

Golden-green, with white spotted; beneath, whitish, black waved. Philippine Islands.

Guira Cuckow. C. Guira. Vieill. Gal. 44. Freycinet. Voy. t. 66. One cannot say why M. Vieillot has made a *Crotophaga* of this species.

Yellowish-white; crested; head, neck, and wing-coverts, brown and yellow, varied; tail, brown, white tipped.*

The Couas. Vaill.

Differ from the cuckows proper, only in having elevated tarsi. They build in the clefts of trees, and do not deposit their eggs in the nests of other birds. This at least is true of such species whose habits are known.

M. Vieillot has made his genus Coccyzus of this division, the Macropus of Spix.

Great Madagascar Cuckow. C. Madagascariensis. Enl. 815.

Olive-brown, waved; throat, yellowish; chestand belly, yellow; outer tail-feather, white tipped Madagascar.

De Laland's Cuckow. C. Lalandii. Col. 440.

Purple-black; throat and beneath, white; vent and thighs, reddish; tail, white tipped. Africa.

Madagascar Crested Cuckow. C. Cristatus. Enl. 589. Vaill. 217.

Crested, greenish-ash; belly, red-white; outer tail-feather, white. Madagascar.

Blue Cuckow. C. Cæruleus. 295, f. 2, Vaill. 218.

Blue; tail, rounded. Madagascar.

* To this genus have also been referred; *C. Panayus*, Gm.; *C. Indicus*, Lath. from General Hardwick's drawings; *C. Pisanus* and *C. Radiatus*, Gm.; *C. Poliocephalus*, and *C. Sonnerattii*, Lath.; *C. Flindersii*, and *C. Orientalis*, of which Messrs. Vigors and Horsfield have constituted their genus Eudynamis.

Spotted Cuckow. *C. Nævius.* Enl. 812.

Ferrugineous-brown; throat, brown, striped; tail, reddish, tipped. Cayenne.

Cayenne Cuckow. *C. Cayanus.* Enl. 211.

Purplish-chesnut; beneath, ash; tail-feathers, white tipped.

Chesnut Cuckow. *C. Brachypterus.* Tem. or *Macropus Caixana.* Spix. 43.

Small; chesnut; bill, arched.

Mangrove Cuckow. *C. Seniculus.* Enl. 813.

Ash; beneath, reddish; throat, white; tail, short, wedge-shaped. Cayenne. Length, twelve inches.

Pheasant Cuckow. *Macropus Phasianellus.* Spix. 42.

Slender; olive-brown, beneath; and top of tail, white; crest, reddish; chest, black spotted.*

We may separate from these an American species with a long bill, bent only at the end.

It is on this distinction that Vieillot has made his genus SAUROTHERA, Gal. 38.

Long-billed Cuckow. *C. Vetula.* Enl. 772.

Brown; beneath, lighter; eyelashes, red; tail, wedge-shaped. Jamaica. Length, fifteen inches.†

* *Coccyzus Minutus*, Vieill. the *Cuculus Cayanus Minor* of Lath.; *Coccyzus Lathami*, of Vieill; *Cuculus Cornutus*, Lin.; *C. Rubibandus*, Lath.; *C. Punctulatus*, Gm.; *C. Brasiliensis*, Lin.; and *Coccyzus Geoffroyii*, of Tem. Col., seem referable to this division.

† Add *C. Plurialis*, which is probably the female.

M. Levaillant has properly separated from the other cuckows

The Coucals.* Centropus. Illig.

Species of Africa and India, which have the thumb nail long, straight, and pointed, like that of the larks. Such as are known belong to the old world. They build also in the clefts of trees.

Egyptian Cuckow. C. Ægyptius and *Senegalensis.* Enl. 332. Vaill. Af. 219.

Grey; beneath, white; crown and tail, blackish. Senegal and Egypt.

Philippine Cuckow. C. Philippensis. Cuv. Enl. 824. *C. Bubutus.* Horsf.

Black; wings, red.

Red-brown Cuckow. C. Nigrorufus. Cuv. Vaill. Af. 220.

Dingy; wings, brickdust colour. Africa.

Long-heeled Cuckow. C. Tolu. Enl. 295, f. 1. Vaill. 219.

Greenish-black; wings, chesnut; head, neck, and back, blackish, red streaked. Madagascar.

Lark-heeled Cuckow. (C. Bengalensis.) Brown's Ill. t. 13.

Ferrugineous brown, white, and black, streaked; belly, yellowish-brown; quills and under tail-feathers, red, black banded. Bengal.

* A compound signifying *Lark Cuckows. Centropus,* spur-footed. M. Vieillot has changed this name into Coridonix; Dr. Leach into Podophilus.

Rufous Cuckow. C. Rufinus. Cuv. Vaill. 221.

Rufous ; stem of the feathers lighter. Africa.

Æthiopian Cuckow. C. Æthiops. Cuv. Vaill. 222.

Black ; bluish over the eyes. Africa.

Giant Cuckow. C. Gigas. Cuv. Vaill. 223.

Rufous, above ; dingy, underneath ; tail, dingy, with numerous black bars. Africa.

Black and White Cuckow. C. Ateralbus. Lesson. Voy. de la Coquille, t. 33.

Forehead, black ; neck and chest, brown-white ; back, belly, tail, and wing, blue-black ; wing, with a white speculum. New Zealand.*

We ought also, with this naturalist, to distinguish

The COUROLS,† or VOUROUDRIONS, of Madagascar,

With the bill thick, pointed, strongly compressed, slightly arched at the end of the upper mandible, with the nostrils pierced obliquely in the middle on each side. Their tail has twelve quills. They nestle like the preceding, and live in woods. They are said to be principally frugivorous.

African Courol. C. Afer. Enl. 387, the male with the bill badly represented ; and 558, the female, where it is better. Vaill. 226 and 227.

Golden-green ; beneath, grey ; head and neck, ash ;

* Add *Centropus Membiki.* Lesson, Voy. Cog. t. 34. Black and green. New Guinea.

† Compounded of *Cuckow* and *Roller*, the genus LEPTOSOMUS of Vieill.

crown, splendid-black; tail, beneath, black. Madagascar and Cape of Good Hope.*

The INDICATORS, Vaill.

Are other African species, celebrated by feeding on honey. They serve as guides in the discovery of the hives of wild honey, which they seek making a cry. Their bill is short, high, and nearly as conical as that of the sparrow. Their tail has twelve quills, and is both slightly cuneiformed and forked. Their skin is particularly hard, and resists the stings of the bees; but these insects attack them in the eyes, and sometimes kill them.

Honey Cuckow. C. Indicator. Vaill. Af. 241.

Ferrugineous-grey; beneath, whitish; shoulder, yellow; base of the tail furcated with a black spot. Egypt and Senegal.

Lesser Honey Cuckow. C. Minor. Cuv. id. 242.

The smallest. Like preceding, but beneath, slate-coloured. Africa.

White-billed Cuckow. C. Albirostris. Col. 367.

The largest. Dusky-brown; ears, throat, and beneath, white; tail, black tipped.

M. Vieillot has adopted this genus and its name in his Galerie. 45.†

* This is the *Bucco Africanus* of Shaw, and well figured by Vieill. Gal. t. 40.

† *Indicator Major* of Temminck is *Cuculus Indicator*. Lath., and Temminck has described a species under the name of *Indicator Levaillantii.*

The Barbacous.*

With the bill conical, elongated, a little compressed, slightly arched at the end, and furnished at its base with loose feathers or stiff hairs, which gives them some relation to the genus Bucco.

Wax-bill Barbet. C. Tranquillus. Enl. 512. Spix. 41, 2.

Black; beneath, ash; wing-coverts, white edged; bend of the wing with small white spines.

White-rumped Black Cuckow. C. Tenebrosus. Enl. 505, and Col. 323. f. 1.

Black; belly and thighs, red; rump and vent, white. Cayenne.

Red and White Cuckow. C. Rufalbinus. Col. 323. f. 2.

Bill, with very long ragged hairs; tail rounded, graduated, short; plumage, red; belly white.

Masked Cuckow. Monasa Personata. Vieill. Gal. 36. *Bucco Albifrons.* Spix. 41.

Lead-coloured black; forehead and throat, yellowish white.

It must be observed that the *C. Paradisæus*, Briss., is the *Lanius Malabaricus*, and that the

* Compounded of Barbu, (*Bucco,*) and Cuckow. The genus Monasa of Vieill. Gal. 36. Lypornix of Wagler.

Temminck refers these birds to the second section of the genus Capito.

Also have been named—

Buc. Rufus. Spix. t. 40. f. 1. *Buc. Nigrifrons.* Spix. 43. f. 2. (the *Lypornix Unicolor* of Wagler. *B. Rubicauda.* Spix. which is *B. Rufalbinus* of Tem.) *B. Leucops.* Ill. which is *Monasa Personata.* Vieill. *B. Striatus.* Spix. *B. Torquatus,* and *B. Fuscus.* Gm.

Cuc. Sinensis, is the *Corvus Erythrorynchos*. These two remarks are made by M. Levaillant, who has done the most towards clearing the history of the cuckows.*

The MALCOHAS. Vaill.

(M. Vieillot calls them PHÆNICOPHÆUS†) have the bill very thick, round at its base, arched towards the end, and a large naked space round the eyes.

Some have the nostrils round towards the base of the bill.

Red and Green Malcoha. Le Malcoha Rouverdin. Vaill. Afr. 225.

Above, and middle tail-feathers, deep green; beneath, marron; tail, long, cuneiform. Africa.

Others have them straight near the edge of the bill.

Red-headed Malcoha. Vaill. 224. *Cuc. Pyrrhocephalus.* Forster. Ind. Zool. t. 6. Vieill. Gal. 37.

Tail, very long, black; chest and belly, white; cheeks and crown, scarlet.

And Latham, in his history, has described several varieties of these species, which may eventually prove distinct.

These birds, which are inhabitants of Ceylon, are said to live principally on fruits.‡

* *C. Clamosus*. Lath., is said to be the same as C. Capensis. *C. Orientalis* is the male of *C. Mindanensis*, *C. Sinensis*, Lin. is a magpie. *C. Dominicus*, Lath., seems to be the female of *C. Americanus*.

† The *Phænicophæus Viridis* of Vieillot. Gal. t. 37.

‡ The *Phænicophæus Leucogaster* is mentioned by Dumeril.

We should, probably, also distinguish from the rest, the species with a bill less thick, and which have scarcely any nudity round the eyes.

Java Malkoha. Malcoha à bec peint, Phœnicophæus Calorhenchus. Col. 347.

Black; head and neck, red; crown, iron-grey; bill, yellow, red, and black.

Java Phœnicophæus. Phœnicophæus Javanicus. Horsf.

Dark greenish-gray; bill, throat, breast, vent, and thighs, ferruginous; tail, tipt white.*

The Scythrops. Lath.

Have a bill still larger than the preceding, and thicker than the last, with two slight ridges longitudinally on each side. Round the eyes is naked, and the nostrils are round. Their bill approximates them to the Toucans, but their tongue is not ciliated. Only one species is known, which is proper to New Holland. It is as big as a crow, whitish, with a grey mantle.

Psittaceous Hornbill. Scythrops Novæ Hollandiæ. Lath. *Scy. Australasiæ.* Shaw.

Pale-green; back, wing, and tail, lead-coloured.

Sh. Philips, t. at p. 165, and White 142, two bad figures. There are better in Pl. Col. 290, and Vieill. Gal. 39.

* Lesson mentions a bird as the *Phœnicophæus Superciliosus* of Cuvier, but Cuvier does not notice it in print.

The BARBETS. BUCCO.* Lin.

Have a thick conical bill, convex on the sides at the base, and furnished with five facetts of stiff bristles, directed forward, one behind the nostril, one on each side the base of the lower jaw, and the fifth under the symphysis. Their wings are short, their proportions and their flight heavy. They live on insects, and attack little birds, nevertheless they also eat fruit. They build in holes in trees.

They may be divided into three subgenera.

The BARBICANS.† Buff. POGONIAS. Ill.

Which have one or two strong teeth on each side the upper mandible, the bend of which is blunt and arched; their quills are very strong. They are found in Africa and India. They are more frugivorous than the other species.

Groove-billed Barbican, Bucco Dubius. Gm. *Pogonias Sulcirostris.* Leach. Zool. Mis. 76. Vieill. Gal. 32. Vaill. Ois de Par. 2. 19.

Black, above; beneath, red, with black band on breast; middle of back, white; bill, channelled. Barbary.

Black-bellied Barbican. Pog. Erythromelas. Vieill. Gal. 32.

Black; beneath, red; chest, band, and vent, black.

* This name was given by Brisson on account of the convexity at the base of the mandible, from *bucca*, cheek.

† On account of their similarity to the Toucans. Pogonias, from πωγων, beard; but Lacepède has long since applied this name to a genus of fish.

Smooth-billed Barbican. P. Levirostris. Leach. 77. Vaill. pl. K. *Le Barb., à Ventre Rosé.* Vieill. loc. cit. pl. A. is the young.

Like *B. dubius*, but the bill is not channelled; top of head, crimson. Africa.

Burchell's or the Masked Barbican. P. Personatus. Tem. Col. 201.

Top of head, cheeks, and throat, bright red, behind the red, black; back, greenish-ash; beneath, yellowish. Africa.

Black-throated Barbican. P. Niger. Tem. Enl. 688. Vail. 29, 30, 31.

Crown, rump, throat, neck, middle of back, and wing-coverts, black; forehead, red; curve over eye, yellow and white. Philippine Islands.

Red Barbican. P. Rubicon. Vaill. pl. D.

Forepart of body, red; beneath, pale-yellow; sides, streaked with black.*

Bucco† properly so called. (Cuv.)

With a bill simply conical, slightly compressed, the crest blunt, a little elevated in the middle. There are some of them in the two continents, many of which

* *Bucco Saltii*, Lath. (Abyssinian Barbican of Salt's Travels. *P. Rubrifrons* of Swainson. Zool. Illus. and *P. Hæmatops*, Wagler.) *P. Leuconotus*, Vieill. (which is *P. Lævirostris* of Leach; and the young is *P. Levaillantii* of Leach.) *P. Stephensii*, Lath. (the *P. Niger* of Tem. *P. Senegalensis*, Lich. *P. Vieillottii*, Leach. *Bucco Fuscescens*, Vieill. which seems to be the *B. Rubescens* of Vaill.

† Vieillot has changed this name into Capito.

have brilliant colours. They live in pairs in the breeding season, and in small bands the rest of the year.

Great Barbet. B. Grandis. Enl. 871.

Changeable green, with the quills varied with black; head and neck, blue; vent, red. China and India.

Green Barbet. B. Viridis. Enl. 870.

Green, with head and neck greyish brown; white spot near the eye. India.

Yellow-fronted Barbet. B. Flavifrons. Cuv. Vaill. Ois. de Par. 55.

Top of head and back of neck, olive-green; about the cheeks and throat, blue; body and wings, green; paler, beneath. Ceylon.

Blue-throated Barbet. B. Cyanops. Cuv. id. *ib.* 21, or *Capito Cyanocollis.* Vieill. Gal. 35.

Head, red, streaked with black; throat and crop, blue; occiput and body above, green; beneath, pale-green. India.

Latham's Barbet. B. Lathami. Lath. Syn. pl. 22.

Greenish; quills, dark; face and chin, brown.

Yellow-throated Barbet. B. Philippensis. Enl. 333.

Green; beneath, yellow, spotted with olive; throat and sides of the head, yellow; top of the head, red. Philippine Islands.

Red-crowned Barbet. B. Rubricapillus. Brown. Ill. 14.

Green, with the crown and throat scarlet; beneath, white, and white spot on shoulder. Ceylon.

Red-necked Barbet. B. Rubricollis. Cuv. Vaill. 35.

If, indeed, these are not three varieties.

Black-throated Barbet. B. Torquatus. Cuv. Vaill. 37.

Back, wings, and tail, brown; beneath, yellow; forehead, cheeks, and throat, red; patch on throat, black. Brazil.

Rosy Barbet. B. Roseus. Cuv. Vaill. 33.

Shining-green, above; greenish, beneath; sides white, with black stripes; face and throat, rose-coloured.

Black-throated Barbet. B. Niger. Enl. 688. Vieill. Gal. 33.

Above, black, with yellow spots; beneath, white. Philippine Islands, and Cape.

*Beautiful Barbet. B. Maynanensis.** Lath.

Green; head, and throat, red, margined with blue; breast, yellow. South America.

The Elegant Barbet. The *B. Elegans.* Gm. Enl. 688.

Appears to be nearly allied to the last, if not the same.

Bearded Barbet. B. Barbiculus. Cuv. Vaill. 56.

Beard longer than the bill; forehead and throat, red;

* *Capito Elegans*, Vieill.

post ocular spot, yellow; general colour, deep-green, varied with yellow; quills, red.

*Little Barbet. B. Parvus.** Vaill. 32. Pl. Enl. 746.

Fulvous brown; beneath, whitish, striated with brown. Senegal.

Red-rumped Barbet. B. Erythronotos. Cuv. Vaill. 57.

Rump and upper tail-coverts, bright red; dark vinous on the upper parts, with yellow streaks on the cheeks; throat, yellow; beneath, whitish. Africa.

Yellow-cheeked Barbet. B. Zeylanicus. Brown.

Green, with head and neck pale-fulvous; wing-coverts, with white spots. Ceylon, Java, &c.

Cayenne Barbet. B. Cayanensis. Enl. 206.

Black; beneath, whitish yellow; forehead and throat, red. Cayenne.

Peruvian Barbet. B. Peruvianus. Cuv. Vaill. 27.

Forehead and throat, orange; above, from the gape, nearly black, with yellow patches; belly, yellow, with brown spots. Peru.

Black-breasted Barbet. B. Nigrothorax. Vaill. 28.

These again may be three varieties.

White-breasted Barbet. B. Fuscus. Vaill. 43.

Brown, with triangular white spot on the breast. Cayenne.

* *Capito Rubrifrons,* Vieill.

Red-collared Barbet. B. Armillaris. Tem. Col. 89.

Green; forehead, and half collar, red; back of neck, blue. Java.

Blue-throated Barbet. B. Gularis. Tem. Col. 89. 2.

Green, top of head, throat and edge of wings, blue; under eye, and half collar, yellow; black between the yellow and blue of the throat. Java.

Yellow Mustache Barbet. B. Chrysopogon. Tem. Col. 285.

Green; top of head varied with white, red, blue, and brown; throat, blue; yellow stripe from the gape. Sumatra.

Party-coloured Barbet. B. Versicolor. Tem. Col. 309.

Green; top of the head, and patch on the side of the breast, red; over the eye, and throat, bright blue, waved; the eye in a black stripe, in which is also a marron spot. Sumatra.

Blue semi-collared Barbet. B. Mystacophanes. Tem. Col. 315. Vaill. pl. C.

This is also green, with the top of the head and throat, deep red; the eye in a blue patch; half collar, blue.

Olive Barbet. B. Auro-virens. Cuv. Vaill. pl. E.

Above, olive-yellow; beneath, orange-yellow; top of head and nape, red; belly, pale. Brazil.*

* To these may be added:—

B. Rubicula, *B. Rufus*, and *B. Striátus*, from Spix's Brazil. *B. Marginatus*, Ruppel, and *B. Nigrescens*, Gray's MSS.

B. Cinereus, *B. Calcaratus*, L. *B. Leucops*, Ill. *B. Albifrons*, and *B. Nigrifrons* of Spix. and *B. Tenebrosus* of Ill. are all Barbicans.

The TAMATIAS* TAMATIA. Cuv.

Whose bill is a little more elongated, and more compressed, and bent downward at the extremity of the upper mandible. Their thick head, short tail, and large bill, give them a stupid appearance. All the species are American, and live only on insects. Their disposition is sad and solitary.

Pied Tamatia. Bucco Macrorhynchos. Enl. 689.

Black; with the forehead, throat, front of the neck, abdomen, and tip of tail, white. Cayenne.

Lesser Pied Tamatia. B. Melanoleucos. Enl. 688.

Black; forehead, throat, spot on scapulars, streak behind the eyes, abdomen, and tip of tail, white. Cayenne.

Collared Barbet. B. Collaris. Enl. 395.†

Rufous; above, striated with black; beneath, whitish, with a broad black band on the breast. Guiana.

Spotted-bellied Tamatia. B. Tamatia. Enl. 746. Vieill. Gal. 34. *Tamatia Maculata.* Cuv.

Reddish brown; beneath, lighter, spotted with black; dark collar on neck. Cayenne.

Black-eared Tamatia. (*Capito Melanotis.*)‡ Col. 94.

Back, wings, and top of head, brown, with small black

* The name of one of these birds in Brazil, according to Margrave. They are called *Chacurus*, in Paraguay, according to Azara. It is to these that Temminck has applied the name CAPITO.

† Probably *B. Capensis* of Gm. and *B. Collaris* of Lath.

‡ The *B. Chacuru* of Vieill. from the Chacuru of Azara, and *B. Strigillatus* of Lich

bars; beneath, and round the breast and neck, brownish-white; a large black patch below the eye. South America.

Black-banded Tamatia. (*Cyphos Macrodactylus*). Spix. 39.

Crown, chesnut; chin, nape, and chest, white; eye-streak, and chest band, black; back, wings, and tail, dark-brown; belly and rump, brown, lineated with blackish. Brazil.*

The COUROUCOUI.† TROGON. L.

Have, together with the brush of hair of *Bucco*, the bill short, larger than high, bent from its base, with the upper ridge arched and blunt. Their small feet furnished with feathers to near the toes, their tail long and large, their plumage fine, light, and plentiful, give them a different appearance to Bucco. There is, in general, some part of their plumage which shines with a metallic lustre; the rest is more or less brightly coloured. They build in holes in trees, feed on insects, remain silent and solitary on the lower branches, in thick humid woods, and fly only during morning and evening.

Some of them are found in both continents.

The American species have the edges of the mandibles indented, while those of the old world have them more entire.

* Also have been named *B. Somnolentus*, Lich. The *Tamatia* of Marcq. *Capito Maculatus*, Wagler. *Alcedo Maculatus*. Gm.

† The expression of their cry, and their name, in Brazil. Trogon was given them by Mœhring.

Red-bellied Trogon. (*T. Curucui.*)* Enl. 452. Vaill. Couroue. 1, 2.

Green, above; fulvous-red, beneath; throat, black; coverts and tail, striped with black. West Indies.

Lesser red-bellied Trogon. (*T. Rosalba.*)† Vaill. 6, or *Variegatus.* Spix. 38.

Belly, red; but the species smaller than the last. Cayenne.

Yellow-bellied Trogon. (*T. Viridis.*) Enl. 195. Vaill. 3, 4. Spix. 36.

Green; beneath, yellow; throat, black; lateral tail-feathers, striped black and white. Cayenne.

Violet-headed Trogon. (*T. Violaceus.*) Nov. Com. Petrop. II. 166. 16.

Violet; back, green; wings, spotted with white; tail, barred. Cayenne.

Cinereous Trogon. (*T. Strigilatus.*) Enl. 765.

Cinereous; abdomen, fulvous; wings and tail, banded white. Cayenne.

Rufous Trogon. (*T. Rufus.*) Enl. 736. Vaill. 9.

Rufous; beneath, yellow; quills, striated black and white. Cayenne.

Black-headed Trogon. (*T. Atricollis.*) Vieill. Gal. 31. or *Oranga.* Vaill. 7, 8. or *Sulfuraceus.* Spix. 38.

Forehead, cheeks, and throat, black; golden-green, above; bluish-green, beneath. Cayenne.

* Pl. Enl. 437, appears to be the female.

† *T. Castaneus*, Spix. and *T. Collaris*, appear to be the female.

Domingo Trogon. (*T.. Domicellus.*) Vàill. 13.

Bright bluish-green, above; coverts, barred, black, green, and white; beneath, greenish-white; vent, red. St. Domingo.

White-bellied Trogon. (*T. Albiventer.*) Vaill. 5.

Bluish-green, above; quills, black; beneath, white. Cayenne.

Fasciated Trogon. (*T. Fasciatus.*) Ind. Zool. pl. 5.

Ferrugineous, above; fulvous-red, beneath; wings, striated black and white. Ceylon.

Mountain Trogon. (*T. Oreskios.*) Col. 181.

Head and neck, green; back and middle tail-feathers, marron; beneath, orange-yellow; edge of tail, white. Java.

Reinwart's Trogon. (*T. Reinwartii.*) Col. 124.

Green; throat and belly, yellow; tail, ash with darker broad bands. Java.

Duvaucel's Trogon. (*T. Duvaucelii.*) Col. 291. Vaill. 14.

Head and throat, black; behind the mouth and over the eye, blue; back and middle tail-feathers, marron; all beneath and rump, deep-red; wings, barred.

White-collared Trogon. (*T. Condea.*) Col. 321.

Head and throat, black, with narrow white collar behind; back, ochreous; beneath, red; wings, black barred with grey; eye-patch, blue; lunule patch on nape, red. Sumatra.

Narina Trogon. (*T. Narini.*) Vaill. Af. 228 and 229. Cour. 10 and 11.

Green, above; beneath, red; quills, black, with white edges. South Africa.

We may be permitted to doubt whether the *Trogon Maculatus* of Brown be a true Trogon.

There is one remarkable for the cut of its tail. (*T. Temnurus.* Tem.) Col. 326, and another, in which the coverts of the tail are nearly as long as the body. (*T. Pavoninus.*) Tem. Col. 372. Spix. 35. It is celebrated in the mythology of the Mexicans, and sought after by the natives for its fine feathers.*

The Ani,† Crotophaga L.

Are known by their thick bill, compressed, arched, without indentation, elevated, and surmounted by a vertical and trenchant crest.

Two species of them are known, both of the warm and humid parts of America, with strong and elevated tarsi, the tail long and round, and the plumage black. *C. Major*, et *C. Ani.* Enl. 182. fig. 1 and 2. Vieill. Gal. 33.

These birds live on insects and grains; fly in flocks; lay and sit many pairs together, in a nest placed on branches, and of a size proportioned to the number of

* *T. Temminckii*, and *T. Cinnamoneus*, Vaill. *T. Ardens*, Tem. and *T Aurantius* and *T. Variegatus*, Spix.

† *Ani Anno*, the name of these birds in Guiana or Brazil. Crotophagus was used by Brown in his History of Jamaica, because in that Island these birds fly on the cattle to pick from their backs the *Tabanni* and ticks.

couples which build it. They are easily tamed, and even learn to speak; but their flesh has a bad scent.

The Toucans.* Ramphastos. L.

May be known from all other birds by their enormous bill, almost as thick and long as their body, light and cellular within, arched towards the end, and irregularly indented at the edges. The tongue is long, narrow, and furnished on each side with barbs like a feather. They are found only in the warm parts of America, and live in small troops on fruit and insects; but they devour also, during the laying season, eggs, and little birds newly hatched. The structure of their bill obliges them to swallow their food without mastication; when they have seized it, they throw it in the air, the more easily to swallow it. Their feet are short; the wings little extended, and the tail tolerably long. They build in the trunks of trees

The Toucans, properly so called, have the bill thicker than the head, they are generally black with bright colours on the throat, breast, and rump. These parts of their plumage were employed formerly in making a sort of embroidery.

Toco Toucan. (*Ramphastos Toco.*) Enl. 82. Vaill. 2.

Black; throat and rump, white; vent, red; bill, red, with black tip. Cayenne.

* Toucan is from their Brazilian name, *Tuca.* Ramphastos was applied by Linnæus, from ῥάμφος, bill, on account of the size of that part.

Carinated Toucan. (*R. Carinatus.*) Wagler. Edw. 329.

Black; bill, large, green; culmen, green; throat, golden-yellow; chest and vent, scarlet; rump, white.

Yellow-breasted Toucan. (*R. Tucanus.*) Enl. 307.

Black; throat, yellow; spot on breast and vent, red. South America.

Brazilian Toucan. (*R. Piscivorus.*) L. or *Callorhynchos.* Wagler. Edw. 64.

Black; rump, white; spot on breast and vent, red. Brazil.

Great Toucan. (*R. Maximus.*) Cuv. Vaill. Toucans, pl. 6.

Black; bill, black; base, yellow; sides of head and throat, pale-yellow; chest, belly, rump, and vent, deep-red.

Red-breasted Toucan. (*R. Pectoralis.*) Shaw. or *R. Tucai.* Lich. Enl. 269.

Black; throat, yellow; band on breast, vent, and rump, red. South America.

Aldrovandine Toucan. (*R. Aldrovandi.*)* Sh. Alb. II. 25.

Black; beneath, red; breast, yellow. Guiana and Brazil.

Red-billed Toucan. (*R. Erythrorynchos.*) Sh. Enl. 262. Vaill. 3.

Black; throat, white; rump, yellow; bar on breast and vent, red. South America.

* *R. Erythrosema,* Wagler.

Vaillant's Toucan. (*R. Vaillantii.*) Vaill. 4. Swainson. Zool. Ill. 56.

Black; bill, red; culmen, black; base, yellow; throat, dirty-white; chest and vent, ochreous. Brazil.

Black and Yellow Toucan. (*R. Dicolorus.*) Wagler, or *Chlororhynchos.* Tem. Vaill. 8.

Black; cheeks and throat, pale-yellow, with a central orange spot; chest and belly, scarlet. Brazil.*

The Aracari. Buff. Pteroglossus. Ill.

Have the bill not so thick as the head, and furnished with a more solid horn; their size is less; and the ground colour of their plumage generally green, with red or yellow on the throat and breast.

Green Toucan. (*R. Viridis.*)† Enl. 727, 8. Vaill. 10. 17.

Dark-green; head and neck, black; rump, red; beneath, yellow. Cayenne.

Aracari Toucan. (*R. Aracari.*) Enl. 166. Vaill. 10 and 11. Vieill. Gal. 30.

Dark-green; beneath, yellow; bar across the abdomen and the rump, red. South America.

* *R. Cuvieri. R. Stors,* and *R. Temminckii,* Wagler, and *R. Ambiguus.* Swainson. Ill.

† *R. Aldrovandi* of Shaw.

Piperine Toucan. (*R. Piperivorus.*) L. or *Pt. Culik.* Wagler. Enl. 577. 729. Vaill. 13, 14.

Olivaceous; head, black; crescent on the neck, orange. Cayenne.

Grooved-bill Aracari. (*R. Sulcatus.*) Swainson. Zool. Ill. 44. Col. 356.

Green; edge of wings, black; top of head, and triangular patch about the eye, blue; throat, dirty-white. Peru.

Azara's Toucan. (*R. Azaræ.*) Vaill. Sup. A.

Top of the head, green-black; neck and throat, marron; back and coverts, olive-green; beneath, yellow, with reddish spots and bars. Brazil.

Black-green Toucan. (*R. Inscriptus.*) Swain. Zool. Ill. 90.

Black-green; head and neck, black; beneath, pale-yellow; thighs, reddish; rump, scarlet. Brazil.

Baillon's Toucan. (*R. Bailloni.*) Vaill. 18.

Olive-green, above; forehead, with transverse yellow band; beneath, yellow.

Spotted-billed Toucan. (*R. Maculirostris.*) Vaill. and 15, Sup. A. A.

Brilliant black; above, and on the breast, a spot of golden yellow; orange-yellow band on neck; back, olive-green; lower tail-coverts, red; bill, black, with red at the base.

* *Pteroglossus Humboldtii,* Wagler. *Pterog Reinwardtii,* Wagler. *P. Langsdorffii,* Wagler.

The Parrots (Psittacus, L.)

Have the bill thick, hard, solid, round in all parts, surrounded at the base with a membrane in which the nostrils are pierced; the tongue is thick, fleshy, and round, circumstances which give them the greatest facility in imitating the human voice. Their lower larynx is complicated, and furnished on each side with three muscles, contributing further to this facility. Their vigorous jaws are put in motion by muscles, which are stronger than in other birds. They have very long intestines, but are without a cæcum. Their food consists of fruit of all sorts. They climb branches by means of their bill and feet; build in holes in trees; have a voice naturally hard and noisy; and are, almost all, of very bright colours: hence they are only found in the Torrid Zone. But they exist in both continents, though the species proper to each are different; even every great island has its own species, as the short wings of these birds do not allow them to traverse much space of sea. The parrots are very numerous, and are subdivided by the shape of the tail, and other characters.

Among those with a long wedged tail we may first distinguish

The Aras, Maccaw (Ara, Kuhl.),

Whose cheeks are denuded of feathers. They are American species; for the most part, very large, and of a very brilliant plumage: for which reason many of them are brought into Europe.

Red and Blue Maccaw. P. Macao. L. Vaill. 1.

Scarlet, with blue wings; yellow on wing-coverts; cheeks naked, white. Size of a fowl. South America.

Red, Yellow, and Blue Maccaw. P. Aracanga. Enl. 12. Vaill. 2.

Rather smaller than the last. Scarlet; rump blue; scapulars, yellowish, tipped with blue; quills, blue and green, most of them black tipped; middle tail-feather, blue; the next blue and red; and the four outer, blue. Guiana.

Varied Maccaw. P. Tricolor. Vaill. 5.

Top of head, cheeks, and under parts, reddish; nape, yellow; quills, blue; coverts, red-brown. South America.

Hyacinthine Maccaw. P. Hyacinthinus. Lath., or *Anadorhynchus Maximiliani.* Spix. t. 11.

Deep blue; bill, and legs, black; orbits, and base of lower mandible, yellow. South America.

The Ararauna, or Blue and Yellow Maccaw. P. Ararauna. Enl. 36.

Blue; beneath yellow; cheeks, naked, with black lines.

Military Maccaw. P. Militaris. Vaill. 4.

Green; forehead, red; quills, and rump, blue; tail, red, tipped with blue.

Brazilian Green Maccaw. P. Severus. Vaill. 8, 9, and 10.

Green; front, purple-brown; wings, blue; tail-feathers, blue above, red beneath.

Parrot Maccaw. P. Macawuanna. Enl. 864. Vaill. 7.

Green; beneath, refescent; crown, and wings, bluish.

Purple-backed Maccaw. Arara Purpureo-dorsalis. Spix. 24.

Large; yellow-green; back, middle of the belly, and forehead, purple; crown, primaries, and tip of tail, blue; base of tail, red. Brazil.*

The others with a long tail have in common the name

PARRAKEET (CONURUS, Kuhl.);

M. Le Vaillant divides them into

MACCAW PARRAKEETS,

Which have a naked space round the eye; they come from America, like the Maccaws.

Pavouane Parrakeet. P. Guyanensis. Enl. 167, 407. Vaill. 14, 15.

Green; orbits, naked, whitish; quill, and tail feathers, yellowish; underneath shoulders, and under wing-coverts, red. Guiana.

* Add from others, *Psittacus ambiguus,* Becht. *Psitt. Illigeri.* Kuhl.

Scaly-breasted Parrakeet. P. Squamosus. Shaw, Mis. 1061.

Top of head, and neck, purple-brown; green on the upper part; brown, beneath; the feathers margined with white. Less than a blackbird.

Banded Parrakeet. P. Vittatus. Vaill. 17.

All above, and sides of belly, green; chest, yellowish-ash, with darker bands; quills of wings, blue above; tail, green. Brazil.

Wave-breasted Parrakeet. P. Versicolor. Enl. 144. Vaill. 16.

Shining-green; throat, brown, each feather with a yellow margin. Cayenne.

Angola-yellow Parrakeet. P. Solstitialis. Vaill. 16, 19, or *Aratinga Chrysocephalus.* Spix. 14. His *Aratinga luteus*, 14, is a variety of it.

Orange-yellow, with green spots on the back and wings; sides, and thighs, red. Angola.*

Parrots with arrow-shaped tails, in which the middle tail-feathers greatly exceed the others in length.

It is of this division that Messrs. Vigors and Horsfield have made their genus PALÆORNIS.

Collared Parrakeet. P. Torquatus. Briss. Enl. 551.

Green, tinged with yellow; beneath, a broad band of yellow, striped with black at the back of the head. Philippine Islands.

* Some place here *P. Auricapilus, P. Patagonus, P Leucotis, P. Inornatus.*

Alexandrine Parrakeet. P. Alexandri. L. Enl. 642.
Vaill. 30. Edw. 292. The young of which, according to Kuhl, is *P. Eupatria.* L. Vaill. 73. Enl. 239.

Bright-green, with a red collar on the neck, and a black spot under the throat.

Yellow-collared Parrakeet. P. Annulatus. Bech. Vaill. 75, 76.

Above, brilliant green; beneath, yellow-green; head of male, blue; collar, yellow; intermediate tail quills, long blue, tipped with yellowish-white. Pondicherry.

Blossom-headed Parrakeet. P. Erythrocephalus. L. *Gingianus.* Lath. Vaill. 45. Edw. 233.

Head, red; bluish, behind; chin, black; with a pale green collar; rest of the plumage green. Gingi.

Malacca Parrakeet. P. Malaccensis. Gm.

Crown and plumage in general, green; head, red, rose-coloured. Malacca.

P. Barbulatus, Bech. Enl. 888. Vaill. 72. *P. Barrabandi.* Swain. 59.

Green, with forehead and throat, golden yellow; chest band, red. New Holland.

Bengal Parrakeet. P. Bengalensis. Gm. Enl. 888. Vaill. 74.

Like *P. Erythrocephalus,* with the hind head blue.

Papuan Parrakeet. *P. Papuensis.* Sonner. Nov. Guin. t. 3.

Scarlet; wings and tail, green; sides of body and tail, yellow; hind head, blue. Papua.

Sincialo Parrakeet. *P. Rufirostris.* Enl. 580.

Green; shoulders, yellowish; tail, long, tipt with blue. St. Domingo.

Red-breasted Parrakeet. *P. Hæmatopus.* Enl. 61, or *Cyanocephalus.* Enl. 192, or *Moluccanus.* Enl. 743, or *Cyanogaster.* Sh. Gen. Zool., viii. t. 59. and White p. 140, all varieties of age.

Green; breast, red; face, blue; neck spot, yellowish.

Vigors and Horsfield having remarked that the tongue of this species has soft bristles under its point, have made their genus TRICHOGLOSSUS of it. It would be interesting to examine whether many other parrots have not the same character.

The character of the tail of this division is observable in the first species known in Europe, into which it was brought by Alexander. This is *P. Alexandri*, before mentioned.*

Parrakeets, with the tail enlarged towards the end.

Black Parrakeet. *P. Niger.* Enl. 500. Edw. 5.

Glossy blue-black; tail, red, beneath. Brazil.

* Others place here, *P. Bitorquatus*.

Large Black Parrot. P. Vasa. Vaill. 51.

Black, with grey and brown reflections. Much larger than the last. South Africa.

Masked Parrot. P. Mascarinus. 3. Enl. 5. Vaill. 139.

Large, brown, lighter underneath, with a black mask; rest of head and neck, ashy-grey; lateral tail, white at the base, the rest brown. Madagascar.

Crimson-winged Parrakeet. P. Erythropterus. Nat. Mis. 653.

Crown, cheeks, and ridge of wings, green; back, black; rump blue; beneath, yellowish-green. New Holland.

Nonpareil Parrakeet. P. Eximius. Vaill. 28, 9. Nat. Mis. 93.

Head, neck, and breast, scarlet; wings and tail, blue; back, black, undulated with green; beneath, yellow. New Holland.

Pennant's Parrakeet. P. Pennanti. Lath. White. t. at p. 174 and 175, or *Elegans*, Gm. Vaill. 78, 79, or *P. Gloriosus.* Sh. 53.

Scarlet, with blue throat, wings, and tail; back varied with black. New Holland.

Brown's Parrot. P. Brownii. Kuhl. Vaill. 80.

Crown, black; cheeks, white; above, black; the feathers margined with yellow; beneath, yellowish-white; shoulders and wing-coverts, blue. New Holland.

Tabrian Parrot. P. Scapulatus. Bech. Vaill. 55, 56. Enl. 240.

Head front of neck, sides and belly, brilliant deep-red, with a few blue spots under the tail; demi-collar, blue; back, green. New Holland.

Tabuan Parrakeet. P. Tabuensis. Lath. or *Atropurpureus.* Sh. Lev. Mus. t. 34.

Dark-crimson with green; back, wings, and tail-quills, and outer tail-feathers, blue. Friendly Islands.

Amboina Red Parrakeet. P. Amboinensis. Gm. Enl. 240. and White, p. 168, and 169.

Head, neck, and all beneath, scarlet; back, rump, and quills of tail, blue. Amboina.

It is of this division that Vigors and Horsfield have made their genus PLATYCERCUS.*

In common Parrots, with the tail wedged almost equally.

Brazil Yellow Parrakeet. P. Guaruba. Kuhl, or *P. Luteus.* Lath. Vaill. 20. or *Aratinga Carolinæ.* Spix. 12.

Body, yellow; wings and tail, yellowish green with blue. Brazil.

* Others place here—*P. Spurius, P. Venustus, P. Cyanomelas, P. Icterotis, P. Multicolor.*

Pavouane Parrot. P. Guyanensis. Gm. or *Macrognathos.* Spix. 25.

Plumage, deep green; paler, beneath; cheeks, spotted with red; edge of wing and under lesser coverts, scarlet. Guiana.

Carolina Parrakeet. P. Ludovicianus. Enl. 499. or *Carolinensis.* Wilson III. 26. f. 1.

Fore part of head, orange; back and throat, yellow; body, green; edge of wing at the bend, orange; coverts, green. Carolina.

Illinois Parrakeet. P. Pertinax. Enl. 528. Vaill. 34 to 37.

Green, with fulvous; cheeks, quills, and tail-feathers, grey. Brazil, &c.

Golden-crowned Parrakeet. P. Aureus. L. Vaill. 41. Edw. 235.

Green; crown, orange; blue bar across the wings; edges of tail-feathers, blue. Brazil.

Red and blue-headed Parrakeet. P. Canicularis. Enl. 767. Vaill. 40.

Green; beneath, inclining to yellow; forehead, red; crown, blue. South America.

Brown-throated Parrakeet. P. Æruginosus. Edw. 177.

Green; crown and quills, blue; cheeks and throat, grey-brown. South America.

Grey-cheeked Parrakeet. P. Buccalis. Vaill. 67.

Green, above; yellowish-green, underneath; lower wing-coverts towards the base, blue; about the eyes, throat and forehead, greyish. Cayenne.

Yellow-winged Parrakeet. P. Virescens. Enl. 359. Vaill. 59.

Pale-green; quills, blue; secondaries, white, yellow edged; coverts, tipt with yellow. Cayenne.

Cayenne Parrakeet. P. Sosova. Enl. 456. Vaill. 58, 9 and *P. Tovi.* Enl. 190.

Deep green, above; lighter, underneath; quills of wings and middle of tail, blue; spot on wings, yellow. Cayenne.

Grey-breasted Parrakeet. P. Murinus. Enl. 768. Vaill. 38.

Olive; quills, green; face, throat, and breast, grey. South America.

Mustache Parrakeet. P. Pondicerianus. Enl. 517. Vaill. 31.

Green; crown, grey; frontal bar, and behind the bill, black; breast, red. India.

Ternate Blue-headed Parrot. P. Xanthosomus. Bech. Vaill. 61.

Bright-green; head, neck, and large quills, bright-blue; bill, red. Ternate.

Bridled Parrakeet. *P. Capistratus.* Bech. Edw. 232. Vaill. 47.

Deep-green, above; yellowish green, underneath; face, surrounded with a blue band; collar, yellow. India.

Lory Parrakeet. *P. Ornatus.* Enl. 552. Vaill. 52. Edw. 174.

Green, varied with yellow; crown, blue; hind head, red; throat and breast, undulated blackish-green. India.

Lace-winged Parrakeet. *P. Marginatus.* Vaill. 60. or *P. Olivaceus.* Enl. 287.

Green; crown, blue; wing-coverts, blue, edged with yellow. Lugonia.

Great-bellied Parrot. *P. Macrorhynchus.* Enl. 713. Vaill. 83.

Green; back and wings, bluish; coverts, black, edged with orange. New Guinea.

Grand Lory. *P. Grandis.* Enl. 518, and 683. Vitter. 1, Vaill. t. 126, 127, 128.

Scarlet; bluish band round the body; edges of shoulders and quills, blue; tail, yellow beneath. Ceylon.

Red-winged Parrakeet. *P. Incarnatus.* Vaill 46.

Green; throat and coverts, red; bill and legs, flesh-coloured. India.

Bornean Lory. P. Borneus. Vaill. 44.

Red; quills and tail-feathers, tipt with green; blue spot on wings. Borneo.

Black Parrakeet. P. Novæ Guineæ. Vaill. 49.

Glossy blue-black; tail, red; beneath. New Guinea.

Crimson-fronted Parrakeet. P. Concinnus. Vaill. 48.

Green; forehead, temples, and sides of rump, scarlet.

Small Parrot of White. P. Pusillus. Vaill. 63.

Green; frontlet and base of tail-feathers, red. About seven inches long. New Holland.

Red-banded Parrot. P. Humeralis. Vaill. 50.

Above, yellowish green; more yellow on the sides; forehead, band, and throat, red. New Holland.

Red-shouldered Parrakeet. P. Discolor. Vaill. 62.

Green; margins and under parts of shoulders, red; coverts, blue; tail-feathers, ferruginous at base. New Holland.

Undulated Parrakeet. P. Undulatus. Sh. 673.

Yellowish-green; undulated above with brown; tail, blue, with a yellow bar across. New Holland.

Golden Parrakeet. P. Chrysostomus. Kuhl. pl. 1.

Olive-green, above; throat, and chest, bright-green; belly and about the eye, yellow; forehead, wing-coverts, and tail, with a narrow blue band. New Holland.

Turcosine Parrakeet. P. Pulchellus. Vaill. 68.

Green; beneath, yellow; wings and front, blue. New Holland.

Zoned Parrakeet. P. Zonarius. Sh. 657.

Green, with transverse collar behind the neck, and a yellow band on the abdomen; head, face, and quills of wings, black. New Holland.

We may insert here some species with a square tail, with the two middle tail-feathers long, but with barbs only at the end.

Racket-tailed Parrakeet. P. Setarius. Tem. Col. 15.

Green; on the crown, a marron lunule patch; behind it another, much larger, light-blue; back of neck, red and yellow; upper wing-coverts, blue; middle tail-feathers, green; the lateral, blue. New Holland.*

Among the Parrots with a short and equal tail may be distinguished

The Cockatoos,

which M. Vieillot has named Plyctolophus.

These have a crest, formed of long and narrow feathers, ranged on two lines, recumbent or erect at the will of the bird. They live in the remotest parts of India. The plumage of most of them is white. They are very docile, and prefer marshy places.

* Others place here—*P. Viridissimus, P Lichtensteinii, P. Lunatus, P. Ruber, P. Unicolor, P. Guebiensis, P. Domicella, P. Radhea, P. Garrulus, P. Cyanurus, P. Stavorini, P. Coccineus, P. Riciniatus, P. Australis, P. Chlorolepidotus*, &c.

Great White Cockatoo. P. Cristatus. Enl. 265.

White, with large crest, which is red on the under side in the males. Molucca Islands.

Philippine Cockatoo. P. Philippinarum. Enl. 191.

White; vent and under tail, red; crest, bright yellow at base; white at the top. Malacca.

Molucca Cockatoo. P. Moluccensis. Enl. 498.

White, with a slight rosy tint; middle feathers of crest, red; the rest white; under tail and wings, saffron. Molucca and Sumatra.

Lesser Sulphur-crested Cockatoo. P. Sulfureus. Enl. 14.

White, with pointed sulphur-coloured crest, and spot beneath the eye. Size of common grey parrot. Molucca.

Great Sulphur-crested Cockatoo. P. Galeritus. White at p. 237.

Like the last, but as big as a cock. New Holland.

Nascian Cockatoo. Ps. Nasicus. Tem. Col. 331.

White, with obscure waves; round the eye, red; before it, an orange spot; cheeks and throat, with lunule spots; under tail, yellow. New Holland.

Some species of these lately discovered in New Holland have the crest more simple, less mobile, and composed of large and moderately long feathers. They subsist especially on roots.

This division forms the genus Calyptorhynchus of Vigors and Horsfield.

Banksian Cockatoo. P. Banksii. Lath. Syn. Sup. 109. Sh. Mis. 50.

Black; tail, red in the middle, with numerous black bars; crest and wings, spotted. New Holland.

Funereal Cockatoo. P. Funereus. Sh. Mis. 186.

This is like the last, but the spotted bar on the tail is buff colour. New Holland.

Cook's Cockatoo. P. Coccii. Tem. or *Leachii.* Kuhl. pl. 3.

This differs from the preceding only in having the middle of the tail-feathers crimson, without black bars. New Holland.

Rose Cockatoo. P. Roseus. Kuhl. Col 81. Vail. Gal. t.

Head and body, fine rose-colour; quills, wings, and vent, ashen.*

Others have for a crest nothing but a few pendant feathers, barbed only towards the end, which form a sort of tuft.

Red-crowned Parrot. P. Galeatus. Lath. Sup.

Dusky-green; abdomen, undulated red and green; crown, red. New Holland.

But the greatest number of these birds have no

* Add *Psit Tenuirostris, P. Eos, P. Temminckii.*

ornament on the head. The species best known for its facility in speaking is

The *Grey Parrot*, or *Jaco*. *P. Erythacus*. Enl. 311. Edw. 103. Vaill. 99. 103.

Ash-coloured, with a red tail. They live in Africa.

The species with green plumage are the most numerous.

White-breasted Parrot. *P. Melanocephalus*. Enl. 527. Vaill. 119, 120.

Green; beneath, yellowish with white crest; quills tipped with blue. Mexico.

Marked Parrot. *P. Signatus*. Vaill. 105.

Above, green shaded with blue; yellowish-green, beneath; upper part of the tail, red. Brazil.

Blue-headed Parrot. *P. Menstruus*. Enl. 384. Vaill. 114, or *Flavirostris*, Spix. 31.

Green; head, neck, and breast, violet-blue; vent, and under tail, red. Cayenne.

Purple Parrot. *P. Purpureus*. Enl. 408. Vaill. 115.

Above, blackish-brown; below, lilac-purple; wing and tail-quills, and lower tail-coverts, blackish-blue, Guiana.

Dusky Parrot. P. Sordidus. Vaill. 104.

Top of head and back, dusky; sides and behind, greenish; throat, blue; beneath, cinereous-brown. New Spain.

Amazon, or Common Green Parrot, with blue forehead. P. Amazonicus. Enl. 13, 120, 312. Vaill. 98, 99.

Green; bend of shoulders and patch on wings, red; crown and head, yellow; front, blue. America.

Amazon Parrot. P. Æstivus. Enl. 547 and 897. Vaill. 110, and 110, bis.

Above, dull green; top of head yellow. Underneath, paler, and more yellow green. Guyana.

Blue-faced Parrot. P. Cærulifrons. Sh. Edw. 230. Vaill. 135.

Green; paler beneath; front blue, inclining to violet. Guyana.

Brazil Parrot. P. Cyanotis. Tem., or *Brasiliensis.* Lin. Edw. 161. Vaill. 106.

Grass-green; paler beneath; head yellow; round the base of the bill bright red. Brazil.

Red-banded Parrot. P. Dominicensis. Enl. 792, or *Vinaceus.* Pr. Max. or *Columbinus.* Spix. 27.

On forehead from eye to eye a red band; green, above; stomach, reddish; quills, blue. Guiana.

Dufresne's Parrot. P. Dufresnianus. Kuhl. Vaill. 91.

Body green; top and front of head red; upper part of wings clear green. Cayenne and Brazil.

Autumnal Parrot. P. Autumnalis. Edw. 164. Vaill. 111.

Dark green, above; apple green, beneath; forehead, red; top of head, blue. Guyana.

Blue-fronted, or Havanna Parrot. P. Havanensis. Enl. 360. Vaill. 122.

Green; beneath, subviolaceous; wing-spot, red; vent, yellow. Havanna.

White-fronted Parrot. P. Leucocephalus. L. Enl. 335, 548-9. Vaill. 107-8, 8 bis and 9.

Green, the feathers fringed with brown; front, white; cheeks and crown, blue; and throat, scarlet. Martinique.

White-headed Parrot. P. Albifrons. Mus. Carl. 52.

Like the last, but rather less, and without the blue on the head.

Dusty-green Parrot. P. Pulverulentus. Enl. 861. Vaill. 92.

Dusty-green; yellow spot on the crown; red patch on wings; quills, black, blue towards the tip. Cayenne.

Festive Parrot. P. Festivus. Enl. 840. Vaill. 129.

Green; front, purple; throat and back, red. Guiana.

Hawk-headed Parrot. P. Accipitrinus. Enl. 520. Spix. 32. a.

Green; head and neck, ferruginous, waved with blue. India.

Senegal Parrot. P. Senegallus. Enl. 288. Vaill. 116, 118.

Green; head and quills, ash-colour; beneath, orange. Senegal.

Le Vaillant's Parrot. P. Levaillantii, Lath. *P. Infuscatus.* Sh. Vaill. 130-131.

Olive-brown, varied with green; shoulders and thighs, orange. Eastern parts of Africa.

Amboina Parrot. P. Gramineus. Enl. 862. Vaill. 121.

Green; beneath, olive; crown, blue. Amboina.

Red-sided, or *Green and Red Chinese Parrot. P. Sinensis.* Edw. 231. Enl. 514. Vaill. 132.

Green; sides of body, and under wing-coverts, red; quills, edged with blue. China, Amboina, &c.

Geoffroy's, or *Red-masked Parrot. P. Geoffroyii.* Vaill. 112, 113, or *P. Personatus.* Sh.

Green; face and throat, scarlet; crown, bluish, and blue under wing-coverts. New Holland.

Yellow-faced Parrot. S. Xanthops. Spix. 26.

Pale green; neck, throat, and chest, obscurely cross-banded; top of head, sides of face, and ears, bright yellow; primaries, and tail, blue-green. Brazils.

Bonnetted Parrot. P. Mitratus. Pr. Max. Col. 207, or *Maitaca,* Spix. 29 f. 1. and 30.

Green; top of head, and ears, orange-red; quills, and tail, edge of humerus, blue; female with head green, and blue forehead. Brazils.

Diadem Parrot. P. Diadema. Spix. 32.

Blue-green; forehead, and throat, bluish; crown, yellow-blue; nape, and chest, green; middle of wings, purple and yellow; primaries, green. Brazils.

Such species as have the ground colour of the plumage red, and the tail slightly cuneiform, and which approximate in some degree to certain of the Parakeets, are called Lorys. They are found only in the East Indies.

Unicolor Lory. P. Unicolor. Vaill. 125.

Entirely scarlet, inclined to crimson on back and wing-coverts; legs and feet, dusky. Molucca Islands.

Collared Lory. P. Domicella. Enl. 119. Vaill. 94-5.

Scarlet; wings, green; crown, black; pectoral bar, yellow. Molucca Islands.

Black-capped Lory. P. Lori. Enl. 158. Vaill. 123-124.

Top of head, black; red nuchal demi-collar; above, violet-blue; wings green, red-fringed. Moluccas.

Ceram Lory. *P. Garrulus.* Enl. 216. Vaill. 96.

Scarlet, with shoulders and spot on back, yellow; wings, green; tail, tipped with blue. Molucca Islands, especially Ceram.

Blue-tailed Lory. *P. Cyanurus.* Sh. Vaill. 97.

Scarlet; scapulars and tail, blue; coverts varied with blue; quills, black. Borneo.

Certain small species, with a very short tail, the PSITTACULES of Kuhl, are also, though improperly, sometimes called Parakeets.

Passerine Parrot. *P. Passerinus.* Enl. 455. Nat. Mis. 893. Spix. 33.

Green patch on wings and rump, and all beneath, blue. Little more than four inches long. Brazil.

Tui, or *Gold-headed Parrot.* *P. Tui.* Enl. 456. f. 1. Vaill. 70.

Green; forehead, orange. Size of a starling. Cayenne.

Black-winged Parrot. *P. Melanopterus.* Enl. 591. f. 1. Vaill. 69. Sh. 132.

Head and neck green; mantle, and alar quills, brown-black; breast, and beneath, apple-green; coverts and quills, yellow, tipped with blue; tail violet, banded black. Java.

Pileated Parrot. *P. Pileatus.* Enl. 744. Vaill. 135.

Green; forehead and crown, red; rump, yellow-green; quills, blue-edged; tail, yellow at the end.

Black-headed Parrot. P. Barrabandi. Vaill. 134. Head, black; mustache, yellow; upper wing-coverts, lively red. Brazils.

Grey-headed Parrot. P. Canus. Enl. 791. f. 2. Sh. 425.

Green; head and neck, grey; black bar across the tail. Madagascar.

Swindernian Parrot. P. Swindernianus. Kuhl. Pl. 2.

Top of head, cheeks, and nape, green; black demi-collar; back and wings, obscure green; belly, &c. yellowish green. Africa.

Sapphire-crowned Parrot. P. Galgulus. Enl. 190. f. 2.

Green; rump and breast; red; crown, blue. Philippine Islands.

Philippine Parrot. P. Philippensis, Enl. 520.

Is treated as a variety of the last by Latham and Shaw, without the blue on the crown.

Vernal Parrot. P. Vernalis. Mus. Carl. 29.

Brilliant green above; yellowish green underneath. Java. Timor.

Indian Green Parrot. P. Indicus. Edw. 6.

Head and neck, vivid green; wings and back, obscure green. East Indies.

Collared Parrot. P. Torquatus. Son. Nou. Guin. 393.

Green above; yellow, undulated with black bar at the back of the neck. Philippine Islands.

Plain Parrot. P. Simplex. Kuhl. Son. Ib. 38. f. 1.

General colour, green; clearer underneath. Isle of Luçon.

Guinea Parrot, with Red Head. P. Pullarius. Enl. 60.

Green; face, red; rump, blue; tail, orange, with a black bar across. New Guinea.

Short-winged Parrot. P. Micropterus. Son. 41.

Back, and upper wing-coverts, black; head, neck, and belly, yellowish green. New Guinea.

Otaheite, or *Violet Parrot. P. Taitianus.* Gm. Enl. 455. Vaill. 65, or *P. Porphyrus.* Nat. Mis. 7.

Blue, with a white throat, and a slight crest. Otaheite.

Sparman's Parrot. P. Sparmanni. Mus. Carl. 27. Vaill. 66.

All the plumage deep blue. Otaheite.

Blue-crested Parrot. P. Fringillaceus. Vaill. 71, or *Porphyrocephalus.* Nat. Mis. 1.

Green; crown, slightly crested blue; throat, and bar on abdomen, red. Sandwich Islands.

Blue-headed Parrot. P. Phigy. Vaill. 64.

Head, deep blue, slightly violet; cheeks, throat, chest, and all underneath, fine red; scapulars, red. Friendly Islands.

Yellow Green Parrot. P. Xanthopterigius. Spix. 34. f. 1, 2.

Yellowish green; middle wing-coverts, yellow; lower part of the back, green. Brazils.

Gregarious Parrot. P. Gregarius. Spix. 34. f. 3, 4.

Herb-like green; wing, green, spotless; bill, streaked; tail rather longer than the wings; head of female, yellowish. Brazils.

All these differences of colour and size can scarcely authorize generic distinctions.

There are left only

The PARROTS *with a trunk*, of Vaill., which have characters sufficiently peculiar by which they may be detached from the others.

Their short and square tail and crest, composed of long and narrow feathers, assimilate them to the cockatoos. They have the cheeks naked like the aras; but their bill, with an enormous upper mandible, and the lower very short, which cannot, therefore, be entirely closed,—their cylindrical tongue,—terminated by a little corneous gland cleft at the end, and capable of being protruded out of the mouth—their legs naked a little above the talons—their tarsi short and flat, on which they frequently support themselves in walking—distinguish them from all other parrots. We know but two species, originally from India.

Black Cockatoo. *P. Aterrimus.* Gm. or *P. Gigas.* Lath. Edw. 316.

Black, with a large crest. East Indies.

Giant Cockatoo. *P. Goliath.* Kuhl, or *L'Ara noir à trompe.* Vaill. Per. I. pl. 12 and 13. Crest large; ashen grey in the whole body. East Indies.

The *Ara gris à trompe.* Id. ib. pl. 11, is probably only a variety.

The name Parrot with a Trunk, is not very applicable. The tongue is not hollow; indeed the tongue, properly speaking, is nothing but the little corneous piece attached to the end of this cylinder. See Geoff. St. Hilaire, ap. vj. Gal. 4.

It is of this division that M. Vieillot makes his genus, MICROGLOSSUS. Gal. pl. 50.

We might also perhaps make a sub-genus of

The NIMBLE PARROTS. PEZOPORUS, Ilig.

Whose bill is weaker, the tarsi more elevated, and the talons straighter than in other parrots. They walk on the ground, and seek their food in the grass.

Ground Parrot. *P. Formosus.* Vaill. I. 32. Nat. Mis. 228.

Green above, each feather banded with black and yellow; beneath, yellow, with many dark waved bands; crown and nape, streaked; forehead, orange. New Holland.

New Zealand Parrot. *P. Novæ Zelandiæ,* Lath. Mus. Carl. 28.

Green; paler, beneath; top of head and vent, crimson. New Zealand.

Horned Parrot. *P. Cornutus.* Lath. Syn. Sup. III. pl. 8.

Head and cheeks scarlet, with two feathers, one inch and a half long, rising like horns from the head. Body in general green, yellow about rump and vent. New Caledonia.

There are two African birds usually placed among the Scansores nearly allied to one another, which appear to me to have also some analogy with the gallinaceous birds, and especially the genus *Crax*.

They have the wings and tail of Crax, and remain like them in trees; their bill is short, and the upper mandible swelled out; their feet have a short membrane between the toes in the fore part, but it is true that the external toe is frequently directed backward, like that of the owl's. Their nostrils are also simply pierced in the horn of the bill. The edges of the mandibles are indented, and the sternum, at least in the Touraco, has not the great slopes common in the gallinaceous birds.

These birds, of which two genera have been made, are—

The TOURACOS CORYTHAIX,* or *Plantain-eater*. (Illig.)

Whose bill does not mount up on the forehead, and whose head is furnished with a moveable crest.

Touraco Plantain-eater. *Cuculus Persa*. Lin. Enl. 601. Vaill. Prom. 16 and 17. *Edw. t.* 7.

This bird inhabits the environs of the Cape; is of a bright green, with part of the quills of the wings crimson. It builds in holes in trees, and feeds on fruit.†

Touraco Géant. Vaill. Prom. pl. 19. *Musophaga Gigantea*. Viel. Cheeks and above, generally shining blue; quills, black at the extremity; crest, large and black.

Pauline Plantain-eater. *T. Paulina*. Tem. Col. 23, or *Opæthus Erythrolophus*. Vieill. Gal. 49.

Green; crest, red; cheeks, white. South Africa.

Varied Plantain-eater. *T. Brun Phasianus Africanus*. Lath. Vaill. 20, or *Musophaga Variegata*, Vieill Gal. 48.

Ashy-blue, above; white, underneath; quills of tail. black, except the two middle, which are brown; crest, long, brown, edged with white.‡

* M. Viellot has changed this name to OPÆTHUS; and Wagler has divided them into two genera, under the names of *Cheyarhis* and *Spelectos*.

† The *Opæthus Buffonia*, Vieil. C. *Africanus Shaw*. Vail. t. 17, and Edw. t. 7.

‡ The genus *Phimus* of Wagler, the *Cuculus regius* of Shaw, Mus. Lever, t. 40 bad.

MUSOPHAGA, Isert.

So named, because they live more especially on the fruit of the banana, and have for their character the base of the bill forming a disk, which covers a part of the forehead.

The species known is—

Violet Plantain-eater. Musophaga Violacea, Vieill. Gal. 47. *Touraco Violet,* Vaill. Prom. pl. 18.

Has a naked red skin round the eyes; the plumage, violet; the occiput and the great quills of the wings, crimson; a white stripe passes under the neck round the eyes. It inhabits Guinea and Senegal.

2. Galbula
1. Picus
4 Cuculus
3 Scythrops
6. Pogonias
5. Trogon
8. Crotophaga
7. Ramphastos
10. Psittacus Aras
9. Musophaga

London. Published by Whittaker & C°. Ave Maria Lane, Aug. 1829.

SUPPLEMENT ON THE SCANSORES.

The genera comprehending the present order constitute a part of the Picæ of Linnæus, and there is certainly much analogy between them and those of the foregoing order; accordingly, Mr. Vigors, in the application of the *quinary* system of Mr. Macleay, to the class of birds, has united the Picæ and Passeres of Linnæus into one order, under the denomination of *Insessores*.

The Jacamars, (Galbula,) in the system of Linnæus, constituted a part of the genus Alcedo; but the character derived from the position of their toes must exclude them from that genus. Still, however, there is a considerable approximation, in the sharp and elongated bill, and short feet, with the front toes united for a considerable portion of their length. Willoughby, Klein, &c. placed them with the woodpeckers.

They are all natives of South America, (with the exception of a single species,) where they remain isolated in humid woods, on low branches. They fly lightly, though but to short distances, and are silent, except during the time of their amours, when they utter precipitate cries, which are heard afar off. They are all entomophagous, and they nestle in the holes of trees, in worm-eaten wood. One species (*Galbula Viridis*) delights in the thickest recesses, and is of so indolent a disposition, that it will remain perched for the greatest part of the entire day on the same branch, from which, however, it will occasionally spring, to seize, on their passage, the insects which constitute its prey.

Under the name of Picus, which we may translate by our popular term *Woodpecker*, Naturalists unite a number of birds, which constitute a very natural genus of the order now under review. Their hard tongue, armed with solid corneous papillæ, is a very proper weapon for seizing the insects, and more especially the soft larvæ, which these birds seek under the bark, or in the tender and rotten wood of old trees. The feet, furnished with claws, strong and crooked, assist them in climbing along the trees, which they most frequently ascend in a spiral direction: they are also capable of running along the branches horizontally, and in opposition to their own proper weight.

The cry of the woodpeckers is sharp and piercing; their flight heavy, and by springs. They are easily recognized by the redoubled blows with which they strike the trees, to terrify the insects which are concealed under the bark, or catch them, if the wood be soft enough to yield to their strokes. The woodpeckers are never fat; their flesh is hard, coriaceous, black, and, consequently, in little estimation. Their plumage is exceedingly varied, and they exhibit in the upper parts of it, all colours, blue excepted.

The woodpeckers are continually occupied in hollowing trees: into the holes of these they retire during the night, and also when they lay their eggs, which the female deposits there without making any nest. The father and mother keep their young ones there until the latter are fit for reproduction. During the day, they remain isolated, and their life seems a laborious and active one. The species of this genus are very numerous; even Buffon was acquainted with nine and thirty; but since his time the number has been wonderfully increased. They are extended over the globe, through every latitude; two-thirds of them are found between the tropics, but they abound in the greatest numbers in the humid forests of America. It is, however, remarkable enough, that none are to be found in New Holland.

Though the woodpeckers constitute a very natural genus, and all appear, as though they were formed on one type, yet the manners of some species vary considerably. There are some which do not climb, although their organization might lead us to believe they did; but, on the contrary, they live on the ground, or in the rocks.

The *Picus Viridis* is the most common species in Europe, where it is well known, under various names, derived from its colour, cry, or habits. It usually abides in the forests, which it causes to re-echo with its harsh and piercing cries, resembling the words *tiacacan*, *tiacacan*, which are heard at a considerable distance and which it particularly utters when flying. Beside this, its usual cry, it has a love-note, which in some sort resembles a noisy and continued burst of laughter, repeated thirty or forty times in succession. It has also another very different and plaintive cry, which the peasants in some places imagine to announce the approach of rain.

Its flight is by springs and bounds. It plunges, rises, and traces undulating semicircles in the air. It can, however, sustain itself for a long time, for it will cross considerable spaces of open land, to pass from one forest to another, and it never fails to mark its arrival by its habitual cry. In spring and summer, and seldom but in these seasons, it is found on the ground, a habit which the other European woodpeckers do not possess, and which arises from its taste for ants, on which it then feeds; it awaits them on their passage, couching its long tongue in the little path which is nearest to the ant hill, and which they are accustomed to pursue in file: when it finds its tongue covered with these insects, it retires to swallow them. If the cold or the rain keeps them, in a state of lethargy or repose, in their retreat, this woodpecker goes to the ant hill itself, opens it with his feet and bill, and presently devours the ants at his ease: he also swallows the chrysalids. In other seasons he continually climbs trees, striking them with redoubled blows of his bill, which may be heard at a very considerable distance, and easily counted. This is the time in which the bird may be approached most easily:

but he withdraws himself from the view of the fowler by turning round the branch or trunk, and keeping himself always on the opposite side. Many persons imagine that after a few strokes of the bill, he goes to the other side of the tree to see if he has pierced it; but if he does make a circuit of this kind, it is rather for the purpose of seizing the insects which he has awakened and put in motion. The sound returned to the strokes of his bill seem to inform him where the hollow parts are, in which the worms lodge, or where there is a cavity where he may lodge himself and arrange his nest. This appears the more probable, as it is always in the heart of a vitiated and worm-eaten tree, that he takes up his abode. He most frequently selects trees of a tender wood, such as the aspen, birch, &c., but very rarely oaks, and other hard trees. The male and female work alternately to pierce through the sound part, until they arrive at the carious centre, throwing out the chips: they sometimes make a hole so oblique and profound that the light of day cannot penetrate into it; they enter and come out by climbing. The nest is composed of moss and wool; the eggs are from four to six, greenish, and with small black spots. During the time of hatching, the male and female rarely quit each other, go to rest early, and remain in their hole until daylight.

When these birds are on the ground, they do not walk, but jump; this is also their manner of climbing, as, in truth, it is of all birds properly called climbers.

These woodpeckers remain during the winter in very considerable numbers in the great forests of France and Germany. A part, however, migrate, for Sonnini has seen them arrive on the coasts of Egypt, in September, at the same time as other birds of passage. The *Black Bearded Woodpecker* is from South America, and is remarkable by the straight form of its bill. A black band extends round the base of the bill, which is dilated on the upper part of the throat, so as to imitate a beard. The occiput is red, the top of the head

BLACK-BEARDED WOODPECKER.

P. MELANOPOGON LICH.

London, Published by Whittaker & C° Ave Maria Lane, June 1829.

CRAWFURD'S WOODPECKER.

PICUS CRAWFURDII. Gray.

London. Published by G. B. Whittaker, July, 1828.

is covered with a large black band, and the forehead with a white band, which cover the fore part of the neck. The nape, back, and scapulars are black, with greenish reflexions. The rump and coverts of the tail are white, the quills black.

Crawford's Woodpecker is from an Indian drawing brought to this country by Mr. Crawford, jun. The whole upper part, except the crest, is deep dark brown, sprinkled with grey on the sides of the neck; across the breast is a large lunule patch of slate colour, with small dark waves; the belly is yellow, with the like crescent-shaped spots, and the crest is deep red.

There is a close analogy between the Wrynecks, (Yunx,) and the woodpeckers, in the extensibility of the tongue, and the position of the toes; but they differ in the want of the piercing bill. The *Wryneck* (*Yunx Torquilla*) which gives its name to the genus, derives that name from a habit of turning its neck, with a slow undulating motion, like that of a serpent, turning its head towards the back, and closing its eyes; this movement appears to be the result of surprise, terror, or astonishment at the sight of some novel object. It is also an effort which the bird appears to make to disengage itself when it is held; but as it executes it equally in a state of liberty, and as the young, even in the nest, have the same habit, it is clear that it must be the result of a peculiar conformation. The wryneck has also another habit, not less singular; when confined in a cage, it has been observed to bristle and elevate the feathers of the head, whenever any one approached,—to spread forth and raise those of the tail,—to advance, and then retire abruptly, striking the bottom of the cage with its bill, and lowering its little crest. This it perpetually continued until it was left alone.

The wryneck is a solitary bird, which, voyages and lives alone, except during the season of love, in which it is seen in society with its female. It arrives in these countries alone, in the month of May, and departs in the same manner in September. It crooks itself to the trunk of a tree, but does not climb, though its feet seem conformed for that motion, like those of the woodpeckers. It even seldom perches, except for the purpose of going to sleep; it has a curious method in perching of holding its body back. It is most frequently seen on the ground, where it collects its food; it darts its long tongue into an ant-hill, and draws it out loaded with ants, which are retained by the viscous liquid which covers it.

The male is heard in the woods soon after the cuckow; its cry is a sort of sharp and prolonged hissing. The female makes no nest, but lays in the hollows of trees, in the dust of rotten wood: the eggs are from eight to ten in number, of an ivory white. The young, by their habit of hissing often deter the plunderer of the nest, who fears to lay his hand on a serpent.

It is difficult to preserve these birds for any time in cages, for want of the aliment which is suitable to them; they may, however, be fed for a time with ant-eggs.

This species, without being numerous, is extended throughout all Europe, from Greece to Lapland; it is also to be met with in Siberia and Kamtschatka; Kolben tells us it is found at the Cape.

The Cuckow (Cuculus) forms a very large genus, divided by our author into several sub-genera. Of the Cuckows proper, the propagation is said to be the same among the foreign species as among our own; that is to say, they construct no nest, but employ those of other birds, in which to deposit their eggs, and confide their progeny to the care of strangers. The cause of this phenomenon, which is without parallel in the history of birds, is as yet unknown. M. Herissant is inclined to attribute it to the peculiar conformation of the viscera in the cuckow, which opposes itself to the process of incubation; in other birds, he says, the stomach is nearly united to the back, and totally covered by the intestines; in the cuckow, on the contrary, its position is totally different: it is placed in the lower region of the belly, and absolutely covers the intestines. From this position of the stomach, it follows that the process of incubation is as difficult to the cuckow, as it is easy to other birds, in which those parts which immediately cover the eggs and the young are soft, and capable of yielding without danger to the compression which they must experience. In the cuckow, on the other hand, the membranes of the stomach are charged with the weight of the body, and being compressed between the aliments within, and hard bodies without, must experience a painful pressure, and one injurious to digestion. Montbeillard does not regard this difference of conformation as a cause capable of rendering the cuckow unfit for incubation; he considers that the stomach is not too hard, because its parietes are membranous; and that if ever it be hard, it is only accidentally so, in consequence of repletion, a case which rarely occurs to a female when hatching.

There is one point of great importance relative to this phenomenon, which is not yet cleared up. It is certain that the female cuckow deposits her eggs in strange nests, usually one in each; but what may be the means which she employs of doing this, in many cases it is very difficult to conjecture,—in such nests, for instance, as those of many warblers, of the redbreast, the yellow wren warbler, &c., which are far from being proportioned to her size, or capable of sustaining her weight, without being deranged. The nest of the yellow wren warbler presents peculiar difficulties, for it is constructed in the form of an oven, with a very small aperture on the anterior side; yet the cuckow's egg has been found in this very nest, and the nest uninjured, so that she must have introduced it without entering the nest. M. Le Vaillant tells us, that he remarked the female of an African cuckow, which swallowed her egg, and retained it in the œsophagus until she regorged it into the nest which she had chosen. Another naturalist told M. Vieillot, that he saw a female cuckow take her egg in her bill, and deposit it in a strange nest. These facts may serve to elucidate the problem we have been considering.

The cuckow is found in Europe, Asia, Africa, and Australasia; but not in America. The birds of that part of the world which have been so called, have not the disposition and attributes of the cuckow, for they construct nests, incubate, and rear their young.

Cuckows live on insects, and principally on caterpillars. They also eat berries and soft fruits; but they are by no means carnassial in the wild state.

We possess but one species in Europe, which is the *Common Cuckow*, (*Cuculus Canorus*,) and of which we may remark the following particulars.

The adult male has the intestinal tube about twenty inches in length, and two cœcums of unequal length, one generally fourteen, and the other ten lines,—both directed forwards, and adhering in their whole length to the great intestine by a slender and transparent membrane. It has one gall-bladder. The reins are situated on each side of the spine, each being divided into three principal lobes, themselves subdivided into smaller lobules. By this whole apparatus is secreted a whitish matter.

The œsophagus is dilated at its lower part into a kind of glandulous pouch, separated from the ventricle by a contraction. The ventricle is a little muscular in its circumference, membranous in its central part, adhering by fibrous tissues to the muscles of the lower belly, and to the different parts which surround it, and much less gross in the wild bird than in that which is tamed and reared by man; for in the latter, this sac, generally distended by excess of food, equals the volume of a middling-sized hen's egg,—occupies all the anterior part of the cavity of the belly, from the sternum to the anus, extending at times five or six lines under the sternum, and at other times leaving no part of the intestine uncovered. But in the wild cuckows this *viscus* does not extend quite as far as the sternum, and permits two circumvolutions of the intestines to appear between its lower part and the anus, and three in the right side of the abdomen.

The males are in this species more numerous than the females, for in twelve individuals not more than two of the latter are reckoned. From this great disproportion, many mistakes have resulted concerning the female. It has been sometimes taken for the young, sometimes for a variety, and sometimes for a distinct species.

The cuckows arrive here and in France in the month of April, and commence singing a few days after. They inhabit the woods, delighting in those which are situated on hills and mountains. They constantly return to the same district in which they have chosen to pass the summer. They are usually alone, and appear unquiet, changing place every moment, and traversing a considerable extent of ground every day, without taking long flights at a time. To this they are forced, by seeking the aliment which suits them. They roam in all directions; sometimes they are seen on the summit of trees,—sometimes they retire among the thickest bushes. Everywhere they hunt insects, caterpillars, and phalenæ, which constitute the basis of their food. They also eat the eggs of small birds, and discover the most concealed nests with an astonishing facility.

The cuckow is approached with difficulty; and especially, when found in the woods, it sometimes exercises the patience of the hunter for a long time, flying from tree to tree, and never removing to any great distance. It certainly is one of those birds which are best known from its name and song; but not equally so in all its habits and manners, for naturalists are far from being in accordance on these points.

They are in reality so extraordinary, and at the same time so difficult of observation, that it is not surprising that, for the want of attentive examination, conjectures have been given for facts, and vulgar stories adopted, which, however absurd, have not failed to obtain a very general credence.—What absurdity, indeed, is too great to shock human credulity?

Naturalists have varied even as to what becomes of the cuckow during the winter. Some with reason assure us that it migrates to more temperate regions; but these are the fewer number. Others tell us, that the cuckow strips itself of all its feathers, and conceals itself during the bad season in the hollow of a tree, where it lives in the midst of a heap of grain, which it has amassed for its nourishment. Such persons either would not, or could not see that, in its conformation and habits, this bird has nothing granivorous. Others, granting this truth, have metamorphosed the cuckow for the winter into a hawk or falcon, and made it live on carcasses, birds, &c. They designate it as a perfect bird of prey, without examining whether nature has given it the physical capacity, or the means of seizing prey or digesting flesh. Had they examined with attention the interior of its body, they could not have fallen into such an error. The true carnivorous animal has short intestines, is destitute of the double cœcum, and has a membranous stomach, furnished with a gastric juice, calculated for the solution of flesh. The cuckow, on the contrary, has no such juice; its intestines are long, and it has a double cœcum. The external characters of this bird would be sufficient to convince us that it was not carnassial; yet this opinion has been supported, because the young cuckow in captivity is fed with meat, and refuses bread, though those who maintain it allow that it has so strong a taste for insects, that it will instantly abandon meat to feed upon them.

A sufficient proof that flesh is not to the natural taste of this pretended carnivorous bird is, that when it is left to itself it will never touch it, and it is necessary to thrust it into the bill to make the bird swallow it; but if insects are put into its cage, it will readily take and swallow them of its own accord. This is not the mode in which a genuine bird of prey would act. In fine, if the circumstance of a bird's eating meat in a prepared or half masticated state, in default of other food, is to be taken as a proof of its carnivorous character, all our singing insectivorous birds must be ranked as birds of prey as well as the cuckow.

It has been an universally vulgar error that the cuckow is nothing but a small hawk metamorphosed, and that this metamorphosis takes place twice a year, at the same periods, namely, in the month of July, when it ceases to sing, and in the spring, when it is again turned into a cuckow. This absurdity is founded on the circumstance that these two birds are seldom or ever seen together in the same places. When the cuckow begins to sing, the hawk returns into the depth of the forest, and does not reappear in the neighbourhood of inhabited places until the former bird ceases to be heard. Also, there is a very great resemblance in their plumage, size, mode of flight, solitary life, &c.; added to which, the colours of the female cuckow are very analogous to those of the merlin. But the cuckow has neither the tarsus, the beak, the toes, the claws, the courage, nor the strength of a bird of prey.

Many other absurdities have been recounted of the cuckow: such as that it returns in spring on the shoulders of the kite— that it casts a saliva on plants, which is fatal to them, by the larvæ which it engenders, and that these again prove fatal to the bird by stinging it under the wing. The fact is, that this pretended saliva of the cuckow is nothing but the frothy exsudation of a species of cigala. It is also reported, that the female cuckow takes the precaution of laying an egg of the same colour as those in the nest where she deposits it, the better to deceive the mother—that she has the habit of hatching strange eggs, though she will not hatch her own—that she is accustomed to visit the nest where she has deposited her egg from time to time, to drive out or devour the young, that her own offspring may be more at its ease. It is moreover said, that the voracity of the young, even when hardly born, is so great that it will devour the other nestlings, and the mother also. To this mother, too, the character of a most cruel step-mother has been attributed, who will kill and eat her own offspring, that she may devote herself altogether to the stranger, and lavish all her attentions upon it.

In fact, if we consult the ancient, and even some modern naturalists, we shall find stories of greater absurdity than those of which we have just given some specimens. It would seem that every thing the most monstrous in fable, or the most odious and criminal in the history of mankind, had been carefully sought out, and attributed to these inoffensive birds: and this, because men could not discover the secret springs which Nature has employed to give to this species manners, habits, and a mode of life altogether opposite to those of others, and the union of which fixes on the cuckows a distinguishing character from all other known animals.

It is not certainly known whether the cuckows pair, like most others of the passerine or climbing birds—at least, there are no demonstrative proofs on the subject. When the female flies, she is observed to be followed by two or three males, who seem extremely eager to obtain her favour. As the males are much more numerous than the females, they frequently fight desperately for the possession of the latter. Montbeillard tells us, that they evince no choice or predilection in this matter, one female being the same as another, and that they are equally capricious and inconstant in their amours: this must naturally be the case with birds so strangely constituted; as they have no nests to build, eggs to hatch, or young to rear, so they have no need of mutual affection, or common care for their progeny.

The female, according to the naturalist just quoted, can only lay an egg or two at most, because the superflux of nutriment being almost entirely absorbed by the growth of the plumage, can furnish but little to the reproduction of the species. This assertion, however, appears to rest on very slight foundation—in the first place, there is nothing less proved, than that these birds re-appear in Spring, with plumage scarcely renewed, that their wings are so feeble that they can seldom ascend large trees, but are forced to drag themselves along from bush to bush; this privation of the power of raising themselves on a tree, and this unfinished state of their plumage on their arrival in this country, would indicate a bird in the moulting state, and the cuckow must on this principle be supposed to moult anew in spring, for it cannot be imagined that the state of the plumage, thus described, can be the remains of the autumnal moulting; but the absurdity of this supposition is manifest, for how is it possible that birds which arrive from Africa, half-deplumed, and with wings so weak that they cannot raise themselves to the moderate height of a tree, should have been able to perform so long a voyage as they must have undertaken to come and pass the summer in the north of Europe? The naturalist who advanced the assertion on which we are commenting, has not thought proper to enter into any explanation of this phenomenon. It is true enough, that in the early days of their arrival, the cuckows frequent bushes, and are often observed to fix themselves on the ground; but this is not in consequence of the weakness of their wings, for at this season they make very considerable flights: it proceeds from their seeking in the plants then in vegetation, and on the young shrubs which begin to be covered with verdure, for the insects which they cannot find on large trees, which are still in a defoliated state. As to the eggs being only two in number, that is decidedly an error, for the females have been found, on dissection, not only to have eggs ready to come forth, but also to have an ovary furnished with as considerable a number as are found in most other birds.

The true egg of the cuckow is, according to Montbeillard, more bulky than that of the nightingale—less elongated—of a grey colour, nearly whitish, spotted towards the gross end with an obscure violet-brown, and also with a deeper shade of brown, and marked in the middle part with some irregular traits of morone colour. It would appear that these eggs vary in size and colour, for they are small in comparison to the bulk of the bird; so great indeed is this disproportion, that they are usually less bulky than those of the hedge-sparrow, although the latter bird is, at least, five times smaller than the cuckow. There is a considerable resemblance between the eggs of these two birds, in the ground colour, and in the spots—at least, in some of them; others are covered with reddish spots, arranged in no regular order—and some are even seen marked with blackish lines.

The female, compelled by some law of her organization, respecting which, naturalists are far from being agreed, to confide her eggs to strange nurses, usually deposits but one in the same nest. It rarely occurs to find two of them together. Her selection too does not fall indiscriminately on the nests of all birds. She prefers those of the warblers, field-larks, larks, wagtails, of the red-breast, the wren, the nightingale, &c.; also those of the thrush, the blackbird, the cole-titmouse, the turtle, the bunting, the greenfinch, the linnet, and the bulfinch. It is very singular to find in the list of the nurses of the young cuckow many birds like the last three just mentioned, which are purely granivorous. They do not feed their own young with insects: nor is it easy to imagine how some of them, from their peculiar mode of disgorging their aliment, can feed the young cuckow at all. Montbeillard, however, thinks that the vegetable substances macerated in the crop of these birds, may suit the cuckow very well, at a certain age, until it is itself enabled to find caterpillars, spiders, coleopterous and other insects, of which it is fond, and which usually swarm in the neighbourhood of its habitation.

Although the female deposits her eggs in the nests of these birds, she does not do so without frequently encountering a very obstinate resistance on their parts. She is sometimes indeed forced to relinquish the attempt. The female redbreast has been observed to join with the male in prohibiting the entrance of the cuckow into her nest. While one of the opponents was striking the cuckow with its bill repeatedly in the abdomen, the latter bird exhibited a slight tremor in the wings, and opened the mouth so wide that the other redbreast which was attacking in front, frequently popped its head in and concealed it altogether, but always with perfect impunity. The cuckow soon fell exhausted, staggered, lost its equilibrium, and, turning itself on the branch, remained suspended with its feet upwards. Having continued about two minutes in this attitude, with the bill open and the wings extended, and still pressed by the two red-breasts, it quitted the branch, went to perch at some distance, and did not make its appearance any more. The cuckow has also been known to be repulsed in a similar manner by the bunting.

The cuckow's eggs are never found in the nests of quails or partridges—or, at all events, they do not thrive there. The young of these birds can run and eat alone, almost as soon as they are born.

What is most surprising in this subject is the forgetfulness of the bird which adopts the young cuckow, of her own eggs and young, and her complete devotion to the stranger. This sacrifice of all natural feeling, and which is made by birds in favour of the cuckow alone, is evidently commanded by an imperious law of nature, for the majority of these birds will not cover other eggs beside those of the cuckow. Lothinger tried a number of experiments in this way, the result of which proved the fact now stated.

"On the 15th of May, 1772," observes this naturalist, "about four in the morning, I put an egg of the gold-crested wren into a nest of the common warbler, which was concealed in some nettles, pretty near the ground, and in which were five eggs, which the warbler had been hatching for some time. I remained in the neighbourhood, to be certain that nobody should lay hands on the nest; but in about a quarter of an hour or so, I could no longer find the egg which I had placed there. I again stole into this same nest a thrush's egg. About five in the evening this egg occupied the centre of the nest, and it appeared from its position as if it was the intention of the warbler to hatch it; but the following morning it had disappeared. I looked for it, and found it on the ground, open and dry. It had either been broken in its fall, or the bird had opened it to get rid of it more easily. On the same day, in the afternoon, I took an egg from the nest of a blackbird in the neighbourhood, and placed it, still quite warm, in the nest of the warbler, which was then absent, though but for a little time, as her eggs had still a remarkable degree of heat. Them I removed, and, imitating the cuckow, left only the blackbird's egg in the nest. After some minutes I approached, and saw the warbler hatching as usual. The following day, returning to the same spot, I found the nest abandoned, and, as is customary with the warblers on the failure of a brood, the birds were already disposing themselves for the construction of another.

"About the end of June, I removed from the nest of a bunting four eggs, which the bird had been hatching for a long time, and I put one of a blackbird in their place. Two hours afterwards the bunting was on the nest, and the egg had experienced no derangement. The following morning I found things in the same state; but in the evening the nest was abandoned, and the egg was cold.

"Knowing that goldfinches, linnets, greenfinches, and chaffinches, will readily hatch eggs which may be substituted for their own, I was curious to try what would happen by acting with these species of birds, as the cuckow is habituated to do. Having suffered a greenfinch to hatch her eggs for the space of six days, I took them away, and put one of a blackbird in their place. This was in the evening. The following morning the nest was deserted, and continued to remain so."

This gentleman tried many similar experiments with a great number of other birds of the passerine order, and the result was invariably the same. Scarcely had the eggs been removed, and others substituted, than the nest was abandoned, and never returned to. This ornithologist, in giving an account of his thirty-first experiment, communicates some observations—the more interesting from the difficulty of making them—relative to the conduct of the yellow-wren warblers towards the young cuckow, and relative to the nursling itself. We give his details in his own language:—

"Arriving at the place in the morning, I posted myself advantageously to observe the father and mother, who had undertaken the nursing of the young cuckow. They, however, were extremely shy, and did not approach at last without the greatest possible circumspection. They were, however, obliged to shew themselves in consequence of the cries of the nestling, who had been a considerable while without food. I then recognised them to be yellow-wren warblers. Growing more familiarized with me, they appeared very often, and more than once I had an opportunity of seeing the nature of the food they provided for their charge, which was nothing but an insect, of greater or less bulk. I availed myself of an occasion so favourable, to observe whether the real parents, after delivering their eggs and offspring to the care of strangers, took no further concern in their welfare. I soon was enabled to learn the real state of the case. I concealed myself under the foliage, so as not to be perceived, and remaining in silence, I saw a cuckow approach, singing and hovering about the immediate neighbourhood of the young bird. The better to attain my object, I took the young cuckow, placed it in a glade at a little distance from the nest, having first excited it to utter some cries, to draw the attention of the parents. This, however, was to no purpose, as they would not approach any more. Still I could observe that the old cuckow redoubled his song in proportion to the cries of the young, and both parents appeared to pay to it the greatest possible attention."

Of all the birds on which Lothinger makes his experiments, the bunting and the yellow wren warbler continued the longest without abandoning their nests. When deprived of their own eggs, they returned to the nest, and covered the strange egg for twenty-four hours.

From the facts which we have been stating, the following deductions are made by the indefatigable naturalist whom we have thus quoted. 1st. That the vulgar notions concerning the cuckow are totally erroneous, and that even to many naturalists its history seems to have been but imperfectly known. 2d. That every bird which has eggs quits its nest if they be taken away and a *single* one belonging to another species put in their place. 3d. That this abandonment is tolerably prompt, and will take place even when the process of incubation is going on. 4th. That in a most extraordinary manner is this law of nature reversed in favour of the cuckow. 5th. That it is a fact, placed beyond all doubt, that the cuckow neither incubates nor builds a nest, but lays in that of some little bird, from which it has previously ejected the eggs. 6th. That this little bird, thus maltreated, makes no sort of difficulty in returning to its nest, and covering the egg substituted by the cuckow, though but a single egg, and totally unlike its own. Finally, that it is neither in consequence of indifference nor of idleness that the cuckow does not build a nest or hatch its own eggs; but that, from the peculiar nature of its conformation, and probably from some other unknown cause, it requires the co-operation of others for the multiplication of its species; and that all this, however singular, is not to be considered as a disorder or caprice in nature, but as an effect of the supreme and sovereign will which regulates the universe.

Montbeillard has not adopted these results of Lothinger; but there does not appear to be much foundation in his objections.

Among the nests which the female cuckow makes choice of, there are some so small that they cannot contain a young cuckow, and the children of its nurse together. The naturalist we have been citing, declares, that he has had multiplied proofs that this female throws, or pushes out, such eggs as she finds in the nest. Others pretend that she eats them. These facts, however, should not be generalized, for eggs have been found in the same nest with a young cuckow. They might, doubtless, have been laid after the introduction of the stranger; but it is quite certain, in some cases, that they were laid before, for a young cuckow, just broken from the shell, has been seen in a thrush's nest, with two young thrushes, which were beginning to flutter. The same thing has been observed in various nests; while in others, the young cuckow has been found to be the oldest. Lothinger assures us, that the female cuckow can introduce her egg into the nest of the wren, and remove the other eggs to make way for it; and he also declares that she does the same with the nest of the cole-titmouse. It would be interesting to know in what manner she removes the eggs, as this nest is always at the bottom of the hole of a tree, and its entrance is in general extremely narrow. Yet Montbeillard speaks of a nest of this bird, in which there were five eggs of its own, with one of the cuckow; but the former disappeared by degrees. All the small nests in which young cuckows are found, are very much flattened, and scarcely to be recognized as to form, in consequence of the weight and bulk of this bird, who, that it may be more at its ease, will often shove out, of its own accord, any eggs or young birds that happen to be in the nest along with it. Of the mode which it employs to make this displacement, some interesting details are given by Jenners. The young cuckow, assisting itself with its rump and wings, endeavours to slide under the little bird which partakes its cradle, and to place it on its own back, where it retains it by raising the wings; then, drawing itself backwards, to the elevated edge of the nest, it reposes for a moment, and then, making an effort, flings its burden clean out of the nest. After this operation, it is soon observed to grope or feel with the extremities of its wings, as if desirous of convincing itself of the success of its enterprise.

This observer has constantly remarked, that the young cuckows make use of the end of their wings to recognize the eggs, or little ones, which they intend to dislodge. He has often repeated the same observations on a great number of nests, and invariably found the young cuckows pursuing the same manœuvre. Sometimes, in climbing the elevated edges of the nest, the young cuckow will let its burden fall; but it soon commences its work again, and never gives over until its enterprise is atchieved. It is surprising to behold the reiterated efforts which a cuckow will make, for two or three days, when it is lodged with a bird rather too heavy for it to raise. It is then in a continual agitation, and never stops working. But when the cuckow arrives towards twelve days old, it loses the desire of throwing out its companions, and after that period it has not been observed to disturb them. It has been also remarked, that the cuckow would much sooner suffer eggs in the nest along with it than young ones. A cuckow of nine or ten days old has been frequently known to chase out a little bird, which had been put into the nest with it, while it never touched an egg which had been placed there at the same time. The peculiar configuration of the young cuckow renders it very fit for the performance of the operation we have described. Different from that of other birds, the upper part of its body, from nape to rump, is very broad, with a perceptible depression in the middle. It would seem that this depression was made for the very purpose of securing, more effectually, the eggs, or the young birds, which the cuckow is desirous of throwing over; for as soon as it has attained its twelfth day, this cavity becomes completely effaced, and its back does not differ in any respect from that of other birds. The obligation which the young cuckow seems to be under of rejecting the eggs or the little ones of its adoptive parent from the nest, may be alleged as a reason why the female cuckow always takes care to lay in the nests of small sized birds. The same observer, whom we last mentioned, found in the same nest two cuckows, and one warbler, which had been disclosed in the morning. There still remained a warbler's egg. In some hours, the two cuckows began to dispute possession of the nest, and their dispute lasted until the afternoon of the next day, when the cuckow that was a little more bulky than the other, succeeded in flinging out the latter and the young warbler, and the egg along with it. Their dispute was remarkable. The combatants appeared alternately to have the advantage, and each, in succession, carried his antagonist to the edge of the nest, from which he fell back into the bottom, overwhelmed under the weight of his burthen. At last, after many efforts, the strongest succeeded, and was the only one brought up by the old warblers.

Of the authenticity and justice of the foregoing observations, there can be no doubt; and they seem to place the cuckow in a very different point of view from that in which it stands in the fables of vulgar credulity. Instead of being from its birth a bird of prey, and devouring the little ones of its nurse, at a time when it can only open its mouth to receive nourishment, this bird is purely insectivorous, and not at all the monster of ingratitude which it has been represented. Yet so much had this notion taken possession of the popular mind, that to be as *ungrateful as a cuckow* is a proverbial expression among the Germans, and the learned Melancthon has composed an eloquent harangue on the ingratitude of this bird. Neither do the wren, the red-breast, the warblers, and other gentle birds, lose their natural character and affection for their own offspring, when they become the nurses of the cuckow; nor do they sacrifice the least handsome of their young, kill, and tear them piecemeal to satisfy the voracity of their nursling. Yet with such fables has the natural history of the cuckow been filled, and they have been repeatedly copied even into very modern works. Klein tells us of a warbler being destroyed by a cuckow; but the fact appears to be, that the former had forcibly engaged its head within the bars of the cuckow's cage, and the latter, in the extremity of hunger, was found pecking at it. When Klein arrived, being influenced by the prevailing notions on this subject, he easily persuaded himself that the cuckow was endeavouring to swallow the head of the warbler. But deprived, as the cuckow is, of all the powers of a carnivorous animal, it would have been choked by such a morsel, and utterly unable to break the bones.

Certain facts mentioned by Montbeillard should set this question at rest. "On the 27th of June," says this naturalist, "I placed a young cuckow of the existing year, nine inches in length, in an open cage with three young warblers, which had not yet a fourth of their plumage, and could not feed alone. Far from devouring, or even threatening them, the cuckow appeared, as it were, to recognize the obligation under which he was to the species. He suffered those little birds, which did not exhibit the slightest fear of him, to seek an asylum under his wings, just as they would have done under the wings of the mother. On the other hand, a young owl, of the same year, hitherto fed only by the bill, instantly devoured alive a fourth warbler, which was placed along with him."

Some writers, a little shaken in their opinions by such facts as these, content themselves with asserting that the young cuckow devours only the little birds which have just burst the shell. But can any one believe this to be possible, at an age when it is not able to take food of its own accord, and can only receive its nutriment from the bill of another bird?

This prejudice would appear to be founded on the menacing air, like that of a young bird of prey, which the cuckow assumes, when it is approached, and that too for a long time previous to its quitting the nest. It opens its bill, as if to deter the adversary, and defend itself—bristles up its plumes, raises and lowers its head repeatedly, turns itself on its back, and endeavours forcibly to seize whatever is presented to it. The noise which it then makes is very similar to that made by the young hawk. At other times, if disturbed ever so little, it sends forth a sort of puffing sound, and moves its entire body in a heavy manner.

These birds attain their growth very rapidly; but though vigorous and large, it is generally some time before they can provide for their own sustenance, and being confined in cages, they remain some months without being able to eat alone, or at least without attempting it. This, however, may be attributed to the sort of aliment which they receive, such as meat, which is not natural to them. They frequently refuse to eat it, but will greedily seize the caterpillars, or worms, which are offered to them. It is to be presumed that in a state of nature they make a more rapid progress in the use of their natural faculties, through necessity, from the enjoyment of liberty, and the choice and abundance of aliments. As soon as their wings are sufficiently strong, they employ them to follow their nurse over the neighbouring branches when she quits them, or to meet her when she brings them their meal. They are insatiable nurslings, continually holding open their wide bill, and repeating every moment their cry of appeal, accompanied invariably by a movement of the wings. This cry is not less sharp than that of the red-breasts, or warblers. They retain it in captivity, according to Frisch, as far as the 15th or 20th of September, and thus salute those who bring them food. But it then commences to grow more grave by degrees, and they soon lose it altogether, and preserve the most profound silence, for they do not sing the first year. These birds, thus captured, at five or six months old, are rather stupid; they seldom move, remaining many hours in the same position, and are so little voracious, that they must be assisted to swallow their meat,—a certain indication that this food is not to their natural taste.

The substances found in the stomachs of the young cuckows must of necessity be different, since they are reared by birds of different species. In the stomach of one of them, brought up by wagtails, were found flies, scarabæi of different kinds, small snails with their shells entire, grasshoppers, caterpillars, and some vegetable substance not satisfactorily identified. The pipets generally feed them on grasshoppers. The most curious substance, however, found occasionally in the stomach of young cuckows, is a ball, formed of hairs closely intertwisted. Some are of the bulk of a pea, and others as voluminous as a small muscovado nut. These balls appeared to be formed entirely of horse hair, and would seem to have been detached by the young bird from the nest. Little pellets of hair have also often been discovered in the stomach of old cuckows; but they were evidently the remains of the hairy caterpillars which these birds had swallowed.

Various kinds of food, not necessary to be enumerated here, are given to cuckows in a state of captivity. It may be observed, that the omnivorous regimen, properly regulated, is as suitable to insectivorous birds as it is to man.

Many observers have remarked in the cuckows a repugnance to drinking, and that they have even rejected water, which has been put forcibly, or by stratagem, down their throats; this remark, however, does not apply to them all. Sonnini has recited a case of a young cuckow taken from the nest, when just about to fly, whose habit was totally different; it drank readily of its own accord, and always gave signs of satisfaction on being presented with water.

Though wild and solitary, the cuckows are not altogether unsusceptible of education; a cuckow has been known to recognize its master, come at his call, and even follow him to the chase, perched upon his fowling-piece. When this cuckow found a cherry-tree in the path, he flew to it, and picked plentifully; sometimes he would go off, and not return to his master for the entire day, but he would always keep him in sight, hovering from tree to tree. In the house, this bird was left at perfect liberty, and used to pass the night on a perch. Olivier asserts, that the cuckow might be trained to the chase, like the hawk, or the falcon; but this is most assuredly an error, occasioned by the resemblance of the plumage.

To preserve these birds during winter they must be carefully secured from cold, especially at the transition from autumn to that season, which is always a critical, and often a fatal period to them. Some get into a languishing state at this time, and exhibit cutaneous eruptions.

Others perish in the moulting; but before they die they fall into a state of lethargy and torpor. Their moulting is more complete than that of other birds: and it takes place later with those in a state of confinement, which are not stripped of their plumage until October or November. At this period, the wild cuckows emigrate to a milder climate, which they could not do were they despoiled of their feathers to the same extent as the tame ones. We must attach no credit to the story of some of them remaining during the winter in a lethargic state, without their feathers, in hollow trees, or holes in the earth. Neither are we to believe, though their moulting is long and slow, that they return in spring, without their due complement of feathers; this supposition is neither borne out by actual observation, nor by any means compatible with the fact of their having performed the voyage from Africa into our northern climates.

The males give over singing in the first days of July; this silence does not announce their approaching departure, but the commencement of the moulting. The greater numbers migrate from the first to the fifteenth of September; those found to the end of this month and later, are doubtless young ones, which were too weak to accompany the others. The early cold, and the dearth of insects, and soft fruits, (for these birds are fructivorous, in case of necessity,) determine them to pass into warmer climates. They pass twice into Malta and the Greek Islands of the Archipelago, where they arrive at the same time as the turtle doves. As the cuckows are less numerous than the doves (only a single one is usually discovered in the midst of a flight of the others, of which he appears to be the chief,) this circumstance has given occasion to the modern Greeks to call the cuckow *trigono kracti*, which means, leader of the turtles. It is important to observe, (according to the remark of Sonnini,) that the migrating cuckow changes almost all its natural habits as known to us. It is no longer solitary; it is seen with other birds of its own species; and it travels, as has already been remarked, with numerous companies of birds of different species. During this exile, which imperious necessity prescribes, it feels no desire of reproduction, and the song of love, which its name expresses, is never heard. It is doubtless with this species as with the greater number of others; it is the business of propagation which causes it to isolate; for these solitary birds have been observed in the course of July to assemble in little flocks of from ten to twelve, young and old, just at the epoch when the cessation of their song marks the close of the season of their amours.

On the ground, the cuckows proceed only by hops; but they seldom remain there, which must be attributed to the extreme shortness of their legs and thighs; when young they scarcely use their feet at all for walking. They employ their bill in drawing themselves onward on their belly, much in the same way in which the parrots use it in climbing; when they climb, it may be remarked, that the external hinder toe is turned forward, but that it is of less use than the two front ones. In their progressive movements, they agitate the wings, as if to assist their progress.

The ordinary song of the cuckow is too well known to need description. It appertains to the male exclusively, and is never heard but in spring; sometimes when the bird is perched on a dry branch, and sometimes when he is flying. Another and a more sonorous sound is heard when the males and females are reciprocally seeking and pursuing each other.

In the fall of the year the adult cuckows are very fat, and good for eating; on their first arrival among us, they are so thin as to have given rise to a proverb on the subject. The young, taken when just ready to fly, is also said to be a delicate morsel. The ancients held their flesh in high estimation, as also do the modern Italians. In some countries, the cuckow is regarded as a bird of ill-omen, and revered in others as a presage of good, and an oracle of wisdom.

In medicine, the flesh of the cuckow has been described as producing the most salutary effects in cases of epilepsy, gout, stone, intermittent fevers, and colic. Its dung has been supposed to be a most efficacious remedy in delirium and madness. Its fat has also been prescribed as a prevention to the falling off of the hair; and, perhaps, it deserves this reputation, as well as most of the nostrums, of every variety of sonorous appellation, with which fools and coxcombs are gulled by the swindling empirics of this metropolis.

The opposite is a figure of a cuckow found by Captain Flinders, on the North Coast of New Holland, during his voyage of discovery. The bill is stout, and horn-coloured; crown of the head, dusky-clay colour; the under parts of the body, pale-buff; wings, mixed with blackish and buff-colour; over the eye, a broad streak of buff-colour; behind the eye, a streak of black; legs, horn-colour; length, about fifteen inches.

The *Sacred Cuckow* (*Cuculus Honoratus*) owes its name to the compass and melody of its voice. It is held in the highest veneration throughout the Indian Peninsula. It lives most generally in small flocks, and prefers unfrequented and well wooded places. These birds fly by springs, or hovering, but to very short distances. Insects are their usual food.

The Couas (called *Coulicous* by M. Vieillot,) are strongly distinguished from the cuckows by their mode of propagation —which indicates a different internal organization. They construct a nest either in the hollow of a tree, or on the branches. They hatch their eggs, and rear their young ones. There are some, indeed, classed with the last sub-genus, that do the same; but their right to be so classed, is very doubtful. The difference in the tarsi and wings, the first being denuded of feathers, and longer, and the latter being shorter and more rounded in the couas, cannot be considered, in a natural method, as characters of so important a kind as that which belongs to the multiplication of the species.

Some of these birds, distinguished by M Vieillot, under his genus *coccyzus*, belong to Madagascar; one to New Holland; but for the most part they are natives of America.

FLINDERS CUCKOW.

EUDYNAMYS FLINDERSII. Vigors & Horsfield.

London Published by G.B. Whittaker, Oct. 1828.

They frequent large forests, and thickets which border on inhabited places, but very seldom are found in open situations. They conceal themselves in tufted woods, of the most sombre character, and on the most thickly foliated trees, the branches of which they traverse in search of caterpillars and insects, which constitute their principal nutriment: still, in default of these, they will feed on berries, which they swallow entire; at least, this is true of two species, which pass the summer in North America. The couas are lively and alert, and seldom descend upon the ground. Some are more wild and tameless than others. Many have a powerful and sonorous cry. From the peculiar sound uttered by some of them, M. Vieillot has derived his popular name of the genus.

Of the habits of the two next divisions of the text, nothing is known.

The INDICATORS are so called from serving as guides to the natives of Africa in the discovery of honey. They nestle in the hollows of trees, lay four or five eggs, and their aliment is composed of honey, wax, and insects. The nest of the *Great Indicator*, as we are told by Sparmann, is composed of weak filaments of the bark of trees, ingeniously tissued together, and in the shape of a bottle turned upside down. It is suspended by the two ends with a loose cord, so that the birds can perch upon it. M. Le Vaillant, on the contrary, assures us, that this bird makes its nest in the hollows of trees, and climbs on them like the woodpeckers. This account, we own, appears more probable, for it is more in consonance with the general practice of the tribe, and far more intelligible than that of Sparmann. The male partakes the incubation; and the eggs are three or four in number, of a dirty white.

This species is found in the interior of Africa, but is not met in the environs of the Cape of Good Hope. If we are to believe Sparmann, this is because there are no bees there; he says that he never saw any but on the farm of a single colonist, who succeeded in fixing some wild swarms, by presenting them with chests or boxes. This bird probably experiences some difficulty in procuring an aliment of which it appears particularly fond; but instinct appears to have directed it to call man to its assistance, by indicating the nest of the bees with a very sharp cry, which, according to some travellers, is the word which signifies honey in the language of the Hottentots. It utters this cry morning and evening, and seems to call the persons who are hunting for honey in the African deserts. They reply, in a graver tone, as they approach. The moment the bird beholds them, it proceeds to hover over the tree which contains a bee-hive; and if they delay to come, it redoubles its cries, flies towards them, and, by various turns backwards and forwards, indicates the spot in a very marked manner. While they are seizing the contents of the hive, the indicator remains in the neighbourhood, and awaits his portion, which is certain always to be left for him. The existence of this bird is valuable to the Hottentots, and they never see it killed but with an evil eye.

The Barbacous and Malcohas we must pass over, as containing no matter of interest to the reader.

Of Scythrops there is but one species, peculiar to Australasia. It feeds on certain grains and scarabæi. It often extends its tail like a fan, and utters a piercing and almost terrific cry, something like that of the cock when he perceives a bird of prey. These birds are never seen but in the morning and evening, sometimes seven or eight in number, but more usually in pairs. Their appearance and cries, according to the natives, constitute a certain index of approaching wind or storm. They are of a wild and fierce character, and cannot be tamed. They refuse all food, and peck violently at those who approach them.

Of the family of the Barbets, including their three subdivisions, as given in the text, our notice must be general, and very brief. The first are natives of Africa and India: of the second, some are found in both continents; and the third

COLLARED BARBET.

B. COLLARIS.

London Published by G.B. Whittaker Nov. 1828.

MARGINATED BARBET.

BUCCO MARGINATUS. Ruppel.

British Museum.

London, Published by Whittaker & Co. Ave Maria Lane, Octr. 1829.

are exclusively peculiar to America. Their habits, manners, and physiognomy are all the same—sad, sombre, and serious. Their figure is massive, and ill put together. Their disposition stupid, taciturn, solitary, and lazy. They invariably prefer covered retreats, and shun the open plains. They neither go in flocks, nor in pairs. Their flight is short and heavy. They place themselves on low branches only, and experience much difficulty in putting themselves in motion. Once fixed, they remain a long while in the same situation, and accordingly are approached with facility. These birds feed on fruits, scarabæi, and other large insects. They make their nest in a hollow tree, and lay from two to four eggs.

The figure of *Bucco Marginatus* of Ruppel, is from a bird in the British Museum. The opposite is a figure of the *Collared Barbet*, a description of which will be found in the text.

The TROGONS may dispute the palm of beauty with the humming-birds. Their plumage in certain parts shines with metallic brilliancy, and exhibits all the colours of the rainbow. On other parts, the tints, though opake, are not less rich and splendid ; but a very short neck, feet disproportioned to their figure and bulk, and a long and broad tail, injure the harmony of their form, and give them a heavy port and aspect. Their long attenuated feathers, with barbs disarranged and luxuriant, make them appear more bulky than they really are. These too are so feebly implanted, that they fall at the slightest agitation. Their skin is so delicate, that it will tear at the slightest tension.

These birds are solitary, and extremely jealous of their freedom. They never frequent inhabited or open tracts. They delight in the silence of deserts, where they even fly the society of their consimilars. The interior of the thickest forests is their chosen abode for the entire year. They are sometimes seen on the summit of trees, but in general they prefer the centre, where they remain a portion of the day without descending to the ground, or even to the lower branches. Here they lie in ambush for the insects which pass within their reach, and seize them with address and dexterity. Their flight is lively, short, vertical, and undulating. Though they thus conceal themselves in the thick foliage, it is not through distrust; for when they are in an open space, they may be approached so nearly as to be struck with a stick. They are rarely heard to utter any cries except during the season of reproduction, and then their voice is strong, sonorous, monotonous, and melancholy. They have many cries, from the sound of one of which their name is derived.

All those whose habits are known nestle in the holes of worm-eaten trees, which they enlarge with their bills so as to form a comfortable and roomy residence. The number of eggs is from two to four, and the young are born totally naked; but their feathers begin to start two or three days after their birth.

The Trogons have several broods in the course of the year. The occupation of the male during incubation consists in watching for the safety of his companion, bringing her food, and amusing her with a song, which, though we should call it insipid, is, to her, without doubt, the expression of sensibility. Some of the couroucous express the syllable *pio* repeated many times in succession with a powerful, yet plaintive tone. Their accent almost reminds one of the wailings of a child who has lost its way, and it is thus that they cry to each other amidst the silence of the forests. As soon as the young are able to provide for themselves, they separate from their parents to enjoy that solitude and isolation which appear to constitute the supreme happiness of the species. Their aliments are composed of larvæ, small worms, caterpillars, coleoptera, and berries, which they swallow entire.

The male, at various ages, the female, and the young, differ in their plumage, which has given rise to the institution of more species than are in reality in existence.

THE INDENTED COUROUCOU.

TROGON TEMNURUS. Tem.

London, Published by G. B. Whittaker; July 1828.

H. Kearsley del.t
Mus. Brit.

PAVONIAN TROGON.

TROGON PAVONINUS. *Spix var?*

London. Published by G.B. Whittaker, March 1829.

Our figure of the *indented Trogon* is from a specimen in the Museum of Paris. The top of the head and quills are deep blue, but on the latter are many white patches; the nape, shoulders, upper wing-coverts, and upper side of the tail, are green; beneath, the bird is bluish white, except the belly and vent, which are red. The tail-feathers are very peculiar, each being indented at the extremity, as if the skin had been divided, and the barbs separated. This species is from Cuba.

We have also inserted the figure of a most magnificent Trogon in the British Museum, which, though it differs in some particulars from the *T. Pavoninus* of Spix and of Temminck, in the *pl. col.*, belongs, in all probability, to the same species. The specimens in the British Museum have the entire head covered with an erect, but extremely soft and silky crest; whereas M. Temminck's figure, from a bird in Mr. Leadbeater's extensive and valuable collection, has but a slight indication of such a crest, and that merely on the front; the wing-coverts, moreover, like the long feathers of the tail, fall in a very elegant manner over the quills. The whole plumage, but especially the pendant feathers of the wings and tail, are of silky softness, and very brilliant. The head, body, and tail are of the brightest green; the belly and vent red, and the quills black. The species is from Brazil.

The birds of the genus Ani are indigenous to the warmest climates of the New Continent. Their wings are feeble, and their flight extremely limited. They cannot bear the wind, and numbers of them perish in the tremendous hurricanes which take place in the countries they inhabit. Their social instinct is truly admirable. They are always found in flocks, the least numerous of which are from eight to ten, and many are five-and-twenty or thirty. They scarcely ever separate, but remain almost continually together, whether when flying or at rest; and even when they perch, it is as nearly as possible to each other. This mutual amity, which constitutes a peaceable and durable community, commences from the very birth of these inoffensive birds. Born in common, they also live in common. Their society is never disturbed. Love itself, that active element of discord, comes among them in peace, unaccompanied by the jealousies, fears, and altercations which usually constitute its train, and turn to offence and bitterness the flowers and fruits which spring beneath its footsteps.

From the month of February, the season of reproduction commences with the anis, whose disposition is not less amorous than that of the sparrows. During the whole of this period they are much more lively and gay than at any other; but the good intelligence which reigns amongst them, suffers no diminution. There are no quarrels—no combats. The males and females work together at the construction of the nest, which serves for several females at the same time. She that is most pressed to lay, does not wait for the others, who increase the size of the nest, while she is hatching her eggs. This common incubation takes place in the most perfect harmony. The females arrange themselves beside each other, and should the eggs happen to be mixed or joined together, a single female hatches the strange eggs along with her own. She draws them together, and surrounds them with leaves, so that the heat may be equally spread throughout the mass, and not be dissipated. The same good understanding prevails when the young are disclosed. If the mothers have hatched together, they feed all the young family in succession. The males assist in the furnishing of provisions. But when the females have hatched separately, they bring up their young ones apart, but still without jealous interference, or discord. They bring them their food in turn, and the little ones receive it from all the mothers. Thus mildness and peace are invariably the happy attributes of these birds, and the inseparable qualities of their association, their household, and their family. If the ancients, who better knew than we how to derive moral precepts from the various productions of nature, because they studied their relations more than their forms,—if they, who sent the sluggard to the *ant* "to learn her ways and be wise," had been acquainted with this peaceful and amiable community of birds, they would not have failed to have held them up as an example to the mischievous, the envious, and the quarrelsome—to the turbulent and intermeddling crowds, which torment, divide, and undermine the societies of men.

The anis construct their nests very solidly, but somewhat inartificially, with little branches of shrubs, bound together with the filaments of plants. This nest is very wide, and raised at the edges. It is sometimes eighteen inches in diameter, and its capacity is proportioned to the number of females that intend to lay their eggs there. The small number of those that hatch apart, form a separation in the nest with blades of grass to contain their eggs. They all cover them with leaves or grass in proportion as they lay them, and continue to do so during the time of incubation, when obliged to quit them to seek for food. These females, which are a little smaller than the males, and have a more opake and sombre plumage, have several broods in the year, and many eggs each time.

The food of these birds is both animal and vegetable. Small serpents, however, lizards, and other reptiles, caterpillars, large ants, and insects in general, seem to be the aliments which they prefer. They also perch on oxen, to pick out the vermin which lodge in the hair and skin of these animals, whence their ornithological name of *crotophaga* (eaters of vermin). In default of animal nutriment, they feed on different species of grain, such as maize, millet, rice, wild oats, &c., but never in such quantities as to be hurtful to the crops; nay, they must be considered as serviceable to these, from the number of insects they destroy.

An usual attitude with the anis is to draw back the neck, and press the head close against the body. They are neither fearful nor wild, and never fly to any great distance. They are not even much alarmed by the report of fire-arms. It is easy to bring down many of them in succession; but they are in no great request, for their flesh is uneatable: and even when alive they have a disagreeable odour. They are, as might be expected, easily tamed; and it is said that when taken young they can be as well educated as the parrot, and even taught to speak. They are found in great numbers in Brazil, Guiana, Mexico, St. Domingo, and always in comparatively open places, a little shady, but never in thick woods.

The most particularly striking character in the genus Toucan is the bulk and length of the bill. In its whole extent it is wider than the head, and in some species it is as long as the entire body. This enormous bill is a cavernous body, filled with empty cells, separated by partitions of an osseous substance, as thin as a sheet of paper, and covered again by an expansion of corneous substance, of so little solidity that it opposes no resistance to the slightest pressure of the finger.

The upper mandible is curved downwards, in the form of a scythe. The lower is shorter, more narrow, and less curved. Both are dentelated on the edges, but the dentelations of the upper are much more marked than those of the under mandible. These dentelations, though equal in number on each side of the mandibles, not only do not correspond above and below, but even do not stand in a relative position on the right and left.

The tongue of the toucan is still more extraordinary than the bill. It is rather a feather than a tongue, the middle or stalk of which is of a cartilaginous substance, about two lines broad, accompanied on each side by closely serrated cartilaginous barbs, conformed exactly like those of an ordinary feather. These barbs are directed forward, and are more long in proportion as they are situated nearer the extremity of the tongue.

The birds classed in this genus are found only in the warmest regions of America. They live on fruits—usually go in little flocks of from six to ten—and fly heavily, and apparently with trouble to themselves. They can, however, elevate themselves to the summit of the highest trees, where they are fond of perching, and are almost in a continual state of agitation. They make their nests in the hollows of trees, and the female lays but two eggs. The young are easily tamed and reared, for they will eat any thing which is given to them—fruits, bread, flesh, or fish. They seize the morsels which are presented them with the point of their bill, throw them upwards, and receive them in their large gullet. If they seek them on the ground, they usually take them on sideways, and fling them up in the air in the same sort of style. The toucans are so sensible to cold, that they dread the freshness of the night, even in those burning climates. Their skin is generally bluish, and their flesh, though hard and black, is yet eatable.

To M. d'Azara we are indebted for some novel observations on the toucans, which serve to complete their natural history. The toucans, according to that most meritorious naturalist—contrary to what might be supposed—destroy a great number of birds, their large and bulky bill rendering them formidable to most species. They attack them, chase them from their nests, and, even in their presence, devour their eggs and young ones, which they either draw out of holes by the aid of their long bill, or bring to the ground along with the nests. Witnesses worthy of credit affirm, that the toucans do not even respect the nests of the *aras* and *caracaras*, and that if the young ones are too strong to allow themselves to be carried away from the nest, their adversaries strike them to the ground, as if their disposition led them, not only to devour, but to destroy. Even the solid nest of the *rufous bee-eater*, which resists time, and other causes of destruction, is not safe from the attacks of the toucans, who wait until the clay of which it is composed is softened by the rain, to batter it with strokes of their bill, that they may devour the eggs and young. During the season of hatching, the toucans have scarcely any other aliment; but at other times they live on fruits, and sometimes on insects and the tender buds of plants; they then leave the other winged tribes in peace.

Notwithstanding the disproportioned bill of the toucans, it offers no more resistance in flying, than those of other birds, whose head and surface are equivalent in extent to theirs, because they always present its point to the wind. Moreover, we must take into consideration the specific lightness which results from the peculiar conformation of this large bill, and which prevents it from retarding the flight of the bird. In a state of repose, the toucan carries its bill a little more raised than the horizontal line which passes by the eyes, and, on observing it pretty closely, this bill actually appears factitious, because the base exceeds this level of the head, which is embossed in it as in a case. According to M. d'Azara, the tongue of the toucan is inflexible, and can be of no use for the direction of the aliments, or the formation of the cry, which, in the two species of Paraguay, is nothing more than may be expressed by the syllable *rac*.

The toucans fly to a moderate height, and in a straight and horizontal line. They beat their wings at intervals, and with some noise. Their flight is swifter than the small extent of their wings would lead us to suppose. They jump from branch to branch, and change position quickly; but do not climb, after the fashion of the woodpeckers. These birds are strong, and exceedingly attentive to all that passes around them. They advance with distrust, and but rarely settle on the ground. They hop obliquely and ungracefully, and with their legs nearly a palm asunder. When they take little birds in the nest, as well as morsels of meat or fruits, they dart them

Griffith sc.

ARACARI TOUCAN.

RAMPHASTOS ARACARI. Gm.

London Published by G.B. Whittaker March 1828.

into the air, and, by a slight movement of the bill, they direct them, so as to be swallowed conveniently; then, by another motion, they receive them into their wide gullet. But, if the morsel should be larger than the aperture of the latter, they abandon it, without attempting to divide it.

The opposite is a figure of the *Aracari Toucan;* the specific description of which is in the text. Its manners are the same as those of the genus, such as we have now described them.

We now come to the most interesting family of this, or perhaps of any other order in the class Aves, we mean the PARROTS. This family is so numerous, and in many points so singularly characterized, that a naturalist might well be tempted to form it into a sub-order. Its importance demands that we should be a little more extensive and detailed in our observations on it, than we are usually wont to be; and, for sake of method, we shall divide those observations as follows:

First. We shall consider the physical peculiarities of parrots in general.

Secondly. We shall review their intellectual, and particularly their *imitative* faculties, their disposition, habits, and manners.

Thirdly. We shall say a few words concerning the modes of classification adopted for them by the most eminent naturalists; and,

Lastly, notice whatever may be interesting respecting any of the individual species, consistently with the limits and the plan which we have generally pursued in these supplemental essays.

The birds of this genus possess, in an eminent degree, the character of the order in which they are placed. They are climbers, in the fullest sense of the word. Their toes, constantly four in number, are opposed, two to two, and armed with solid and crooked claws—less so, however, than the claws of the birds of prey. The two anterior toes are united at their base by a little membrane; the hinder are completely separated. The tarsi, in the majority of species, are very short, but in some they are elongated in a proportion nearly equal to that in the passeres in general. Their skin is scaly, as is that of the toes; the wings are usually short; the tail is more or less long, and assumes a variety of forms; the colours of the plumage are almost always brilliant.

The head of these birds is voluminous, and of a rounded form; the bill varies in its bulk, relatively to that of the head and body, in the different species: thus, in the long-tailed parrakeets, which the French naturalists term *perruches*, the bill is rather small, being, measured in its curve, but a third of the length of the head. In the parrots proper, it is nearly one-half that length; while in the maccaws and the microglossa, it is as long as the head itself. The upper mandible is always the most powerful, and it often entirely conceals the lower. It is articulated on the forehead, so as to admit of considerable mobility; for it forms with it, very sensibly, a re-entering angle, when the parrot opens its bill and gapes. In general, the back of this mandible is rounded, though in one species it is carinated, or keel-formed. Its point is very sharp, though less so than in the beak of the accipitrine birds, and prolonged, more or less, underneath. Its edges are trenchant, and occasionally furnished with a sinuosity, or a tooth, which reminds us of the beak of the falcon. The under facet of this same part is vaulted, slightly arched from back to front, and its superficies is furnished with numerous *striæ*, parallel to each other, and formed like a V, or a chevron, with the point forwards. The use of these appears to be, to render the surface against which the aliments rest, less slippery, while they are divided by the under mandible. This under mandible is short, sometimes joining only the base of the upper, and incapable of closing the bill entirely. It is also rounded, very slightly compressed, and trenchant at the end, which alone is used for the division of the food. These two portions of the bill, formed of a very hard and thick horny substance, are put in action by more numerous muscles than are to be found in other birds. Their colours vary. In general, those of the base, or the end, are deeper than those of the middle ; and the under mandible more obscure than the upper.

The bill is surrounded at the base with a naked skin, or cere, less apparent than that of the birds of prey, and variously coloured. In this the nostrils are pierced, which are smooth, orbicular, and pretty large.

The tongue is thick, fleshy, soft, and extremely mobile, in the parrots proper, and the skin which covers it is often very fine, and dry, and furnished with papillæ. These papillæ, according to M. de Blainville, are longitudinally arranged, on a sort of anterior disk, supported by a corneous half-ring, which is at the lower part of the tongue, and they are covered by a kind of deposit, or pigment, above which is the epidermis, which is very slender. In the parrots which Levaillant has named *Aras à trompe* (some of our cockatoos, &c.) the tongue forms a small cylinder, is flesh-coloured and solid, tolerably long, not flexible, and terminated by a small black gland, rather corneous, hollowed in its centre. The true tongue consists in this little corneous gland, while the cylindrical part, which sustains it, is a dependency of the hyoïd apparatus, susceptible of being more or less extended from the bill, at the will of the animal, by a mechanism analagous to that which elongates the tongue of the woodpeckers. This tongue is at once an organ of sense, and an instrument of touch and prehension, for the purposes of deglutition.

In some species of parrots belonging to New Holland and the South Sea Islands, the tongue is terminated by a crown-formed bundle of hairs, or rather cartilaginous filaments, which M. de Blainville considers as papillæ, in consequence of the bulk of the nerves which communicate with them.

The eyes of the parrots are moderately large, and situated laterally. The upper and lower lids form a rounded orifice, edged with small tubercles, supporting the lashes in its entire circumference. The upper is evidently mobile; the third lid, or nictitating membrane, is very small, and the parrots are never seen to make use of it. The pupil is round, and not situated exactly at the centre of the iris, but more inward; so that the iris is a little broader on its external than on its internal side. The colour of this last varies according to the species; but it is generally remarked to grow deeper with increasing age. A peculiar character in the parrots is the ability of contracting the pupil, more or less, independently of the action of the light, when they turn their attention to any object—when they feel any sudden internal movement, such as fear or anger—or even when they are in a sportive mood. These birds are evidently diurnal.

The aperture of the ear is oval and small, especially if compared with that of the ululæ; it is directed obliquely forward, and constantly covered with the feathers.

In certain birds of this genus the cheeks are naked of feathers, and covered with a white farinaceous powder, as is remarked in the maccaws; or the skin is coloured, as in the microglossi. In others the circumference round the eye is more or less divested of feathers, and also covered with a sort of farina. This appears to be an epidermic production, and is very abundant on other parts of the skin of these birds, whose plumage, when they shake it, gives out a considerable quantity of white dust. In no species are fleshy caruncles to be found. Some, as the cockatoos, many psittaculi, &c., have the head ornamented with long and slender plumes, which can be elevated in the form of a tuft, or crest, according to the inclination of the bird, but which, in general, are reclined along the neck.

The neck, in general, is but moderately long; sometimes it is even short, and tolerably thick; still, when the parrots wish to attain an object without changing place, they have the power of elongating this neck to a certain extent.

The body varies in degrees of robustness or elegance, according to the species. In the parrots proper it seems thicker than in the others, which perhaps is only the effect produced by the shortness, and the strength and solidity of the tibiæ, toes, and tarsi. Some long-tailed parrakeets, on the contrary, are distinguished by the fineness of their form, and elegance of their general proportions. The breast of these birds is usually broad and rounded.

The wings are short, and their point rarely exceeds one-half the length of the tail, even in species in which the tail is shortest. The first three remiges are the longest of all, and pretty nearly equal with each other.

There are differences in the tail, relatively to the greater or less extent of the various quills which compose it, and which are twelve in number. As to its total size, it is either shorter than, equal in length to, or longer than the body, comprising the head and neck. In form it is sometimes straight or squared, when all the quills are of equal length; sometimes round, sometimes graduated, sometimes arrow or spear-formed; sometimes it is peculiarly broad at the end, and some species have the caudal quills sharp at their termination; sometimes the tail is very short, and at the same time graduated, which is the case with many psittaculi. In some species of this last division the upper tail-coverts elongate into a point, in an extraordinary manner, and conclude by reaching almost to the extremity of the caudal quills; but this character is rarely observed in those whose tail is almost straight.

Though the feet of these birds, as we have said, are robust, and the toes well adapted for climbing, yet there are some exceptions to this. In the pezopori, for instance, the tarsi are elongated, and the claws but little crooked; they accordingly remain constantly on the ground, where they walk with swiftness, which the other species cannot do. The legs of the parrots are usually feathered to the heel; but in two species, the microglossa, the bottom of the leg is as naked as it is in all the grallæ. The colour of the feet is usually grey, but it is in some roseous, brown, or black.

The colours of the plumage of the parrots are exceedingly varied, and almost always pure and brilliant. The adult females often differ in this respect from the males; whilst the young, in their first or second livery, and even after the third moulting, present characters peculiar to themselves. Green is, in general, the predominating colour; then comes red, then blue, and finally yellow. This last colour appears among the parrots to be the general substitute for the white observed in other birds; and it is remarkable that, in many of their species, there are varieties uniformly yellow, as among birds in general we behold Albin's varieties. Very often, when the feathers are plucked, red and yellow ones will shoot forth, whatever may have been the colour of the former. In the countries inhabited by the parrots, the people give the name of *tapirés* to such of them as have their plumage mingled with those reproduced red and yellow feathers, in the places which have been plucked. It has been pretended, that the blood of a certain species of frog introduced into the little wound caused by the plucking of the feather, will make the red plume be reproduced; but this is without foundation. There are some species violet, purple, brown, or lilac-coloured. Some are known whose plumage is entirely grey; some have it black, and some, in fine, entirely white.

Some rules are observable in the distribution of the colours. Thus, the wing-quills are generally grey, brown or black, at their lower face and on their interior barbs, which are concealed; while their external ones are brightly coloured in their visible part. The tail-quills, in general, have the lower face more obscure than the upper. The most external of these lateral quills, and the two intermediate ones, are often of a different colour from the others.

The epaulette of the wing, or the edge of this part towards the carpus, is often of a different colour from the upper part of the wing, and this colour is usually red or yellow. Almost always the lower tail-coverts have a different tint from the upper and from the rump. When the mantle is green, as well as the back, it is seldom that the wing-quills, in their visible part, and the lateral and intermediate quills of the tail, are not of an aquamarine blue, or present some shades of blue, more or less deep. Very often the forehead is marked with a blue, red, or yellow band, which contrasts with the colour of the top of the head. On this last is sometimes a tuft, or cap, also coloured differently from the rest of the head, and bounded by the eyes and occiput. Many parrots or parrakeets have received specific names, derived from the existence of mustachios, or of spots situated at the base of the bill, on the cheeks, or on the *lorum*, which is the space comprised between the bill and the eye, or from having a complete collar, single or double, according as it is formed of one tint or of two; also from the existence of demi-collars, placed sometimes behind the nape, and sometimes under the neck. The plumage of the parrots is never spotted or striated, like that of certain passeres, and some birds of prey.

A disposition of colour frequent amongst the parrots, and in all probability peculiar to young individuals, is that which gives rise to the meshed, or scaled plumage. This takes place when the feathers of the body, and especially of the lower parts, are edged with a border of a different colour from that of their ground. Then, these feathers are so disposed, one over the other, as to resemble the effect produced by the scales of fish; from which circumstance, this sort of plumage is termed, in French *écaillé*.

In young individuals, scattered feathers are sometimes observed of a colour different from those in the midst of which they are found; such feathers belong to the next succeeding coat, which have appeared sooner than their usual time, and indicate the colour which shall be proper to the peculiar parts on which they are found. These feathers constitute an excellent criterion, by which to refer these young birds to their peculiar species.

When the plumage of the female does not differ from that of the male, it is yet observable that the tints are less pure and less lively; there is also, at times, a variation in the colour of the bill.

The following peculiarities are worthy of remark in the internal structure of the parrots. The head is strong, and the cranium rounded; the *os furcatum* is a little pointed towards the sternum, and formed like a V. The sternum is furnished with a powerful keel, or median crest; there is neither lateral nor posterior emargination; its body, on the contrary, is very wide, and only provided with an oval foramen, moderately large, and closed by a membrane near the abdomen, similar to that in the birds of prey, and palmipedes. The lower larynx is complicated, and provided with three peculiar muscles; a circumstance which, united to the mobility and conformation of the tongue, may produce the facility with which these birds imitate the human voice. Their gizzard resembles that of frugivorous and granivorous birds; their intestines are very long, and destitute of cœcum; the liver is of middling size, and divided into two lobes, nearly equal. The spleen is small and round. The heart is of moderate size, and rounded at the end.

The usual habitat of these birds is under the torrid zone, both in the old and new continent, and in the Oceanic islands. The greater number of species are found under those parallels which are nearest the equator; but some are extended in both hemispheres, to very high latitudes. Thus, in the northern hemispheres, is the parrot of Carolina found as high as 42° in the western states; and, in the southern hemisphere, the parrots and cockatoos which inhabit New Zealand and the Macquarrie islands, are placed under the 52d degree of latitude, as is also the emerald parrakeet of the Straits of Magellan. The first mentioned species, therefore, exists in a latitude equal to that of the central parts of Spain and the Neapolitan territories; the second, very nearly in the same latitude as that of London. But as the temperature of the Austral hemisphere is much less elevated than that of the northern, and as about ten degrees are allowed for comparative temperature, at any given points, in these different parts of the world, it follows, that the parrots of the Macquarrie islands, and the lands of Magellan, are placed in a situation nearly analogous to that of Stockholm.

In America, Brazil and Guiana are the countries which contain the greatest number of species of parrots, all of them appertaining to the division of parrakeets; that of the parrots proper, and that of the psittaculi. The maccaws are exclusively confined to these countries. It does not appear that any birds of this genus are found on the chain of the Cordilleras; they are not very numerous even in Paraguay; but a single one has been noticed in Patagonia; and but a single species has been marked by Buffon as peculiar to the Magellanic lands. Some species belong to the islands in the Gulf of Mexico; and it is not improbable that some may exist in the Floridas, though they have not been indicated. On the other side of the Andes, from Chili to California, none have been noticed; but many exist in Chili, on the shores of the southern ocean.

Many birds of this genus belong to the African continent, from Senegal as far as the forests which neighbour on the Cape of Good Hope. They are, however, fewer in number than those of India and America. The Barbary coast, from

Morocco, as far as Egypt, that is, the entire chain of Atlas, and the northern reverse of that chain, are destitute of them. There are some in Madagascar; but none in the Canary Islands.

In Asia, parrots are found only in the countries to the south and east of the table-land of Thibet—that is, in Hindostan and its dependent islands, in Cochinchina, in China, and in the eastern archipelago. There the handsomest and largest species, and those most remarkable for their forms, are in abundance.

In Polynesia this genus is considerably extended. New Holland has species peculiar to itself. These birds are also numerous in New Zealand, the Macquarrie Islands, and in the groups of the Friendly and Society Islands. The habitat of the cockatoos is limited to the Indian archipelago, to New Holland, to New Zealand, and the Macquarrie Islands. The lories are peculiar to the Philippines, and New Guinea, and the psittaculi, with the tongue terminated by a pencil of cartilaginous filaments, belong to the countries which extend from New Holland to the Friendly Islands. Two species only of this genus are known in the Sandwich Islands.

Europe, all the northern and central regions of Asia, the polar countries, Greenland, Iceland, the northern and temperate parts of America, Kerguelin's Land, and the South Shetlands, are almost the only portions of the globe in which the genus, or rather family, of the parrots has no representatives.

The ancients were acquainted with several parrots, among which the most celebrated was that sent from India, by Alexander, in the course of his expedition to that country. Mr. Vigors, who has written on a group of psittacidæ known to the ancients, and has treated this subject with his accustomed elegance of style, methodical discrimination, and profound classical research, tells us, that "the ancient writers are unanimous in informing us, that the parrots known to their times came exclusively from India. In that country, these birds were ever held in the highest estimation. We are informed by Ælian that they were the favourite inmates of the palaces of the princes: and were looked up to as objects of sacred reverence by the religious feelings of the people. From thence they were introduced into Europe at the time of the Macedonian conquest, and the specific name of *Alexandri*, applied by modern science to the type of the group, in honour of the first European discoverer of it, serves to perpetuate the name of a warrior, who is said to have valued the conquests that extended the boundaries of his empire, chiefly as they served to extend the boundaries of science. It was not until the times of Nero, that the *parrots* of Africa became known to the Romans. Some of these birds were among the discoveries made in the course of an expedition sent out by that prince. They came apparently from the neighbourhood of the Red Sea, and it is probable that as that country became more known, numbers of the same race were imported from it into Rome, and formed the chief part of those victims of the *parrot* tribes which, in after times, are said to have supplied the inordinate luxury and wantonness of Heliogabalus."*

We cannot resist the pleasure of transferring to our pages some further observations by this gentleman, suggested by the subject before us, nor will we do him the injustice of using any language but his own.

"But there is another point of view in which the interest of such researches is strongly apparent. In general, we are acquainted with the ancients chiefly through the records of their most splendid actions. The dignity of history, and the elevation of poetry, to which we are almost exclusively indebted for our knowledge of ancient manners, confine the representations which are transmitted to us of them, for the

* Zoological Journal, vol. II. p. 44, &c. *Sketches in Ornithology.* By N. A. Vigors, Esq.

most part, to those which are most important and heroic. We are presented with little beyond the achievements or the apothegms of the warrior, the statesman, or the philosopher. All the minor occurrences of domestic life, all the more endearing traits of private feeling, are cast into the shade. We see the ancients almost always in full dress, almost always in the stately attitude, and on the exalted pedestals of life. It is only by scattered references that we are enabled to enter their homes and their bosoms, and investigate—the most attractive of all subjects—the windings and variations of the human heart. Natural history affords us an occasional insight into feelings of this nature. Through its means we possess a subject of common interest, by which we find ourselves, as it were, on familiar terms with those who are removed from us, not merely by time, but by that imposing dignity which time never fails to confer. When our feelings are called forth in admiration of a bird or insect, which is known to have equally excited the admiration of an Alexander or an Aristotle, we become almost unconscious of the lapse of time which has separated us from such characters: we feel ourselves attracted to them by a community of sentiment, and rejoice in that sympathy which brings us in contact with the patron of science, and the man of genius of the days that are gone by. Science, it is said, levels all distinctions of rank and station, and unites all the adventitious differences in society under the powerful influence of genius and of knowledge: but science goes still farther in the present case, for it appears to level all the distinctions of time and space. In pursuing such researches into antiquity, we find not merely that external nature was the same two thousand years ago as it is at the present time, but that human nature itself has undergone but little variation."

The group of psittacidæ, thus known to the ancients, constitutes a distinct and detached division, which Mr. Vigors has characterized under the name of "*Palæornis*." The generic characters are strongly marked in every species belonging to it, and widely distinguished from those which are peculiar to those of modern discovery. He considers them perfectly identified with the Indian parrots, which are so much in request among ourselves. The rose-coloured collar, the emerald body, and the ruby bill, which mark these birds, have been very distinctly described, by several writers of antiquity.

We have now to treat of the habits and manners of these singular birds, which merit, without contradiction, to be placed at the head of the feathered race, on the score of intelligence. We shall first notice their peculiar modes of locomotion.

The parrots, as we have said before, are eminently climbing birds, as the form, the arrangement, and the strength of their toes clearly evince. When they walk on the ground, it is with a slowness, which is owing to a vacillating motion of the body, occasioned by the shortness and separation of their feet, in which the base of sustentation is very wide. They frequently place the point or upper part of their bill on the ground, which thus serves them as a point of support. In climbing, its hooked form is still more useful to them; and often when they hold any object in this bill, they rest upon the branches by the under part of their lower mandible. When they descend, they sustain themselves by the upper. This is a common habit with the majority of the parrot-tribe. Still, there are some species, which, having more elevated legs, toes less long and less crooked, can walk on the ground with tolerable swiftness, and which never perch. These have been formed by Illiger into a separate genus under the name of *Pezoporus*. Others, again, have the tarsi short and flat, on which they rest in walking.

The wings of the parrots being generally short, and their bodies bulky, they have some difficulty in rising to a certain point of elevation. But that once attained, they fly very well, and often with much rapidity, and through a considerble extent of space. The majority confine themselves to lofty and thickly tufted woods, frequently on the borders of cultivated lands, the productions of which they plunder and destroy. Their ordinary mode of flight is from one branch to another; and it frequently happens, that they will not fly continuously, except when pursued. Many of them emigrate according to the season, and, in particular, the Carolina parrots. Such travel every year some hundreds of leagues, differing in this respect from the habits of the others; but they are comparatively few in number. The difficulty of flight, with many, is the cause of their restriction within narrow limits; and their concentration in certain islands, while they are not found in others, which border closely on the former This is peculiarly the case in many of the island groups of Polynesia.

The food of the parrots consists principally of the pulps of fruits, such as those of the banana, the coffee-tree, the palm, the lemon, &c. They are especially fond of almonds; for the most part they attack the pulp only to get at the kernel; this, when once seized, is fixed on the under-wrinkled surface of the upper mandible; they turn it repeatedly until it is placed by the tongue, in a proper direction for the introduction of the trenchant edge of the lower mandible; then the bird soon forcibly separates the valves of the almond-shell, and, getting the almond into its bill, soon divides it, so that all its envelopes are rejected. The fragments are finally swallowed in succession. Some cockatoos of New Holland are said to live on roots, and the pezopori seek their aliment in herbs.

In domestication the parrots, maccaws, parakeets, and cockatoos, shew the same partiality for vegetable seeds, and, in general, are fed very well on hemp-seed, the skins or husks of which they detach with wonderful address. Some that receive bones to gnaw, are known to acquire a very determined taste for animal substances, but especially for the tendons, ligaments, and other less succulent parts. From feeding thus, some parrots contract the habit of plucking out their own feathers, that they may suck the stem; and this becomes so imperious a want with them, that they strip their bodies absolutely naked, not leaving a vestige of down wherever the bill can reach. They spare, however, the quills of the wings and tail, the plucking out of which would cause them too much pain. M. Desmarest mentions an instance of one of these birds belonging to M. Latreille, the body of which thus became as naked as that of a pullet plucked for roasting. This bird, notwithstanding, supported the rigour of two very severe winters, without the slightest alteration of health or appetite. M. Vieillot observes, that this habit of deplumation is produced, in many parrots, by an itching of the skin, and not in consequence of their being accustomed to eat animal substances.

The parrots drink little, but often, and do it raising up the head, but less strongly than other birds. The major portion of them may be accustomed, in domestication, to drink wine, or, at all events, to eat bread which has been steeped in wine. They all use, with great dexterity, one of their feet, to carry their food to their bills. while they stand perched on the other.

These birds sojourn much on the borders of streams and rivers, and in marshy places. They are fond of the water, and seem to take the greatest delight in bathing themselves, an operation which they perform several times a-day, when in a state of nature. When they have bathed, they shake their plumage, until the greatest portion of the water is expelled, and then expose themselves to the sun, until their feathers are completely dried. In captivity, and even during the most rigorous seasons, they seek to bathe; and, at all events, plunge the head repeatedly into water.

With the exception of the time of incubation, the parrots live in flocks, more or less numerous; go to sleep at the setting, and awake at the rising of the sun. In sleep, they turn the head upon the back. Their sleep is light; and it is not unfrequent to hear them utter some cries during the night. In a state of domestication, after they go to rest, is said to be the most suitable time for repeating to them such words as they are intended to learn, because they then experience no distraction.

Their life is very long; and the mean duration of it, among the parrots, properly so called, is calculated at forty years. Instances, however, have been known, of individuals who lived in a state of domestication ninety, and even a hundred years and more. The parrakeets generally live about five and twenty years.

One effect of captivity, on some species, according to M. Le Vaillant, is to change the colours of the plumage; and to this cause he attributes the frequent varieties observable among these birds. Those that are termed *tapirés*, which we have already alluded to, are regarded by M. Virey as natural varieties; but he considers them to be produced by a state of weakness, or malady.

The birds of this genus are monogamous. They make their nests in the trunks of rotten trees, or in the cavities of rocks; and compose them, in the first case, of the *detritus*, or dust of the worm-eaten wood, and of dry leaves in the second. The eggs are not numerous; usually only three or four each time; but the broods take place several times in the year. The young, when born, are totally naked; and the head is so large, that the body seems to be merely an appendage to it. They remain sometime without having sufficient strength to move it. They are subsequently covered with down; but are not completely invested with feathers for two or three months. They remain with their parents till after the first moulting, and then leave them for the purpose of pairing. The eggs are ovoid, short, as thick at one end as the other, and those which are known are of a white colour; some of them are nearly equal in size to those of a pigeon.

It was for a long time imagined that these birds could procreate in their native country only. Many parrots, however, were born in Europe, as far back as 1740 and 1741. In 1801, some Amazon's parrots were born at Rome. M. Lamouroux has given us considerable details respecting the broods of two blue maccaws, that were at Caen some years ago. These birds, in four years and a half, from the month of March 1818 to the end of August 1822, laid sixty-two eggs, in nineteen broods. Of this number, twenty-five eggs produced young ones, of which ten only died. The others lived, and became perfectly accustomed to the climate. They laid eggs at all seasons; and the broods became more frequent and more productive, in the course of time; and in the end much fewer were lost. The number of eggs in the nest used to vary, six having been together at a time; and these maccaws were seen to bring up four young ones at once. These eggs took from twenty to twenty-five days to be hatched, like those of our common hens. Their form was that of a pear, a little flatted, and their length equal to that of a pigeon's egg. It was only between the fifteenth and five and twentieth day that the young ones became covered with a very thick down; soft, and of a whitish slate-grey. The feathers did not begin to make their appearance until towards the thirtieth day, and took two months to acquire their full growth. It was a dozen or fifteen months before the young arrived to the size of their parents, but their plumage had all its beauty from six months old. At three months old they abandoned the nest, and could eat alone; up to this period they had been fed by the father and mother, which disgorged the food from their bill, in the same manner as pigeons do.

In all probability, the success of this education was owing to the care which was taken in providing these birds with a suitable nest. This consisted of a small barrel, pierced, towards the third of its height, with a hole of about six inches in diameter, and the bottom of which contained a bed of sawdust three inches thick, on which the eggs were laid and hatched. Since these observations were published by M. Lamouroux, collared parrakeets of Senegal, and pavonian parrakeets, have been born in Paris, in hollows made in large billets of wood, where the parents had fixed their nest.

Of all animals in the creation, there are none so calculated to attract the attention and admiration of man, as those which appear to approximate to his own nature, and to partake of some of the attributes of humanity. This is the case with the apes among the mammalia, and the parrots in the class of birds. Both exhibit some of the physical peculiarities of man, and both present a very striking analogy with each other.

The ape, from his external form, so like the human, his gestures and gait, the rude resemblance of his face, to that of man, from the analogous arrangement of all his organs with ours, but especially by the use of his hands, a certain air of intelligence, and from actions imitative of ours, has been regarded as a species of imperfect and wild man. Had he received the gift of speech, like the parrot, he would have passed for a genuine man in the eyes of the multitude, who judge always rather from external appearances than calm and reflective examination. The parrot is in the order of birds what the ape is in that of viviparous quadrupeds. It would appear, on first view, to be still more closely connected with us, than the latter, because the communion of speech is more intimate than that of mere sign and gesture. Besides, speech is the expression of thought, while gesture is nothing but the demonstration of physical wants. The latter is altogether corporeal, the former appertains to the mind.

We must not, however, consider the articulated voice of the parrot as a proof of the superiority of his intelligence over that of other animals, or of its analogy with our own. It is certainly true, that the parrots exhibit the most perfect brain to be found among any of the feathered races. The anterior lobes of its hemispheres are more prolonged than they are in rapacious birds, and the encephalon is wider, and more flatted than long; but as to the intelligence of the bird, compared with ours, it can only be considered that there is a point of contact between them, as it were, but no resemblance. The parrot's imitation seems purely mechanical; it articulates words, indeed, but this cannot be deemed a true language. In the same manner as an air is taught to a linnet with a bird-organ, so a word is taught to a parrot, which he repeats without knowing wherefore. He does not comprehend its signification, and though he may repeat it on certain occasions, because he has learned it, he sees no reason for doing so like man. He utters, indifferently, a prayer or an insult, and those involuntary substitutions, which really prove his want of intelligence, pass, with unreflecting persons, for a mark of wit, of irony, or of some other quality of mind of which the animal is utterly destitute and incapable of acquiring.

There are two kinds of imitations: one which is altogether physical, and dependant on similitude of organization; the other, the fruit of reflection, volition, and intelligence; the first is possessed by the ape and the parrot—the second by man alone; one requires nothing but memory, and an aptitude of organic functions—the other demands a profound study, like that of comedians and tragedians. A mere imitation of the exterior, such as a brute can give, is insufficient. The mind and soul must be moulded, as it were, on the model imitated; this requires a certain equiponderance of mental faculties, which cannot exist between man and brute of any species.

The imitations of which we have been speaking differ again, in a most essential point. It is thus; the imitation which the animal can acquire being totally physical, perishes with the individual, and is not, and cannot be, transmitted by education. A dog, ever so well trained and educated, never, of his own accord, teaches to his whelps what he has acquired from the intelligence of man; he dies, and all perishes with him; nothing remains but the natural qnalities inherent in the species. This is a point which the advocates for the identity of animal intelligence with human can never get over. The successive generations of brutes are indebted for nothing to their predecessors; the superintendance of man is indispensable to their improvement; but the case is quite different with him. His moral existence is extended and embellished by the accumulated acquisitions of past and of contemporary ages. He lives not isolatedly and individually—he co-exists by his acquaintance, and his multiplied relations, with his entire species. The generations of mankind do not pass away, and be as though they had never been. Posterity is the inheritor of the fruits of their labours; to it their intellectual estates descend, and, under favourable circumstances, are bequeathed to remoter ages, cultivated, improved, and enlarged. The instruction of the species becomes that of the individual; and the tide of our moral existence, if we may so express ourselves, is swelled by a myriad of tributary streams, whose sources are hidden in the night of time, and the impenetrable recesses of antiquity.

One main cause of this moral perfection, or rather perfectibility, is to be found in the long duration of the infancy of man. The animal, yet scarcely endued with sufficient strength, abandons his family. He either becomes a solitary individual, or joins in flocks and herds where family relations are utterly unknown, and there is no bond of union, but what spring from the mere necessities of subsistence and procreation. There are no moral ties to connect the communities of the brute; for though animals are often gregarious, they are never, in the true sense of the word, *social*. But in the human species, the wants, resulting from a long incapacity of living solitarily, multiply the moral relations, and increase the intellectual lights of every individual.

Thus we find that the imitative powers of the parrot do not, of themselves, entitle this bird to any marked superiority over others, and still less to the possession of any thing approaching to the genuine character of human intelligence. Its imitation is nothing but an organic mimicry, depending on the conformation of the voice, and perhaps on some peculiar aptitude of the ear. Besides, as we have already seen in the course of this work, the capacity of articulating words is not exclusively confined to the genus. Pies, jays, blackbirds, stares, and other, even small birds, can imitate human speech more or less, from organic facility, rather than the possession of any very superior intelligence. The ear of such animals, though different from ours, is not without a certain musical justness, and a delicate apprehension of sounds. This correctness of ear is observable among certain individuals of our own species, in a much higher degree than in others; and it is almost superfluous to observe, that neither that, nor a facility of mimicing sounds in general, is always accompanied, in such individuals, with a marked superiority in point of intellect. The reverse, indeed, is often strikingly remarkable. We may remark here by the way, that the species of birds which articulate words the best, shew less aptitude in the rendering of modulated sounds.

Parrots, parrakeets, &c. which are imported into Europe, are generally taken young in the nest, and brought up in their native country. Some are taken adult; they are caught when inebriated by eating the seed of the cotton-tree, which they are very subject to become, or they are brought down by arrows, which, having a button on the end, stun without killing them. The natives of Paraguay, according to M. d'Azara, take them in a manner which appears very singular, if not incredible. They attach one or two pieces of wood to a tree frequented by these birds for the sake of its fruits. They put a stick or two across from those pieces of wood as far as the tree, and construct with palm-leaves a sort of cabin, sufficiently large to conceal the fowler. He has got with him a tame parrot, which, by its cries, attracts the wild ones of the forest, and the last never fail to come at the voice of the prisoner. The hunter, without loss of time, passes round their necks a running knot, attached to the end of a long wand, which he moves from within his cabin. If he has five or six of these wands, he can take as many parrots, for they will not attempt to escape, unless the cord presses tightly on their necks.

They are all susceptible of education, but the young ones are more so than the old. The means employed consist in imposing certain punishments upon them, such as immerging them in very cold water, of which they are greatly afraid, or puffing at them with tobacco smoke. Rewards are also used, as well as punishments, and when they perform what is desired, such things are given to them as they are fondest of; more especially, sugar and sweet wine. They are tamed and kept obedient, by taking them with boldness, and speaking to them with authority, and in a loud tone of voice.

They may be thus taught to perform various gestures, and assume different postures; some will lie down on their backs, and not rise but at the command of their master; others will perform exercises with a stick, or dance in a manner more or less grotesque. They are taught by constantly repeating, close to them, such words as they are meant to learn. Success, however, does not always attend such endeavours. Some species are better disposed than others for this kind of education, and the same is the case with different individuals of a species. The grey parrots and the amazons, are those which speak most distinctly, and imitate, most naturally, the cries of animals, and other noises, which they are in the habit of hearing. Some may be taught to whistle entire airs; but they seldom go through them, sometimes whistling only the middle, sometimes the beginning, and sometimes the end. Their natural voices are shrill and disagreeable, and they constitute the only sounds uttered by them in a state of nature. It is not unfrequent to hear an entire flock of them thus crying and chattering at the rising of the sun.

In fine, though we cannot allow such a degree of intelligence to the parrots, as to suppose them capable of understanding the signification of the words which they repeat—though we can yield no sort of credence to the various absurd stories promulgated on this head, yet we cannot refuse to them a great superiority over birds in general, in their relations with man. They attach themselves to those who tend them, and exhibit an aversion to those who have ill-treated them, and that with a very marked discrimination. Many, however, exhibit antipathies of a capricious kind, and cannot be corrected but by the inspiration of fear. It has been said, that the males attach themselves to women in preference to men, and exhibit much ill-temper towards the latter: while exactly the reverse takes place with the females. This assertion M. Vieillot declares to be well founded; he instances the case of a male ash-coloured parrot, in his own possession, which he never could approach without being provided with thick leather gloves; which, however, was perfectly obedient in all respects to Madame Vieillot, and would exhibit the greatest fondness towards that lady; while, on the other hand, a female of the same species showed the greatest attachment to him. But such facts cannot very safely be generalized, and we ourselves have observed the contrary. Many species exhibit a capricious temper more than others; and, on the whole, the parrots are birds that must not be trusted with implicit confidence, without a very intimate acquaintance with them. The surest mode of taming them is by firmness and the exercise of authority. Like too many of the human race, they are best imposed on and subdued by the tones of menace and audacity. It will always, however, prove very advisable to put on thick gloves in dealing with them, to secure the hand from their bite. By degrees, they will become docile to those whom at once they fear, and from whom they yet receive occasional good treatment. All these birds, when taken adult, are very fierce; and yet the savages tame them very speedily, by means of the tobacco-smoke already mentioned: this vapour produces in them inebriation and swooning; they may then be touched without danger, and generally, when the effect of the smoke is over, they are not so violent as before. But if their temper is not so easily overcome, the same operation is recommended, and reiterated until it succeeds, which, in the end, it never fails to do, at least to a sufficient extent.

All these birds are eminently destructive in their disposition; it seems as if breaking and tearing every thing with their bills, that happens to be within reach, constituted a physical want with them; this is especially to be remarked among the larger species. In a state of liberty they devastate the trees, cut their branches, and despoil them of their leaves and fruits. In domestication, they will damage furniture and every thing that is near them. If they are shut up, and kept chained on the perch, to hinder this mischief, they compensate themselves for the constraint, by redoubled cries, by breaking their cage, and destroying their perches with their bills.

In the same way as the simiæ of the New Continent are not found in the old, so the American parrots do not inhabit any part of the ancient world. It is also to be observed, that each species remains within its own particular district, without mingling with others, or suffering them to mingle with it. The same is true of the ape genus; each recognizes its own peculiar livery, unites with its compatriots, and permits no intrusion of foreigners into the republic; there is no more possibility of usurping the rights of a citizen in their society, than there was at Lacedemon; each of them, however, will occasionally traverse the adjacent countries to levy tribute, as the wandering hordes of Tartars sweep successively through the deserts, to find subsistence for themselves, and pasturage for their cattle.

It will now be necessary to take a brief view of the different systems of classification adopted for this numerous and important family of birds, by the more distinguished writers on ornithology.

The species of Psittacus are extremely numerous; and characters, derived from the length and form of the tail—from the presence or the want of a tuft of feathers on the head—from naked or feathered cheeks, have, in general, constituted the basis of such divisions as have been made among them. These divisions, however, must not be considered as decidedly and strongly distinguished from each other; like the other subdivisions of the animal kingdom, they pass on insensibly one to the other, by the gradation of characters in the various species.

By Linnæus, Psittacus is formed into a genus of that ill-defined order, the picæ. He had but forty-seven species.

Buffon divided the parrots,—first, into parrots of the Old Continent; second, into parrots of the new. The first are subdivided thus:—

1. Cockatoos, with short and square tail, and mobile tuft.

2. Parrots proper, short and equal tail, and head destitute of tuft.

3. Lories, with small bill, curved and sharp: red the predominant colour in the plumage; voice, sharp; and motion, quick. Some, or the lories properly so called, have the tail moderately long, and rather angular, or corner-like.

Others, the lory-parrakeets, have the tail longer, and more resembling that of the parrakeets.

4. Parrakeets, with long tails, subdivided into those which have the tail equally graduated, and those which have the two intermediate quills much longer than the others.

5. Parrakeets, with short tails.

The second subdivision is composed of—

1. Aras or Maccaws, with long graduated tails, and naked cheeks.

2. Amazons, with tail short and equal; green plumage; red on the carpus of the wing, and yellow on the head.

3. Cricks, like the preceding, but without the red, having it only on the coverts; plumage, duller green, without the pure yellow on the head, and of smaller size.

4. Papegais (for which perhaps the word *popinjay* may be admitted as a translation), smaller than the cricks, and without red on the wing.

5. Parrakeets (*perruches*), subdivided into long-tailed and short.

Dr. Latham has simplified this division, and distinguishes but two groups, without respect to the habitat, for, as he well observes, the uncertainty of the country of many of these birds renders such a division inconvenient. He divides the parrots into—first, those with equal: second, those with unequal tails.

Le Vaillant has in some measure modified the classification of Buffon, without taking the habitat into consideration. He acknowledges the groups of aras and cockatoos, with the characters above cited; he unites the parrots, the amazons, the cricks, and papegais, under the general denomination of parrots (*perroquets*). He places in the division of parrakeets (*perruches*), all that have graduated tails, and feathered cheeks; but still subdivides it into four groups:—

1. Parrakeet Maccaws (*perruches-aras*), in which the circumference round the eye is naked.

2. Parrakeets proper, with cheeks entirely feathered, tail more or less long, but equally graduated and always sharp.

3. Arrow-tailed parrakeets (*perruches à queue en flèche*), in which the two intermediate quills are much the longest.

4. Parrakeets with broad tails, whose quills are not attenuated towards the end, among which are arranged the greater portion of the lories of Buffon.

This arrangement of Le Vaillant has, as the reader may have observed, been completely adopted by the Baron, with the exception of its having two particular divisions for the *ara à trompe* (grey cockatoo) of Latham; genus *microglossum* of Geoffrey; and the long-legged parrakeet *(perruches aras)* Le V. of which Illiger has made the type of his genus *pezoporus*.

The late M. Kuhl, who died in the island of Java, where he formed some very valuable collections, has published a monograph of this family, founded on very extensive observations, which he had ample opportunities of making, in the various public museums of most of the chief towns of Europe, and in the richest and most celebrated private cabinets. This work is distinguished by its striking superiority over all that has been hitherto published on the subject before us. Its character is thus given by Mr. Vigors, in the paper from which we have already done ourselves the pleasure of quoting:—

"In mentioning that work, I cannot allow myself to pass it over with mere simple approbation. It has the merit of being the first instance in which the principles, so successfully developed in the 'Horæ Entomologicæ,' in reference to some departments of the *annulosa*, were applied to a group of the *vertebrated animals;* and where the circular disposition in which the groups of nature return into themselves, and the uninterrupted series of affinities by which they are connected together, have been asserted and satisfactorily demonstrated. Whether the views which M. Kuhl unfolded in his monograph were the result of his own observations on nature, or whether he was originally indebted for them to the 'Horæ Entomologicæ,' it is now impossible to determine. Whatever may be our opinions on this point, it affords a superior example of an attempt at a natural arrangement. The leading divisions, with some slight modification, will be found to accord with those more comprehensive and philosophic views, which, from accurate observation of nature, are now almost universally allowed to offer the most faithful interpretation of her laws."*

M. Kuhl separates into two groups the species of which he treats, which are 209 in number. 1st. The species which he has seen in nature, and the existence of which cannot be disputed. These are 171. 2d. Those mentioned by ornithologists, but not seen by him, or which he considers doubtful. These are 38.

The species which he admits, M. Kuhl separates into six divisions: —

1st. The Aras, *Macrocercus*, with long tail and naked cheeks.

2d. The Parrakeets, *Conurus*, long and graduated tail, and feathered cheeks.

3d. The Psittacules, *Psittaculus*, very short tail, rounded or sharp, and feathered cheeks.

4th. The Parrots, *Psittacus*, equal or squared tail, and without tuft.

5th. The Cockatoos, *Kakadoes*, equal or squared tail, feathered cheeks, and head provided with a mobile tuft of feathers.

6th. Probosciger, *Aras à trompe*, Le Vaillant, *Microglos-*

* Loc. cit.

sum, Geoff., equal or squared tail, naked cheeks, and tuft on head.

Each of these divisions are subdivided, according as the species which they comprehend are of America, of India, of Australia, or of an unknown country. In that of the parakeets, (*Conurus*,) M. Kuhl has admitted the groups proposed by Le Vaillant, but rendered them subordinate to the geographical divisions above mentioned. This is the only part of M. Kuhl's system that seems objectionable.

The uncertain or doubtful species M. Kuhl has divided into two groups, according to the method of Latham. 1st. Those with long tails. 2d. Those with short. *Macrourus* and *Brachyurus*. Each of these divisions is afterwards separated geographically.

We shall now finally notice the new genera introduced by our distinguished countryman, Mr. Vigors, and his able coadjutor, Dr. Horsfield, in the family of the psittacidæ.

1. PALÆORNIS, Vigors. The type of which is *Psittacus Alexandri*.

2. LORIUS, Vigors. The type *P. Domicella.*

3. BROTOGERIS, Vigors. The type *P. Pyrropterus.*

4. PSITTACARA, Vigors. The *parrakeet maccaws.*

5. PLATYCERCUS, Vigors. The chief generic character is the broad and depressed tail.

6. TRICHOGLOSSUS, Horsfield. Filamentous tongue.

7. NANODES, Horsfield. A beautiful group, belonging to Australia, so called from their diminutive size.

8. ANDROGLOSSA, Vigors.

9. CALYPTORHYNCUS, Horsfield.

Besides these genera, Mr. Vigors has adopted, from other writers, the following: PSITTACUS, MICROGLOSSUM, PLYCTOLOPHUS MACROCERCUS, PEŽOPORUS, and PSITTACULA. Of all these genera he makes five sub-families, viz., *Psittacina*, *Plytolophina*, *Macrocercina Palæornina*, and *Psittaculina.*

We regret that neither our limits nor our plan will permit us to enter into an analysis of the merits of this very scientific arrangement. We can only say that, in our humble apprehension, it appears to approximate more nearly to a true natural method, than anything hitherto put forth upon the subject. The characters upon which all its divisions are founded, are important; and the affinities of the groups, the gentle gradations by which they naturally run into each other, are clearly distinguished.*

Having so far extended our observations on this family in general, we must be very brief in our notices of particular species. There is, indeed, but little to add respecting their habits. We shall take them by their sub-genera, except where some one species is deserving of especial mention. And we shall insert several figures of New Holland species, from specimens in the collection of the Linnæan Society.

The ARAS, or MACCAWS, are among the handsomest species of this family. Their plumage glistens with dazzling reflections of azure, of purple, and of gold. Their long tail, and majestic deportment, add greatly to their attractions; and their wonderful docility renders them peculiarly susceptible of domestication. But their voice is exceedingly harsh and croaking, and they are fond of keeping in continual and annoying exertion. Their intelligence appears to be rather

* We cannot dismiss this subject without referring our readers to a most masterly reply of Mr. Vigors, to certain injurious observations respecting his arrangements in the thirty-ninth volume of the "Dictionnaire des Sciences Naturelles." This reply is in the ninth number of "The Zoological Journal," Jan 1827. From this we have formerly done ourselves the pleasure of making a quotation on the subject of nomenclature. (See An. King. vol. 6, pp. 511, 512.) We regret that it is not in our power to condense the arguments of Mr. Vigors in defence of himself and Dr. Horsfield; but can assure our readers that they will peruse his paper with equal pleasure and profit.

more limited than that of other parrots, and their apprehensions less quick. They would seem, if human passions can with any propriety be attributed to the animal races, to be vain of their fine plumage, and to seek for admiration. They are not so capable of affection as parrots or parrakeets. They have not, however, the petulance of the other species, being more remarkable for gravity. They pronounce the word *ara*, from which their name is taken ; but their pronunciation is not so distinct as that of other parrots. They are not distrustful.

These birds are very subject to epilepsy or cramp, for which bleeding in the foot is recommended. They are very destructive to the coffee plantations in America. When young, their flesh is tolerably good.

The aras are not gregarious, like the parrots and parrakeets; they generally associate only in pairs, and seven or eight are rarely seen together. They seldom go to the ground, probably on account of the difficulty they experience in rising from it, in consequence of the length of their wings and shortness of their legs : hence it is easy to catch them when they are met with in that situation, before they can climb a tree, from the elevation of which they can easily spread their wings ; they fly horizontally, at no great height ; the seeds of forest-trees are their favourite food, in preference to cultivated fruits.

They construct their nests in hollow trees, and one species, the Hyacinthine Maccaw, is said to make holes for the purpose in the perpendicular banks of rivers, in which they lay only two eggs ; the male assists his mate in her incubation. The young make no cry for food.

The *Red and Blue Maccaw* of the English writers, *P. Maccao*, is as big as a fowl. It inhabits South America and the Antilles, but is observed to recede from those parts which the colonists bring, by degrees, into cultivation. It is fond of the maccaw plant (*Borassus flabellifer*), and frequents the wet savannahs in search of it; it feeds also, occasionally, and perhaps only in the absence of more grateful aliment, on the manchineel apple, which imparts its poisonous qualities to the flesh, and renders these birds sometimes injurious to the Indians who eat them. The eggs are said to be spotted, and to resemble those of the partridge; but Dr. Latham states, on the authority of a gentleman who kept a tame maccaw, which laid several eggs, that these were white.

Though without the docility and pleasing manners of many parrots, these birds, if taken young, may, nevertheless, be taught to be interesting and amusing, and to repeat several words; their natural voice, however, is rough and disagreeable, and in common with the genus, their disposition is comparatively sedate.

The *Pavouane Parrot*, *Psuyanensis*, Lin., which stands first in our table of the Parrot Maccaws, CONURUS of Kuhl, is extremely common in Guiana and the Antilles, where they are found in flocks, during the day, in the forests, but in the mornings and evenings they resort to the meadows and streams. They do great damage to the coffee plantations, by eating the pulp which surrounds the grain.

Birds of this species are noisy, importunate, and mischievous, but are easily taught to pronounce words; indeed Le Vaillant mentions one which could recite the Lord's prayer, in Dutch, lying at the same time on its back, and joining its toes together as men join their hands in prayer.

The division with arrow-shaped tails, PALÆORNIS of Vigors and Horsfield, we shall illustrate by a figure of the *Blue-banded Parrakeet*, the *P. Venustus* of Temminck; it is olive-green on the upper parts, and beneath yellow; on the forehead is a deep blue band; the rectices blue, with the tips yellow; the quills are black.

Dr. Horsfield has described six species of the TRICHO-

BLUE-BANDED PARAKEET.

P. VENUSTUS.

Mus. Lin. Soc.

One half size.

London. Published by Whittaker & Co. Ave Maria Lane. Aug. 1829.

SPOTTED PARROT.

P. MATONI.

Mus. Lin. Soc.

London. Published, by Whittaker & Co. Ave Maria Lane, Aug. 1829

BLUE-FACED PARROT.

P. CAPISTRATUS.

Mus. Lin. Soc.

London Published by Whittaker & Co. Ave Maria Lane. August 1829.

GLOSSI, or Filamentous-tongued Parrots, of New Holland, in the collection of the Linnæan Society.

Of these the *Spotted* or *Matous Parrot* is green, the breast and belly waved with yellow; and the under wing-coverts red.

And the *Masked* or *Blue-faced Parrot*, green, with the head and throat bright-blue; the breast is scarlet, and the nape is yellow.

The division with the tail enlarged toward the end, the PLATYCERCUS of Vigors, affords some very splendid species, for the most part from Australasia; but of their manners, dispositions, and habits, we know but little. Mr. Caley has indeed given some information on these points, with reference to such of the species as are in the Museum of the Linnæan Society.

He states that *Pennant's Parrot* is called by the natives *Dulang*, and *Julang*, and is found in large flocks among the ripe Indian corn, in company with the Tabuan Parrot (*P. Scapulatus*). It varies in colour in different individuals. It is said to breed principally in the body of the peppermint tree, but not in the boughs; sometimes it enlarges the hole, through which it enters, and year after year the same place is frequented for the purposes of incubation, the nest itself being nothing more than the decayed part of the tree. It has four young ones; the eggs are white. Mr. Caley has met with it in November, in the most mountainous parts of the country; but he apprehends that it migrates in winter.

The *Nonpareil Parrot*, appears at first to have been called the Rosehill Parrot, from the name of the settlement, afterwards called Paramatta. The native name of this bird is *Bundullock*. It is said to breed in dead trees, chiefly on farms, making a nest with feathers in the body of the hollow tree. To whatever depth the tree may be hollow, the bird always descends to the bottom; it has six young at a time; the eggs are white, and without spots.

This may also be frequently seen, in small flocks with the last mentioned and its companions, in fields of Indian corn. Mr. Caley, however, never saw this species take the corn from the stalk, like the other two, and he thinks it merely picks up what falls from the others. I have seen, says Mr. Caley, the most of this species on new-sown wheat, early in the morning, but never in large flocks. I do not recollect ever to have seen the King's Parrot, or Lory (*P. Scapulatus*), pulling up the young wheat like this bird. All three species are caught, and are very good eating. The present species frequent Van Diemen's Land; whether the other two are found there seems uncertain.

Brown's Parrot, which belongs to this division, is yellowish-white, varied with black; the wings and tail are blue, and the vent is red; it is only eleven inches in length. This, which may be considered, from the extreme delicacy of its colours, as the most beautiful of the family, was discovered by Mr. Brown, whose name it justly bears, at Arnheim Bay.

We insert three figures of other species, belonging to this division, which with the last are all in the very interesting collection of New Holland birds belonging to the Linnæan Society.

Bauer's Parrot is about fifteen inches long; the head and part of the neck, are black; the cheeks, throat, and feathers, which fall over the lower mandible, more or less ultramarine-blue; neck behind, dun-yellow, nearly in shape of a crescent; the general colour of the rest of the plumage above, the fore part of the breast, and of the two middle tail-feathers, is green, deepest on the breast. The second quills are blue; the primaries are black, edged with grey; the under wing-

H Kearsley del^t

Mus. Lin. Soc.

BROWN'S PARROT. *Lath* II. 139.

P. BROWNII. *Lin Trans.* XIII.

London. Published by Whittaker & Co Ave Maria Lane. 1829.

BAUER'S PARROT.

P. BAUERI.

Mus. Lin. Soc.

London. Published by Whittaker & Co. Ave Maria Lane. August. 1828.

BARNARD PARROT.

P. BARNARDI.

Mus. Lin. Soc.

London. Published. by Whittaker & C.° Ave Maria Lane. Aug. 1829.

VARIED PARROT.

P. MULTICOLOR.

Mus. Lin. Soc.

London Published, by Whittaker & C° Ave Maria Lane, Aug. 1829.

coverts, verditer-blue. This was presented to the Society, by Mr. Brown, and was taken at Memory Cove, on the South Coast of New Holland.

Barnard's Parrot was presented by Edward Barnard, Esq. to the Society. It came from the interior of New Holland, but appears to be a scarce species in its native country.

It is about fifteen inches long; the forehead is deep crimson; the rest of the head, pale-green; on the nape is a broad brown patch; the back and wing-coverts are deep-blue; the rump and upper tail-coverts, pale-green; and across the wings there is a broad stripe of the same colour; the two middle tail-feathers are green, the others are blue from the base to the middle, the rest is pale-blue, fading almost to white towards the end.

The *Varied Parrot* was procured by Mr. Brown, on the South Coast of New Holland. It is not much more than ten inches long; the plumage is emerald-green; on the forehead, yellow; across the crown, chestnut; on the rump are three shades of colour, first, pale-green, then deeper, and lastly reddish, or chestnut; the belly, thighs, and vent, are orange-yellow; the quills are edged with deep-blue; the tail, green, blue at the end; on the two outmost feathers is a bar of black, and on the third a patch of white.

Of the *Tabuan Parrot*, Mr. Caley says it was seldom he saw a full coloured specimen, viz. red. When the Indian corn is ripe, they may be seen, in large flocks, clinging on the stalks, and doing much mischief; but as it is rare to see a bright red one among them, it may be presumed that the greater part of these flocks are young birds. The natives say that it breeds chiefly in a white gum tree, a species of *Eucalyptus*, making its nest of a little grass, and lining it with feathers. It has as many as twelve young at a brood. The eggs are dirty-white, with black specks. The nest is found by the bird enlarging the hole to creep in at, which process gives the surrounding part a reddish appearance, which, by contrast with the whiteness of the nest, renders the hole conspicuous.

The division of Parrots with the tail wedge-shaped equally, presents us with nothing worthy of remark in relation to manners, habits, &c.; or, rather, this interesting branch of their history has not hitherto been sufficiently cultivated to furnish us with much biographical matter relating to them.

At the end, as it were, of this division, comes the species named by M. Temminck *P. Setarius*, or the *Racket-tailed Parrot*, on account of the peculiarity of the middle tail-feathers, which, being much larger than the rest, and furnished with barbs only at the end, look something like a pair of rackets. The male of this species has the forehead, cheeks, neck, and all the under parts of the plumage, green; a reddish-ash band passes from one eye to the other; and beneath this, on the occiput, is pale blue, which passes to the nape; the smaller wing-coverts at the angle of the wing, are blue, passing into green; all the upper parts are of a deep-green, except a portion of the extremities of the caudal feathers, which are deep-blue; the tail beneath is blue. The young appears in all probability to want the red band on the forehead. This species inhabits the Indian Islands, but nothing is known of its manners.

The New Holland Cockatoos include three species, which are very much assimilated to one another, viz. *The Banksian, Funereal, and Cook's Cockatoo.* Of the first of these, Mr. Caley informs us that its native name is *Geringora*, and that it occurs in various parts of New Holland, especially northward of Paramatta; but that they are not seen many together. The natives say it breeds in the winter in *Mun'niny-trees*, or *Blood-trees* of the colonists, a species of *Eucalyptus*. It

RACKET-TAILED PARROT.

P. SETARIUS. Tem

London. Published by Whittaker & C.º Ave Maria Lane Aug. 1829.

COOK'S COCKATOO. var.

P. COOKII. Lin. Trans. P. BANKSII. Ind. Orn.

London. Published, by Whittaker & C° Ave Maria Lane Aug. 1829.

makes no other nest than the vegetable mould formed by the decay of the tree. It has three young at a time; but its eggs are unknown.

The native name of the Funereal Cockatoo is *Wy'la*, so called from the similitude of that word to the sound which it makes. These also are not gregarious, except in small flocks of not more than about half a dozen; but they appear to be extensively located in this extensive island. These also construct their nests in a species of *Euçalyptus*.

Cook's Cockatoo seems to be the *Carat* of the native New Hollanders; if so, it is a very shy bird. It scrapes the dirt out of hollow boughs, and makes its nest in the manner of the former species, which nest may be found by watching the bird when it enters the hole in the tree where it is deposited. It cuts off the fruit of two species of *Persoonia*, without however eating it before it is ripe, to the great injury and vexation of the natives.

The short specific characters of these species, given in the text, are sufficient to shew a general and very considerable similarity. The little information we have on their habits (more perhaps than we possess of most of the other species,) bespeaks also as great a similarity in this respect. Our figure is from a specimen which was several years ago in Bullock's Museum. In it the broad red band across the tail is not interrupted by small black bars, which exist in the Banksian Cockatoo, properly so called. Both, however, were considered varieties of one species, but M. M. Temminck, Kuhl, Vigors, and Horsfield have treated them as distinct; we have, therefore, inserted a reimpression of this figure, altering the name to that of Cook's Cockatoo, leaving the trivial name of Banksian undivided in the other species.

The other species above mentioned, the Funereal Cockatoo, differs from Cook's and the Banksian principally in having the tail-band buff-coloured, marked with numerous black spots instead of bands.

The *Rose Cockatoo* is palish-ash coloured; the neck, the body beneath, and the lower-coverts, are rose coloured; the crest is of the same colour, but of a lighter tint. This, together with the three species above named, is an inhabitant of New Holland.

In that comprehensive division of parrots with short, even tails, and without a crest on the head, are included the two well known species of the grey or ash-coloured parrot, *P. Erithacus*, from Africa, and the *Amazon*, or *Green Parrot*, and their numerous varieties. Many stories have been told, and repeated *usque ad nauseam*, of the marvellous deeds of these species supposed to be consequential on their mental faculties; indeed, most persons are in possession of anecdotes, more or less wonderful, of particular individuals of the species, which have fallen under their own observation, or that of their friends—anecdotes, which too often increase by repetition, till the true extent and character of the original facts are lost. Parrots will certainly sometimes repeat a word or a sentence, which circumstances may render particularly apt and applicable, as monkeys will sometimes use a gesture or an action strikingly human in its appearance; but a very slight acquaintance with these animals will convince any reasonable person that these imitative or mechanical qualities are not to be attributed to superior reason or sagacity; and, as much has been already said upon the subject, we shall not subjoin any repetition of thrice-told tales, or search for others of a similar character, which, however amusing, may be considered as destitute of instruction, and of equivocal veracity.

At the end of the Scansores, Cuvier has placed two analogous genera, Corythaix and Musophaga, which by certain

ROSE COCKATOO.

P. EOS.

Mus. Lin. Soc.

London Published by Whittaker & C.º Ave Maria Lane, Aug. 1829.

PAULINE PLANTAIN-EATER.

London, Published by Whittaker & Co. Ave Maria Lane, June. 1829.

characters to which he has referred, approximate to the gallinaceous birds on the one hand, and the scansores on the other, and hold therefore an intermediate station between the two. Genera of this description are ever annoying to the systematist, particularly the supporters of a natural method; for whether they be referred to one order or the other, is perhaps arbitrary and indifferent.

The *Touraco Plantain Eater* inhabits various parts of India and South Africa. They are very difficult to be shot, perching only at the extremities of the highest branches of trees, out of gun-shot, and rarely suffering any one to approach. They may, however, be easily caught alive by means of snares baited with fruits, on which they feed. They are good eating.

The *Touracou Pauline* is so named by M. Temminck from Mlle. Pauline de Ranchaup. The crest is red, but some of the feathers are tipt with white. It is composed, as in the rest of the genus, of several thin and delicate feathers, which, rising on each side of the head, meet at the top, and form a crest, not unlike an ancient helmet, passing down to the nape. The feathers which cover the nostrils, the neck, back, wings, the upper side of the tail, the quills, throat, and breast, are of a shining copper-colour; the belly and abdomen, green; the remiges are red; the eyes are situated in a large white patch. It is about as big as a pigeon.

These birds are gentle and familiar. They leap on the ground, and from branch to branch of trees, with quickness and agility. Their voice is sonorous and deep; and they inhabit South Africa.

Our figure of the Blue Curassow, var., corresponds with Dr. Latham's description of a species under that name, which he refers to CRAX. It seems, however, very probable that the bird when better known, will be found to belong to this genus of the plaintain-eaters. A short description of it is inserted at the commencement of our observations on the gallinaceous birds.

END OF VOL. VII.

LONDON:
SHACKELL AND BAYLIS, JOHNSON'S-COURT, FLEET-STREET.

For EU product safety concerns, contact us at Calle de José Abascal, 56–1°,
28003 Madrid, Spain or eugpsr@cambridge.org.

www.ingramcontent.com/pod-product-compliance
Ingram Content Group UK Ltd.
Pitfield, Milton Keynes, MK11 3LW, UK
UKHW042206080726
473066UK00007B/279
* 9 7 8 1 1 0 8 0 4 9 6 0 3 *